Fundamentals and Advancements in Hydraulic Measurements and Experimentation

Proceedings of the Symposium

Sponsored by the
Hydraulics Division of the
American Society Civil Engineers

in cooperation with the
Environmental Engineering Division
Irrigation and Drainage Division
Water Resources Planning and Management Division
Waterway, Port, Coastal and Ocean Division
of the American Society of Civil Engineers
and the
Canadian Society of Civil Engineering

Hydraulics Division Programs Committee
George V. Cotroneo, Chairman
William H. Espey, Jr., Vice Chairman
Steven R. Abt, Secretary
S.T. Su, Past Chairman

Hydraulics Division Executive Committee
Edward R. Holley, Chairman
Arlen D. Feldman, Vice Chairman
David S. Biedenharn, Secretary
Steven R. Abt, Past Chairman
Adnan M. Alsaffar
Catalino B. Cecilio
Linda S. Weiss, News Correspondent

Buffalo, New York
August 1-5, 1994

Edited by Clifford A. Pugh

Published by the
American Society of Civil Engineers
345 East 47th Street

ABSTRACT

This proceedings, *Fundamentals and Advancements in Hydraulic Measurements and Experimentation*, contains papers presented at the Symposium held in Buffalo, New York, August 2-4, 1994. The objective of the Symposium was to provide a forum for timely dissemination of information and for stimulating interaction among engineering and research professionals in the arena of hydraulic measurements and experimentation. Topics included in the proceedings are: 1) Typical problems associated with hydraulic measurements and instrumentation; 2) flow measurement; 3) scour instrumentation and sediment measurement; 4) sediment modeling and measurement techniques; 5) turbulence measurements and studies; 6) visual and optical techniques for sediment measurements; 7) ice measurement techniques; 8) air/water flow measurement techniques; 9) techniques in hydraulic measurements; 10) instruments for velocity measurements; 11) measurements at hydroelectric and pumping plants; and 12) experimental data collection techniques. These papers provide civil engineers working in the hydraulic speciality with guidance concerning the principles, limitations, and the application of measurements and experimental techniques in hydraulic engineering.

Library of Congress Cataloging-in-Publication Data

Fundamentals and advancements in hydraulic measurements
   and experimentation: proceedings of the symposium spon
   sored by the Hydraulics Division of the American Society of
Civil Engineers, in cooperation with the Environmental
   Engineering Division...[et al.] of the American Society of
   Civil Engineers.../edited by Clifford A. Pugh.
      p.   cm.
   Proceedings of the Symposium on Fundamentals and
Advancements in Hydraulic Measurements and
Experimentation, Buffalo, New York, August 1-5, 1994.
   Includes indexes.
   ISBN 0-7844-0036-9
     1. Hydraulic measurements—Congresses. 2.
Hydraulics—Research-Congresses. I. Pugh, C. A. II.
American Society of Civil Engineers. Hydraulics Division.
III. Symposium on Fundamentals and Advancements in
Hydraulic Measurements and Experimentation (1994:
Buffalo, N.Y.)
TC177.F86 1994   94-26048
627'.028'7—dc20   CIP

The Society is not responsible for any statements made or opinions expressed in its publications.

Photocopies. Authorization to photocopy material for internal or personal use under circumstances not falling within the fair use provisions of the Copyright Act is granted by ASCE to libraries and other users registered with the Copyright Clearance Center (CCC) Transactional Reporting Service, provided that the base fee of $2.00 per article plus $.25 per page copied is paid directly to CCC, 27 Congress Street, Salem, MA 01970. The identification for ASCE Books is 0-7844-0036-9/94 $2.00 + $.25. Requests for special permission or bulk copying should be addressed to Permissions & Copyright Dept., ASCE.

Copyright © 1994 by the American Society of Civil Engineers,
All Rights Reserved.
Library of Congress Catalog Card No: 94-26048
ISBN 0-7844-0036-9
Manufactured in the United States of America.

PREFACE

One of the most effective methods for encouraging timely dissemination of information and for stimulating interaction among engineering and research professionals is through conducting conferences and symposia. With this in mind, the ASCE Technical Committee on Hydraulic Measurements and Experimentation proposed a three day Symposium on Fundamentals and Advancements in Hydraulic Measurements and Experimentation.

Sound procedures for making field and laboratory measurements and conducting experimental studies are crucial in hydraulic research, development, and design. Classical methods have been practiced for decades, if not centuries, in hydraulics. Recently, the Hydraulics Division of ASCE has recognized the need to emphasize fundamental concepts and new technologies in the arena of hydraulic experimentation by creating a technical committee to provide guidance to civil engineers on experimental design and instrumentation. This committee, the Technical Committee on Hydraulic Measurements and Experimentation, was formed to establish a focal point within ASCE to study, evaluate, and promote the use of appropriate instrumentation and techniques for the wide range of measurement problems facing civil engineers working in the hydraulic specialty. More specifically, the committee initiates activities to review, summarize and disseminate information on the principles, limitations, and the application of measurements and experimental techniques in hydraulic engineering.

Tremendous advances have been made over the last decade in instrumentation technology designed to aid in hydraulic and hydrologic measurements. Examples include fiber optic sensors, acoustical methods, lasers, and digital data acquisition systems, just to name a few. Not surprisingly, the effort required for individuals to stay abreast of these improvements has grown proportionately. Also, with the abundance of new tools available, it is now feasible to successfully perform measurements of phenomena which have been traditionally very difficult to measure, such as turbulence and shear. However, even with improved technologies and instruments, a thorough understanding of classic experimental methodologies is still crucial to insure that the information gathered is dependable and appropriate for the application.

In addition, many of the principles associated with today's hydraulic measurements are not covered in typical engineering curricula. As the technology continues to develop, additional knowledge is necessary in areas not traditionally covered in engineering courses. There exists a need to provide the civil engineering profession an avenue of education on the subject of measurements and instrumentation; not only to assure proper application of equipment, but to enhance awareness of the tools available to assist in arriving at the most accurate and efficient solution to a problem.

Each of the papers included in the Proceedings has been reviewed and accepted for publication. All papers are eligible for discussion in the Journal of Hydraulic Engineering and all papers are also eligible for ASCE awards.

Many people contributed to the success of the Symposium. The members of the original Technical Committee on Hydraulic Measurements and Experimentation: Charles Almquist, Richard McGee, Dominique Brocard, and the Editor, conceived and proposed the Symposium.

Adnan Alsaffar supported and advocated the proposal as the contact member on the Executive Committee. The Organizing Committee which arranged and conducted the Symposium included: Bobby J. Brown, Vito J. Latkovitch, Robert Ettema, and the Editor. The support and assistance of overall conference chairman, George Cotroneo, was essential to the planning and success of the Symposium. The assistance in organizing, editing, and correspondence of Fran Haefele and Opal Brittain is greatly appreciated. The support of the ASCE staff including Shiela Menaker, Julie Taylor, and others in New York is appreciated. The assistance of the Session Moderators is also appreciated. Ultimately the authors' papers and presentations form the focus of the Symposium. The Proceedings provide a valuable reference for the profession, and the authors' contributions are especially appreciated.

                                                      Clifford A. Pugh
                                                      Editor

CONTENTS

INVITED LECTURES—SESSIONS S-1, S-5, S-12

Measurement of Turbulent Properties in a Natural System (Session S-1)
Jörg Imberger and Roger Head 1
Sediment Measurement Instrumentation—A Personal Perspective (Session S-5)
E.V. Richardson 94
Recent Applications of Acoustic Doppler Current Profilers (Session S-12)
Kevin A. Oberg and David S. Mueller 341

SESSION S-1
PLENARY SESSION—VIGANDER LECTURE
Moderator: Clifford A. Pugh

Measurement of Turbulent Properties in a Natural System
Jörg Imberger and Roger Head 1

SESSION S-2
FLOW MEASUREMENT—I
Moderator: Kathleen H. Frizell

Using Drop Structures for Stream Gaging
Bobby J. Brown 21
A Computerized Open Channel Flow Measurement Device
Stan E. Malinky 27
Demonstration Project for Scour Instrumentation
J.D. Schall, W.R. Ivarson, and T. Krylowski 37
Laboratory and Field Evaluation of Acoustic Velocity Meters
Tracy Vermeyen 43

SESSION S-3
FLOW MEASUREMENT—II
Moderator: Tracy B. Vermeyen

Mapping 2-D and 3-D Velocity Components in Circular Conduits Using an
Electromagnetic Current Meter and a 5-Hole Pitot Probe
Joseph J. Orlins and Lawrence J. Swenson 53
Velocity and Turbulence Measurement from the Illinois and Mississippi Rivers
Ta Wei Soong and Nani G. Bhowmik 62
Velocity Measurements by the "One-Orange" Method
B.A. Christensen 76
Experimental Design and Measurement Techniques for Investigation of Two-Phase Flow
Andreas J. Kuck 86

SESSION S-4
PANEL DISCUSSION
Moderators: Robert Ettema and Vito Latkovitch

Instrumentation Needs and Possibilities: A Dialogue Between Suppliers and Users
Robert Ettema and Vitto Latkovitch 539

SESSION S-5
SCOUR INSTRUMENTATION AND SEDIMENT MEASUREMENT
Moderator: Vito Latkovitch

Sediment Measurement Instrumentation—A Personal Perspective
  E.V. Richardson 94
Real-Time Data Collection of Scour at Bridges
  David S. Mueller and Mark N. Landers 104
Low Cost Bridge Scour Measurements
  J.D. Schall, J.R. Richardson, and G.R. Price 114

SESSION S-6
SEDIMENT MODELING AND MEASUREMENT TECHNIQUES
Moderator: E.V. Richardson

Calibration of Movable Bed Model for Armant Area
  F.S. El-Gamal and A.F. Ahmed 119
Using the SedBed Monitor to Measure Bed Load
  Roger A. Kuhnle and Robert W. Derrow, II 129
Vertical Sorting Within Dune Structure
  Mohamed Abdel-Motaleb 139
On Measurements of Particle Spinning Motion
  Hong-Yuan Lee and In-Song Hsu 149

SESSION S-7
TURBULENCE MEASUREMENTS AND STUDIES
Moderator: K. Warren Frizell

An Investigation of Turbulence in Open Channel Flow via Three-Component Laser Doppler Anemometry
  Mahalingam Balakrishnan and Clinton C. Dancey 159
Hot-Film Response in Three-Dimensional Highly Turbulent Flows
  P. Prinos 176
Experimental Study on Turbulent Structures in Unsteady Open-Channel Flows
  Iehisa Nezu, Akihiro Kadota, and Hiroji Nakagawa 185
Response of Velocity and Turbulence to Abrupt Changes From Smooth to Rough Beds in Open-Channel Flows
  Iehisa Nezu and Akihiro Tominaga 195

SESSION S-8
VISUAL AND OPTICAL TECHNIQUES FOR SEDIMENT MEASUREMENTS
Moderator: Gregory Gartell

Development of a Visual Method to Track the Movement of Hydrogen Bubbles in a Laboratory Flume
  Anthanasios N. Papanicolaou and Panayiotis Diplas 205
High-Speed Video Analysis of Sediment-Turbulence Interaction
  Yarko Niño, Fabián López, and Marcelo García 213
Optical Methods for Sediment-Laden Flows
  R.N. Parthasarathy and M. Muste 223
Visual Investigation of Field Bed-Load Sampling
  Moustafa T.K. Gaweesh 233

SESSION S-9
ICE MEASUREMENT TECHNIQUES
Moderator: Robert Ettema

Estimation of Mean Velocity for Flow Under Ice Cover
   Martin J. Teal and Robert Ettema 242
Effects of Simulated Ice on the Performance of Price Type-AA Current Meter Rotors
   Janice M. Fulford 251
Innovative Instrumentation for a Physical Model of River Ice Transport
   Johannes Larsen, Jon E. Zufelt, and Randy D. Crissman 259

SESSION S-10
AIR/WATER FLOW MEASUREMENT TECHNIQUES
Moderators: Tony Rizk and Perry Johnson

Developing Air Concentration and Velocity Probes for Measuring Highly-Aerated, High-Velocity Flow
   Kathleen H. Frizell, David G. Ehler, and Brent W. Mefford 268
Void Fraction Measurement Techniques for Gas-Liquid Bubbly Flows in Closed Conduits; A Literature Review
   Mahmood Naghash 278
Measuring Air Concentration in Flowing Air-Water Mixtures
   Boualem Hadjerioua, Tony A. Rizk, Emmett M. Laursen, and
   Margaret S. Petersen 289
Application of a Needle Probe in Measuring Local Parameters in Air-Water Flow
   A.R. Zarrati and J.D. Hardwick 296

SESSION S-11
TECHNIQUES IN HYDRAULIC MEASUREMENTS
Moderator: Charles Almquist

A New Basic Principle for a New Series of Hydraulic Measurements. Erosion by Abrasion, Corrosion, Cavitation, and Sediment Concentration
   Lucien Chincholle 301
Anomalous Measurements in a Compound Duct
   David G. Rhodes and Donald W. Knight 311
Hydrodynamic Forces in Hydraulic Jump Stilling Basins
   António N. Pinheiro, António C. Quintela, and Carlos M. Ramos 321
Measurements of the Hydrodynamic Lift and Drag Forces Acting on Riprap Side Slope
   A.F. Ahmed and F.S. El-Gamal 331

SESSION S-12
INSTRUMENTS FOR VELOCITY MEASUREMENT
Moderator: Janice Fulford

Recent Applications of Acoustic Doppler Current Profilers
   Kevin A. Oberg and David S. Mueller 341
Acoustic-Doppler Velocimeter (ADV) for Laboratory Use
   Atle Lohrmann, Ramon Cabrera, and Nicholas C. Kraus 351
Field Performance of an Acoustic Scour-Depth Monitoring System
   Robert R. Mason, Jr. and D. Max Sheppard 366

SESSION S-13
VELOCITY AND FLOW MEASUREMENT TECHNIQUES
Moderator: Christopher R. Ellis

Comparison of Current Meters Used for Stream Gaging
   Janice M. Fulford, Kirk G. Thibodeaux, and William R. Kaehrle 376

LDV System for Towing Tank Applications
  R.N. Parthasarathy and F. Stern 386
Experimental Study on Unsteady Flow in Open Channels With Flood Plains
  A. Tominaga, M. Nagao, and I. Nezu 396
Coherent Structures in Compound Open-Channel Flows by Making Use of
Particle-Tracking Visualization Technique
  Iehisa Nezu, Hiroji Nakagawa, and Ken-ichi Saeki 406

SESSION S-14
MEASUREMENTS AND TECHNIQUES AT HYDROELECTRIC PLANTS AND PUMPING PLANTS
Moderator: Bobby J. Brown

Strain Measurement on the Runner of a Hydroelectric Turbine
  K. Warren Frizell 416
Testing Turbine Aeration for Dissolved Oxygen Enhancement
  Tony L. Wahl, Jerry Miller, and Doug Young 425
Eliminating Water Column Separation and Limiting Backspin at a 12,000-Horsepower
Pumping Plant
  Paul Otter, David Hoisington, and David Raffel 435
Pressure-Time Flow Rate in Low Head Hydro Plants
  Charles W. Almquist, David B. Hansen, Gerald A. Schohl, and Patrick A. March 445

SESSION S-15
EXPERIMENTAL DATA COLLECTION TECHNIQUES—I
Moderator: Tony L. Wahl

The Use of Piezoelectric Film in Cavitation Research
  Saurav Paul, Christopher R. Ellis, and Roger E.A. Arndt 454
Design and Execution of Hydrodynamic Field Data Collection Using Acoustic Doppler
Current Profiling Equipment
  Timothy L. Fagerburg and Thad C. Pratt 530
Design and Operation of a System to Monitor Sediment Deposition for Protection of
an Endangered Mussel
  Michael S. Griffin and David S. Mueller 472
Model-Prototype Conformance of a Submerged Vortex in the Intake of a Vertical
Turbine Pump
  K. Warren Frizell 482

SESSION S-16
EXPERIMENTAL DATA COLLECTION TECHNIQUES—II
Moderator: Iehisa Nezu

Simultaneous Flow Visualization and Hot-Film Measurements
  Fabián López, Yarko Niño, and Marcelo García 490
Non-Intrusive Experimental Setup for Inflatable Dam Models
  T.A. Economides and D.A. Walker 500
A Computer-Controlled, Precision Pressure Standard
  Othon K. Rediniotis 509
Observations on the Growth of an Internal Boundary Layer With a Lidar Technique
  Chia R. Chu, Marc Parlange, William Eichinger, and Gabriel Katul 519

Subject Index 541

Author Index 545

MEASUREMENT OF TURBULENT PROPERTIES IN A NATURAL SYSTEM

Jörg Imberger[1] Member ASCE and Roger Head[1]

ABSTRACT

Instruments designed to measure flow properties in stratified estuaries, lakes and oceans are reviewed. Emphasis is given to the measurement of turbulent fluxes and recent developments in laser and acoustic technologies.

INTRODUCTION

Over the last 10 years there has been a radical change in the type of instruments which are available for use by the limnologist, and the rate of improvement in our instrument capabilities will further increase markedly in the next 10 years. Instruments have become totally portable so that they can be deployed easily from fishing vessels, largely alleviating the need to invest in expensive research vessels. Four major factors have lead to these changes. First, the introduction of specialised signal processing chips (DSP chips) has allowed the construction of intelligent instruments capable of processing say FFT's at astounding rates. Such chips are already finding application in microstructure profilers and many other data intensive instruments. Second, the reduction of the cost and power utilization of both general purpose microprocessors and memory has made it possible to produce loggers with memory cards each containing 40 MBytes of RAM; this is enough storage to hold data from 20 thermistors acquiring data at 10 second intervals for 3 months! Third, the introduction to the field of both laser and acoustic velocity and particle concentration measurement techniques has been shown to be feasible (Imberger & Head 1994; Farmer et al., 1987; Agrawal, 1984). Broadband coherent Doppler's are already commercially available and once the cost falls on these sensors we will see a major increase in field capability. Fourth, new miniaturized sensors, with greatly improved time response have recently been introduced

---

[1] Professor, Centre for Water Research, The University of Western Australia, Nedlands 6009, Australia
[2] Electronic Engineer, Centre for Water Research, The University of Western Australia, Nedlands 6009, Australia.

for dissolved oxygen (Oldham, 1993), pressure (Moum, 1990). A new field where electronic pick-up and control are arranged on the same chip as mechanic pump devices and chemical reactors is presently emerging (Zdeblick, 1994; Redwood, 1993). Once this technology is perfected then standard instruments such as auto-analysers and spectrophotometers, as well as standard chemical tests such as the Winkler method for oxygen, will all be available in miniature enabling us to profile the water column quickly and with spatial resolution down to the Kolmogorov scale.

Remote sensing is being increasingly used (Joyce, 1985) to document the horizontal surface variability of parameters such as temperature, turbidity, salinity, chlorophyll, depth and other variables. In general these images are collected from satellites independently to the field data and reconciliation takes place during data analysis. With the improvement of lightweight spectral cameras it will become possible to interlink remote sensing of a lake's surface and deployment of time series and portable equipment. What is required is a real time link between the aeroplane and the research vessel; this should become available shortly. When this happens, then it will become feasible to carry out detailed event experiments where inflows, frontogenesis upwelling etc. are documented as processes.

In general, the most rapid progress in identifying particular problems is to carry out field monitoring followed by field process oriented research. Once the problem has been identified, laboratory experiments, theoretical analysis and numerical analysis of the individual processes allows the underlying physics of the particular mechanisms to be identified. Once the building blocks or individual processes have been established, numerical models can be used as design tools for water quality related management strategies. The circle is closed when the application of these strategies leads to discrepancies with respect to observations; this then requires further process field research and we cycle through the whole loop again.

As discussed below, field instrumentation has undergone a revolution in the last 10 years. The same is true for laboratory and numerical experimentation; the reasons are identical and originate from the introduction of very fast computer and DSP chips. In the laboratory it is now routine (Cheremisinoff, 1986) to measure velocity fields not point by point, but rather over whole sheets and even three dimensionally with a resolution of about 1 mm. This is done by particle tracking. When combined with very fast response temperature and conductivity profiling sensors (Cheremisinoff, 1986) or laser induced fluorescence measurements, the turbulent transport can be measured directly with a resolution high enough to resolve the Kolmogorov scale.

Further very fast work stations are now able to solve the full Navier-Stokes' equations for Reynolds numbers up to 10^3 for 3D and 10^4 for 2D. These are high enough to meaningfully study such important topics as internal wave dynamics, inflow intrusions, surface layer dynamics and others.

In this presentation we review a number of field instruments which have now become standard and may be purchased or constructed from commercially available components but we shall emphasise the newest developments in laser Doppler anemometry. The paper is divided into a discussion of time series instruments used mainly to document time changes of basin scale phenomena, fine scale profilers which allow transects to be performed with great accuracy, and microstructure profilers which are designed to measure the turbulent transport as well as many of the characteristics of the turbulent field.

The reader is referred to the review by Imberger (1994) for a full description of the application of these new measurement techniques to limnology and to the article by Imberger and Head (1994) for more details on the Portable Flux Probe.

TIME-SERIES MEASUREMENTS

Time series instruments, as the name implies, are instruments which are either moored in the water column, fixed to the bottom of the lake, on islands, on headlands, or are designed to profile the water column unattended. The word *in situ* is also commonly used for such instruments

Thermistor Chains

Thermistor chains have been available for a very long time, but it is only recently that modular construction has made it possible to assemble a thermistor chain which can measure the water temperature with an accuracy better than ± 0.005 °C easily and cheaply. If the correct sampling interval is used then an array of thermistor chains will allow the identification of the three dimensional internal wave spectra and it should be possible to extract the local dissipation in the water column (Imberger, 1994). Given the dissipation it is possible to estimate the vertical mass flux. We may therefore expect the use of arrays of fast sampling thermistor chains to greatly increase.

Chains consisting of conductivity devices and current meters are also available, however, the *in situ* fixed point current meter is being rapidly replaced by acoustic profiling current meters, especially in the low velocity environment of a lake. On the other hand, conductivity measurements are difficult to perform over long periods due to fouling.

Acoustic Doppler Current Profilers

The use of acoustics to measure particle and bubble concentrations and water velocities is now wide spread (Patterson, 1991; Pinkel, 1986; Stanton, 1987; Whitehouse & Imberger, 1991; Brumley *et al.*, 1991; Gordon, 1989). The concentration measurements rely on measuring the intensity of the backscatter signal which is a strong function of the signal frequency and the acoustic target size of the particles (Urik, 1983).

Velocity measurements fall into two categories. First, we have the incoherent Doppler current meters. In these instruments a pulse of sound is emitted from a transducer and the return signal is binned or gathered into lengths of return signal corresponding to lengths of the water column from where the return signal originates. The pieces of record in each bin are then analysed for frequency content which is compared to the frequency of the emitted signal. The difference in frequency, over the bin length, is attributed to the Doppler shift induced by the fluid motion in the direction of the emission beam. Each returned bin yields one estimate of the velocity. Since relatively long time series are required to estimate the frequency with some statistical confidence, the signal resolution (bin size) tend to be in excess of 5 meters for these instruments. Such instruments are thus constrained to measuring basin scale velocity fields and really only find application in studies of larger lakes or the deeper ocean.

The second type of device is called a coherent Doppler current meter and in these instruments pulses of sound are emitted with phase coherence so that adjacent pulses form wave trains which are phase coherent. By blinding the receiver to the "holes" in the signal it is possible to construct a time series of returns across as many pulses as is required to get a long time series for the purpose of estimating the frequency of the scattered return; in essence this instrument trades decreased temporal resolution for the poorer spatial resolution of the incoherent Doppler. By further coding the pulses bin sizes as small as a few millimetres are possible for high frequency instruments (Stanton, 1987). A typical instrument is shown in Fig. 1

The frequency of the emitted signal, the pulse length and the power are all important parameters and must be varied in order to optimize the strength of the return signal (Whitehouse & Imberger, 1991). However, both the coherent and incoherent sonar's are limited by the range ambiguity constants

$$vR \leq \frac{C^2}{4 f_c}$$

where v is the fluid velocity, R is the maximum range, C is the speed of sound in water and f_c is the carrier frequency (Pinkel, 1986).

There are proprietary techniques available which increase the range ambiguity but as a rule of thumb the present limitation without dealiasing is a range ambiguity of about 2, a bin size of the order of centimetres, and a velocity resolution around 5 mm s^{-1}.

The most important limitation for the use of these instruments is the particle size and concentration in the water column; we cannot expect a return signal if there are no scatterers in the water column. Combined with an array of thermistor chains the acoustic Doppler current profiler (ADCP) will likely completely change field experimentation. It will become feasible to document the velocity field down to horizontal scales of 10's of meters and vertical scales down to 10's of centimetres using cheaper ADCP's. This information can then be combined with estimates of the mass transport from the thermistor chains and microstructure instruments. An example of a two day deployment of an acoustic Doppler profiler in a tidal estuary, where the water was about 5m deep, is shown in Fig. 2

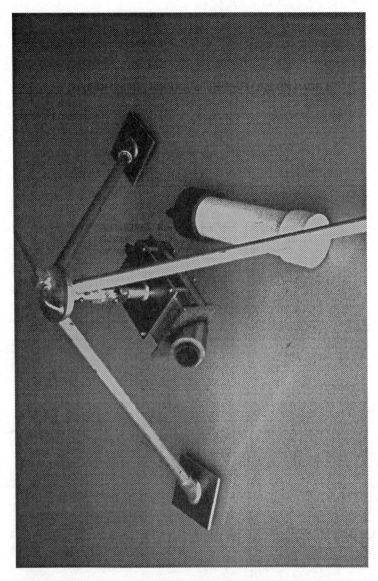

Figure 1. Acoustic Doppler coherent current meter with data logger tube. The current meter is suspended from a gimbol arrangement to align it vertically.

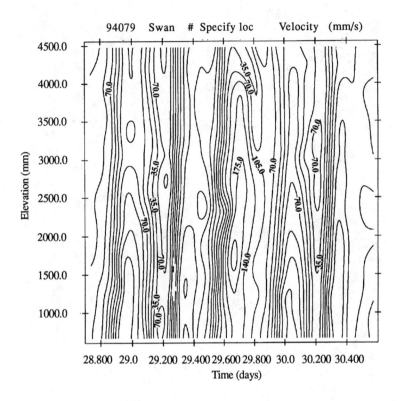

Figure 2. Contour plot of the ADCP horizontal velocity showing position of good velocity estimates.

FINE SCALE PROFILERS

In general the flow field in a lake may be divided into basin scale motions ($\sim 10^4$ m), synoptic features ($\sim 10^3$ m), internal wave and intrusions (10^2 m), entraining motions (1 m) and turbulent mixing (10^{-3} m). The term 'fine scale' brackets the internal wave, intrusion and entraining motions. In order to document these motions the instrument must be extremely accurate (temperature $< 10^{-3}$ °C; conductivity $< 10^{-4}$ Sm^{-1}, depth ; 10^{-2} m), easy to deploy, robust and preferably equipped with a GPS unit for quick position fixing.

There are many manufacturers of such instruments, the great majority being originally developed either for deep oceanographic application or for water quality spot measurements in lakes and estuaries. Shown in Fig. 3 is a free-falling version, able to profile with a resolution of 2×10^{-2} m falling at 1 m s^{-1} (13). It is common to digitally enhance the signals from the sensors in order to carefully match their response times (Fozdar *et al.*, 1985). This is important so that quantities such as salinity or oxygen derived from two sensors do not contain artificial spikes.

When combined with a modern laptop computer it is possible to gather profiles, tows or yo-yo's in order to build up three dimensional contours of the variables under question. Recently, these instruments have also been equipped with an ever increasing array of water quality sensors making them a valuable tool for almost any limnological investigation.

MICROSTRUCTURE INSTRUMENTS

Microstructure instruments are designed to document the turbulent and entraining motions in the water column. This means that they must resolve the parameter variability from about 2 meters to 10^{-3} meters for velocity (Kolmogorov scale) and down to 10^{-4} meters for conductivity (Batchelor Scale). The variables which can currently be measured at these scales are temperature and conductivity (Gregg *et al.*, 1993(a); Carter & Imberger, 1986; Gregg *et al.*, 1993(a)), pressure (Moum, 1990) and velocity (Imberger & Head, 1994; Moum, 1990; Osborn, 1974; Lemckert & Imberger, 1994); conductivity can, however, only be resolved down to the Kolmogorov scale.

Temperature Microstructure

Consider first the temperature microstructure measurements. These are now usually made with an FP07 fast response thermistor (response time of approximately 10 ms). The fast response temperature sensor is used to measure temperature fluctuations down to scales smaller than the Batchelor scale. Spectral analysis is then used (Caldwell *et al.*, 1980) to estimate the roll-off of the temperature gradient spectrum at high wave numbers; the wave number where the spectrum rolls off is associated with the Batchelor scale. The magnitude of this scale can then be used to estimate the dissipation of turbulent kinetic energy.

Figure 3. Typical fine scale profiler with float for free fall deployment.

This methodology has led to the development of very small portable profilers which traverse the water column at the rate of 0.10 ms^{-1} either rising through the water column and piercing the free surface, or falling vertically and impacting on the lake bottom. The first mode of profiling is used when the surface layer is being investigated and the second is used when the main water column and the benthic boundary layer are under investigation. This type of equipment is now commercially available and can be coupled with a GPS unit, all small enough to be deployed from a small rubber zodiac (see Fig. 4).

Oceanographers usually profile at much higher speeds, close to or greater than 1.0 ms^{-1} (Moum, 1990; Oldham, 1994) sacrificing the high wave number resolution needed to estimate the turbulent kinetic energy dissipation and obtaining the dissipation directly from the velocity gradient signal from an Osborn Shear Probe (Osborn, 1977). This has great advantages in the ocean, as the higher profiling speed is much more practical in the great depths of the ocean. The disadvantage is that the knowledge of the dissipation spectrum is lost. Also in lakes the density gradients are much higher and thus the displacement scale is much smaller, requiring high resolution in temperature and conductivity to even just measure the displacement scale.

Conductivity Microstructure

The conductivity of the water may also be measured relatively conveniently at such small scales using a four electrode micro sensor now manufactured commercially. The new design of this sensor has four spherical platinum sensors spaced about 1 mm apart. An oscillating current is passed through two electrodes so that the induced e.m.f. across the other two is a constant. Such an arrangement leads to a system which is tolerant to contamination by plankton and other particles in the water. The present size of the conductivity electrode is not small enough to allow the measurement of the roll of the conductivity gradient spectra, but it does yield accurate measurements of salinity down to and past the Kolmogorov scale; combined with the temperature measurement this may be used to calculate the displacement scale in even very salty water.

Oxygen Microstructure

Recently Oldham (1993; 1994) has modified a proven design for a micro oxygen electrode so that it can be used in a field situation together with the temperature and conductivity microstructure sensors. The electrode yields a response time of about 100 ms and has a 5 mm tip size. When the signal is digitally enhanced to account for sensor roll-off the effective response time can be improved to about 30 ms, which yields a resolution of around 3 mm. Combined with the information about the turbulent transport such instruments allow a much more complete separation of the variability due to physical and biological factors in specific biological systems (e.g. Oldham, 1994).

Velocity Microstructure

Over the last 10 years numerous groups have developed field versions of laser Doppler anemometers (Imberger & Head, 1994; Agrawal, 1984). The great advantage of this

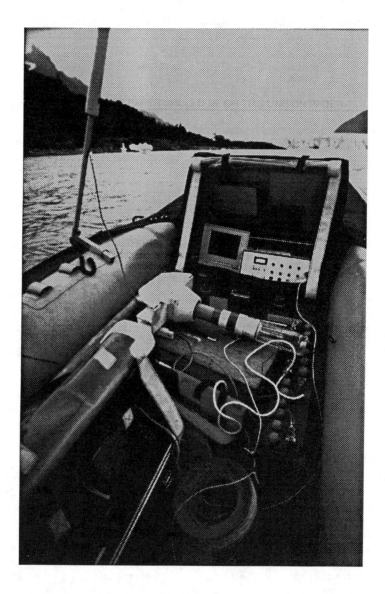

Figure 4. Temperature and conductivity microstructure instrument ready for deployment in Lago Argentino. Instrument in the foreground is a GPS unit.

technique over the shear probe is that the method is immune to vehicle vibration/ accelerations; the signal is only influenced by the vehicle velocities. With proper design, vehicle velocities can be kept much smaller than the fluid velocities, at least for the high wave numbers where the turbulent transport is taking place.

Such a vehicle and laser Doppler system was recently designed and constructed at the Centre for Water Research. It was designed so that the laser Doppler assembly could be coupled with the temperature, conductivity and oxygen microstructure system. When combined, all the parameters, together with the vertical and horizontal components of velocity, can be measured in a volume of about one cubic millimetre.

In order to achieve a high degree of vehicle stability the portable flux profiler was designed to have a compact buoyant body which carries the microstructure and LDA system, separated by a slim metal tube from a smaller pressure vessel containing the vehicle motion sensors and ballast release mechanism (Figs. 5 & 6) A lead mass sufficient to reduce the total vehicle buoyancy to that required for the nominal vertical velocity of $0.1 ms^{-1}$ is attached at the very lower end, and is trimmed on the basis of one or two preliminary casts at the deployment site. A flat drag plate is mounted immediately above the lead mass, and helps stabilize the vertical velocity of the vehicle. Digital signal communication between the LDA and the vehicle motion package is by way of a bidirectional fibre-optic link. A thin Kevlar-reinforced umbilical cable carries microstructure probe power and analogue sensor signals, plus digital data from the LDA signal processor, to the data acquisition system on the surface. A strain relieving construction allows the umbilical to be used to retrieve the vehicle to the surface alongside the deployment vessel, although care must be exercised.

The electrical/optical assemblies are built inside concentric aluminium tubes, with the smaller diameter section used at the upper end so that the distortion of the water column flow field is kept to a minimum. The separation point between the upper pressure housing and main cylinder, and the axial position of the aluminium tubes is such that with the upper housings removed all optical adjustments and alignments can be carried out with the exception of those associated with the He-Ne laser tube. On the detector side, access is available to the preamplifiers, bandpass filters, A/D converters, and a diagnostic connector on the main digital signal processing circuit board.

When the PFP is to be deployed in the falling mode, the buoyancy packs on the sides of the main body are transferred to the vehicle motion package, and replaced by two blocks of lead ballast. The single lead ballast ring on the vehicle motion package is removed, and the vehicle trimmed to be negatively buoyant. At the completion of the descent the vehicle is returned to the surface using the umbilical cable.

Optical Arrangements

Velocity measurement using the Doppler frequency shift technique requires an energy source of known frequency to be directed at the target and the change in frequency of the scattered energy measured.

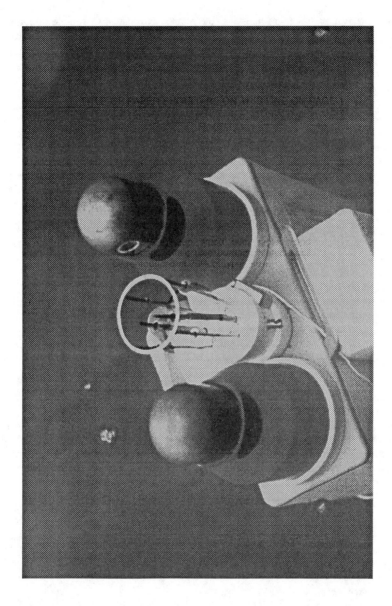

Figure 5. Portable Microstructure Flux Profiler rising through the water column.

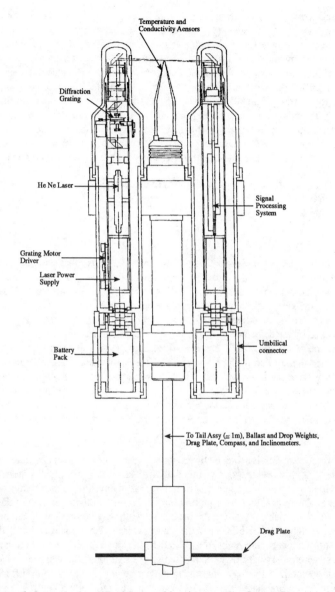

Figure 6. Internal schematic of the portable microstructure flux profiler.

In the heterodyne system, used in the present design, the maximum detector efficiency is obtained when the scattered light and the source sample are co-linear when they impinge on the detector. Although there are various means of achieving this condition, the most convenient is the reference beam method. In this system a reference beam containing only a few percent of the total source energy is directed to intersect the main beam (called the scattering beam), thereby defining the measurement volume. The detector is arranged on the far side of the measurement volume with the reference beam falling directly on it. When a target transits the measurement volume, light from the scattering beam will be scattered throughout a considerable solid angle, but only that portion which travels from the target to the detector along the same path as the reference beam will be sufficiently co-linear to generate significant mixing products.

The velocity derived from a measurement of the Doppler frequency for the configuration used in the PFP is given by

$$V = \frac{\Delta f \cdot \lambda}{2 \sin \theta/2}$$

In the PFP the beam intersection half-angle is 2.53 °, so that with a He-Ne laser of wavelength 632.8 nm the Doppler frequency shift is 139.5 Hz/mms^{-1}, and the design target velocity range of ± 200 mms^{-1} will generate frequency shifts of ± 29 kHz. A pre-shift of ~65 kHz is applied to the reference beams to remove sign ambiguity, so that the frequency range of interest is 36 kHz-94 kHz. It must be noted that the frequency measured by the detector only represents one component of the actual target velocity in the plane of the drawing . By arranging a second pair of beams and detector in an orthogonal plane and which intersect in the same measurement volume, another component of the target velocity can be determined. Similarly, a third pair of beams would allow a total determination of the target motion. In practice, the optical arrangements to measure all three components is complex, but a simple configuration of two reference beams and a single scattering beam will yield two components, and that is what is used in the PFP.

The frequency shift is achieved by passing the beam through a rotating radial diffraction grating. In this case the side beams that arise as a result of normal diffraction grating principles will have a frequency shift imposed on them equal to the product of the beam order and the rate at which the grating lines cross the incident beam, while the main part of the beam which passes directly through the grating (i.e. order n = 0) will be unaltered in frequency. By suitable design of the grating it is possible to control both the distribution of energy between the main (n = 0) and higher-order beams, and the distribution of energy within the high-order beams themselves. Although there are several drawbacks to this method, it does generate the complete suite of beams required for a two-axis forward-scattering heterodyne LDA system in one device at a negligible power cost, and for these reasons was chosen for use in the PFP.

The choice of a He-Ne laser was based on expediency. Previous experience with their use in an LDA system existed, and physically small and robust units were available off the shelf. Their long coherence length means that neither precise equality of optical path lengths for the various beams, nor precise control of the beam waist position, is critical. The need for a small physical size of the laser tube also implies a low optical power output, and the device selected has a rated output of only 0.5 mW. Previous experience with a forward-scattering reference-beam LDA of similar optical configuration indicates that approximately 0.01%-0.05% of the optical energy delivered to the measurement volume will be detected as a Doppler signal, given the short water path and the low turbidity conditions that the PFP is used in. In practice, most of the signal is below 0.02%, but modern microelectronics allows detection with a signal-to-noise ratio high enough to provide satisfactory data for the digital signal processing that follows.

Electrical Arrangements

In the same way that the mechanical design was dictated by physical compatibility requirements with the existing microstructure probes, the electronic systems also had to be integrated with the existing temperature and conductivity microstructure data acquisition equipment. In specific terms, this meant that the combined PFP-microstructure probe had to use the same umbilical cable and data acquisition electronics as the microstructure probes, and to not degrade the excellent low-noise characteristics of those probes.

Because of conductor and bandwidth limitations in the umbilical cable, it was impractical to transmit the raw Doppler signal from the PFP to the surface computers for extraction of the velocity estimates. A highspeed Digital Signal Processing (DSP) system was therefore implemented in the PFP, and a digital data packet containing only the processed velocity estimates, compass heading, and inclinometer data is transmitted over the umbilical every 10 msec.

The electronics in the PFP may be broken down into a number of distinct systems:

- Laser power supply and filter
- Diffraction grating motor drive
- detection of the Doppler signals
- Digital processing of Doppler signals
- Data transmission
- Compass, inclinometer, and weight-release processing
- Power supplies and control

The laser tube used in the PFP is a Melles-Griot 05-LHR-625 He-Ne device which strikes at about 8000 volts and operates at about 900 volts.

The optical frequency shift applied to the two first-order reference beams is directly proportional to the angular velocity of the diffraction grating, and therefore, nominally, to the speed of the motor driving the grating. Without feedback or knowledge of the instantaneous grating velocity, variations in that velocity will be detected in the Doppler signal and erroneously attributed to particle speed variations in the measurement volume. In practice, it is almost impossible to prevent perturbations in the grating velocity due to bearing imperfections and due to the very lightly damped magnetic field of the motor armature. These effects are overcome by tightly controlling the long-term speed variations of the motor electronically, and allowing the doppler signal processor to monitor the short-term variations and make the appropriate corrections to the velocity estimates. In this way it is possible to keep the peak-to-peak velocity noise attributable to the grating to less than 0.5 mms^{-1}.

Selection of the photo-detector was faced with two conflicting requirements besides the more obvious ones of high sensitivity and low noise. The first was the need for a sensitive area significantly larger than the incident beams, so that alignment procedures are eased, and, more importantly, so that beam wander due to refractive index changes in the water path does not result in the beam moving off the detector. The second requirement was for the detector to have the lowest possible capacitance, which would allow the highest gain and bandwidth, and the minimum noise, to be realized while retaining a satisfactory stability margin. These requirements implicitly demand the smallest die size, and a satisfactory compromise requires a careful evaluation of the expected operating conditions. Due consideration of these factors led to the selection of the BPX-65 PIN photodiode for use in the PFP. It is operated in the photoconductive mode, where the reverse bias minimizes the capacitance, and has an active area of 1 mm^2.

Apart from the intrinsic technical suitability of a DSP chip for a particular function, there are many other considerations to be taken into account. These include size, power consumption, support chips required, and technical documentation, as well as software support in the form of assemblers, compilers, simulators, etc. Previous experience with a design using the ADSP2100 from Analogue Devices had been very satisfactory, and its successor, the ADSP2101 appeared to be ideally suited to the requirements of the PFP design. Subsequent experience has only reinforced that opinion.

The standard microstructure probes with which the PFP has to be compatible use a 15-conductor umbilical cable, of which thirteen are used for power and analogue signals, and the remaining two are used as an opto-isolated current loop via which the operator can release the drop-weight. To avoid any changes to the analogue power and signal structures, the PFP intercepts only the weight-release conductors, and uses them for the following functions:

- Data packet request signalling (surface to PFP)
- FSK transmission of data packets (PFP to surface)
- Power on-off control (surface to PFP)
- Weight-release operation (surface to PFP)

An electronic compass and two orthogonally mounted inclinometers are contained in the vehicle motion package and provide data for the decontamination of velocity estimates and the determination of the vehicle attitude and orientation as it moves through the water column. The weight-release motor drive circuit is activated when the interrogate pulses are replaced by a continuously active signal (i.e. the microprocessor is not involved in the drive-circuit operation), and operates the motor until the signal is released. Thus, even if the batteries are exhausted to the point of being unable to power the microprocessor, it may still be possible to release the drop-weight. The time required to release the drop-weight is only of the order of 3-5 seconds, and the absence of interrogation pulses for that time does not affect the operation of the regulated supplies, which only switch off when pulses have been missing for ~20 seconds.

Power is supplied to the vehicle by batteries, housed in the vehicle, for the laser system and weight release, and down the umbilical cable for the temperature and conductivity microstructure system. This arrangement allows independent use of the two systems.

On Board Processing

Considerations of the processing requirements and capabilities indicated that it would be possible to deliver velocity estimates for each of the two channels at about 500 Hz (i.e. every 2 msec), by using overlapped acquisition and processing of the raw data. These estimates are then filtered and transmitted to the surface acquisition computer at a 100 Hz rate.

The A/D sampling rate is set at 256 k samples/sec, and the DMA feature of the serial I/O ports used to acquire blocks of 256 samples from each channel into an internal data buffer. This takes 1 msec, following which processing takes place. The initial processing operations are designed to be completed within the following 1 msec, and are carried out on an 'in-place' basis with the results being delivered to a second area of memory, freeing up the first buffer for acquisition of another 256 data samples. Processing continues on the original data, and is complete prior to the new data becoming available, at which point the cycle begins again. Transmission of processed data packets over the umbilical cable is asynchronous to the raw data acquisition and processing cycle, and is handled under interrupt in response to requests from the surface computer. Communication with the vehicle motion package to acquire new compass and inclinometer data, or to release the drop-weight is scheduled by the requests from the surface computer, but only carried out by the main-line DSP code when the necessary resources (certain registers, and a known amount of free time) become available. The latency is never more than 2 msec. Several possibilities exist for use as master synchronizing signal, such as a timer based on the DSP master clock, or the 100 Hz data-request pulses from the surface computer, but the final selection was to use the clock generated by the encoder on the diffraction grating shaft. This method has benefits in the area of generating velocity estimate corrections.

The individual steps involved in forming a velocity estimate are:
    Raw data acquisition
    Raw data qualification
    Windowing

FFT execution
Unpacking of complex transform results
Peak search and estimation
Correction for instantaneous grating speed
Accept/reject estimate via bandwidth tracking algorithm
Update current tracking parameters
Update current filtered velocity estimate

Applications

Such instruments may be deployed in two distinct ways. First, as vertical up or down profilers. The problem here is that many profiles must be taken in order to form suitable averages (Section 4.4). Second, they can be fixed in space and a time series may be recorded. The averages are then straight forward, but the turbulence may change with time due to the intermittent nature of the flow (81) so once again averaging over similar events becomes necessary. The advantage of the vertical profiling mode is that ℓ_c and e can be obtained so that Frt and Ret may be computed allowing easy classification of the activity of the turbulence; events of similar activity should then be averaged from the various profiles. An example, taken from Lemckert and Imberger (1994), is shown in Fig. 4.4.

Conclusion

As described above, it is clear from the advances made in our ability to measure both mean and turbulent flow field properties that it is now possible to carry out similar measurements in the field as is done in the laboratory. Further, advances in our numerical simulation techniques has also brought these to the point where the influence of turbulence on the mean flow may be estimated. The hydraulic engineer is thus at a most exiting stage of development. He or she is no longer restrained by a lack of tools, but rather the challenge is to exploit the advantages of individual tools and to determine which combination of tools provides the optimum return for the effort invested and the particular question asked.

References

Agrawal YC, A CCD Chirp-Z FFT Doppler signal processor for laser velocimetry, The Institute of Physics 1984; 458-461.
Brumley BH, Cabrera RG, Deines KL, Terray EA, Performance of a broad-band acoustic Doppler current profiler, IEEE J Oceanic Engng, 1991; 16(4):
Caldwell DR, Dillon TM, Brubaker JM, Newberger PA, Paulson CA, The scaling of vertical temperature gradient spectra, J Geoph Res 1980; 85(C4): 1917-1924.
Carter GD, Imberger J, Vertically rising microstructure profiler, J Atmos Oceanic Technol 1986: 3; 462-471.

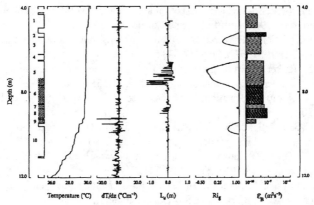

Figure 7(A). A portion of the data collected by the MFP as it ascended 50 m from the operating Argyle bubbler, at 1406 hours on day 90318. The data presented are, (a) stationary segments with the shaded sections indicating the intrusion position, (b) temperature, (c) the corresponding gradient temperature, (d) L_d, (e) Ri_g and (f) ε_B. NB. the strong, one sided gradient signals result from the step structures and the lack of active mixing activity. Also Ri_g values were derived from data smoothed by a 2nd-order Butterworth recursive filter with a 3 db cut off at 1 m length, which corresponds to the corresponds to the typical patch size.

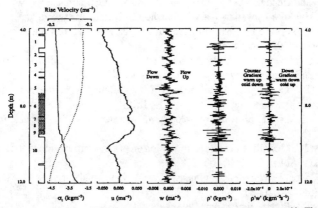

Figure 7(B). Further processed data obtained from the same cast as presented in Figure 7(A). The data displayed is: (a) stationary segments with the shaded sections indicating the intrusion position, (b) (ρ-1000) [solid line] and probe rise velocity [dotted line], (c) horizontal LDV velocity relative to the MFP [u], (d) vertical LDV derived velocity relative to the MFP [w], (e) density anomaly derived by subtracting the monotonised from the recorded density profile [ρ'] and (f) and a point by point estimate of $\rho'w'$.

Cheremisinoff NP (Ed) Encyclopedia of Fluid Mechanics, Vol II, Flow Phenomena and Measurement, 1986, Gulf Publishing Co, Houstan, Texas.

Farmer DM, Clifford SF, Verrall JA, Scintillation structure of a turbulent tidal flow, J Geophys Res 1987; 92(C5): 5369-5382.

Fozdar FM, Parker GJ, Imberger J, Matching temperature and conductivity sensor response characteristics, J Phys Oceanogr 1985; 15 1557-1569.

Gordon RL, Acoustic measurement of river discharge, J Hydr Engng Sci 1989, 115: 825-936.

Gregg MC, Seim HE, Percival DB, Statistics of shear and turbulent dissipation profiles in random internal wave fields, J Phys Oceanogr 1993(a); 23(8): 1777-1799.

Gregg MC, Winkel DP, Sanford TB, Varieties of fully resolved spectra of vertical shear. J Phys Oceanogr, 1993(b); 23(1): 124-141.

Imberger J, Physical Limnology: An Update. In: Limnology Now, A Paradigm of Planetary Problems, R. Margalef (Editor) 1994 (in press).

Imberger J, Head R, Portable flux profiler, to be submitted (1994).

Joyce TM, Gulf stream warm-core ring collection: An introduction, J Geophys Res 1985; 90(C5): 8801-8802.

Lemckert C, Imberger J, Turbulence within inertia-buoyancy balanced axisymmetric intrusions, submitted to J Geophys Res (1994).

Moum JN, Profiler measurements of vertical velocity fluctuations in the ocean, J Atmos Ocean Technol 1990; 7(2): 323-333.

Müller P and Henderson D (Eds) Dynamics of Oceanic Internal Gravity Waves Proce 'Aha Huliko'a Hawaiian Winter Workshop University of Hawaii at Manoa, 1991.

Oldham C, A portable microprofiler with a fast-response oxygen sensor, Limnol Oceanogr. 1994, in press.

Oldham C, The effects of hydrodynamics on the distribution of dissolved oxygen in a lake, PhD Thesis, University of Western Australia, 1993.

Osborn TR, The design and performance of free-fall microstrucutre instruments at the Institute of Oceanography, University of British Columbia, Report No. 30, 1977.

Osborn TR, Vertical profiling of velocity microstructure, J Phys Oceanogr. 1974; 4: 109-115.

Patterson JC, Modelling the effects of motion on primary production in the mixed layer of lakes, Aquat Sci 1991; 53: 218-238.

Pinkel R, Doppler sonar measurements of ocean waves and currents, Mar Technol Soc J 1986; 20(4): 58-67.

Redwood M, Electronic Design, 1993; Nov: 34.

Stanton T, High resolution acoustic Doppler velocity profile measurements, Dept of Oceanography Naval Postgraduate School Monterey CA 93943, 1987.

Urik RJ, Principles of underwater sound, 3rd edition, McGraw Hill Book Company 1983.

Whitehouse I, Imberger J, Current acoustic velocity profiler for coastal monitoring, Proc IABSE Colloquium, Nyborg, Denmark, 1991.

Zdeblick M, Fluidic transistors will change the face of data acquisition, Electronic Design, 1994; Jan: 106.

USING DROP STRUCTURES FOR STREAM GAGING

Bobby J. Brown[1]

Abstract

A long term monitoring program of 15 watersheds, for the purpose of studying channel response to erosion control measures being constructed in north Mississippi, is being conducted by the Hydraulics Laboratory of the U.S. Army Engineer Waterways Experiment Station (WES). As part of the monitoring program, a study is in progress to evaluate the utility of using grade control structures for stream gaging on small ungaged streams. A low-drop structure has been instrumented with continuous stage recording gages to measure the head and tailwater conditions at the structure. Channel thalweg and cross-sectional data have been obtained upstream and downstream of the structure to provide geometric data for calculating hydraulic parameters necessary for developing discharge ratings from the stage data and physical model discharge coefficients. A cableway has been installed upstream of the structure to provide prototype gaging data for comparison with rating curves developed from the physical model discharge coefficients.

Introduction

The design criteria presently being used for the design of low-drop grade control structures in north Mississippi has evolved from field and laboratory studies. The criteria relative to basic dimensions of the low drop structures were developed from model tests at the Agrcultural Research Service (ARS) Sedimentation Laboratory, Oxford, MS (Little and Murphy, 1982), and thus this type of structure is referred to as the ARS-type low drop structure. A low drop is defined as a hydraulic drop with

---

[1]Chief, Hydraulic Analysis Branch, Hydraulics Laboratory, US Army Engineer Waterways Experiment Station, 3909 Halls Ferry Road, Vicksburg, MS 39180-6199.

a difference in elevation between the upstream and downstream channel beds, H; a discharge, Q; and a corresponding critical depth, Y_c, such that the relative drop height, H/Y_c, is equal to or less than 1.0. Conversely a high drop is defined as one with a relative drop height, H/Y_c, greater than 1.0.

A physical model study of an ARS-type low drop structure was conducted at Colorado State University by Water Engineering Technology, Inc. (WET,1990) with the objectives, to evaluate the performance of the structure under flow conditions not investigated by Little and Murphy (1982), and to determine if cost-reduction modifications to the structure were feasible.

Two additional model studies were conducted to develop riprap sizing criteria for the ARS-type low drop structures, one at Colorado State University (Abt et al. 1991) for a drop height of six feet, and one at the U.S. Army Engineer Waterways Experiment Station (Martin et al. 1993) for a drop height of ten feet.

Monitoring Program

Effective evaluation of channel response for the monitoring sites selected in the watersheds is contingent on defining the hydrology that occurs during the evaluation period. Therefore, at least initially, discharge ratings will be estimated by measuring the stages at those sites that have structures and estimating the discharge using model discharge coefficients. A low drop structure on Long Creek, MS, was instrumented in 1992 for the purpose of obtaining field data to correlate with model data, particularly with regard to discharge coefficients.

Low-Drop Structures

Data analysis of the physical model studies indicated that the discharge coefficient was constant up to a submergence of 0.80. Submergence is defined as the ratio of the difference between the tail water elevation and the weir crest elevation (t') and the difference in elevation of the upstream water surface and the weir crest elevation (H_1), i.e. (t'/ H_1). See Figure 1 for a definition sketch.

The discharge coefficient (C_d) is defined by:

$$C_d = \frac{Q}{(B+zH_1)H_t^{3/2}} \quad (1)$$

where:
    Q = discharge. ft³/sec
    B = width of weir, ft

```
z  = lateral side slope of weir, z/1 ft vertical
H₁ = upstream flow depth, ft
Hₜ = total upstream head = H₁ + V²/2g
V  = upstream average flow velocity, ft/sec
t' = difference between tailwater elevation and weir
     crest elevation.
```

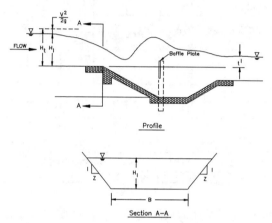

Figure 1: Definition Sketch

Figure 2 shows the relation between the discharge coefficient and submergence for the CSU data. Linear regression curve fitting techniques were applied to the model data with the following results:

$$C_d = 2.5 \qquad\qquad 0 < \frac{t'}{H_1} < 0.8 \qquad (2)$$

$$C_d = 2.354\left(\frac{t'}{H_1}\right)^{-0.328} \qquad 0.8 < \frac{t'}{H_1} < 1.4 \qquad (3)$$

Using the above discharge coefficients with the geometry of the cross-section at the upstream and downstream gage locations, stage-discharge curves (Figure 3) were developed for the head- and tail water cross-sections. Linear regression was applied to the stage-discharge curves to develop equations that were used to translate the stage hydrographs recorded at the site to discharge hydrographs.

HYDRAULIC MEASUREMENTS AND EXPERIMENTATION

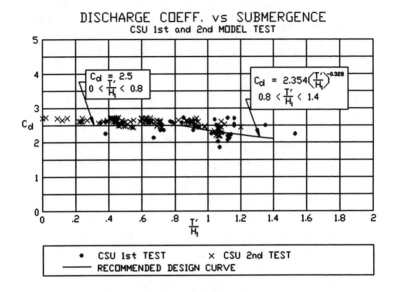

Figure 2: Discharge Coefficients - Model Data

Discharge Measurements

Standard methods of stream gaging are being used at the drop structure to obtain flow rate measurements. Measurements are made by wading at low flows and from a bank-operated cableway described in USGS (1991) at high flows.

Stream gaging data obtained through February 1994 are plotted on Figure 3. Some difficulty in obtaining gaging data has been experienced due to the short duration of runoff events. Furthermore, since the study began, there has not been any tremendously large storm events. Stage data indicate one ungaged event occurred early in the Fall 1991 that was about twice the largest event shown in Figure 3.

Figure 4 shows the comparison between measured discharge and predicted discharge for the 5 events measured at Long Creek.

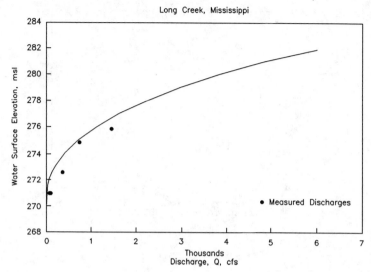

Figure 3. Stage-Discharge Relation - Long Creek, MS.

Summary

A comparison of measured discharge and predicted discharge (Figure 4) from coefficients developed through physical modeling indicates that using the structures as a gaging tool may be a viable alternative for estimating discharge in small ungaged streams. The measured values for the five gaged events are consistently higher than the predicted values indicating that the Long Creek structure is more efficient than the model. However, because the points are few and the range of discharge is small additional data is needed before any conclusions can be made from the study. Stream gaging will be continued at the structure in an attempt to obtain data over the entire range of operating discharges.

Acknowledgements

This work was sponsored by the US Army Engineer District, Vicksburg. Permission was granted by the Chief of Engineers to publish this information.

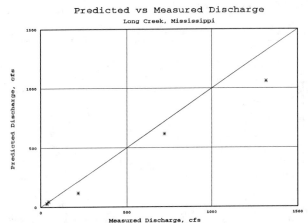

Figure 4: Measured and Predicted Discharge

References

Abt, Steven R., Watson, Chester C., Johns, Derek D., Hamilton, Glenn B., Garton, Andrew D., Florentin, C. Bradley, and Thornton, Christopher I. (1991). "Riprap sizing criteria for ARS-type drop structures," prepared for U.S. Army Engineer Waterways Experiment Station, Vicksburg, MS, under Contract No. DACW39-90-K-0027, by Colorado State University, Fort Collins, CO.

Water Engineering Technology, Inc. (1990). "Value Engineering Of ARS-Type Low Drop Structures," Prepared under Contract No. DACW39-89-d-0001 for U.S. Army Engineer Waterways Experiment Station, Vicksburg, MS.

Martin, Sandra K., Raphelt, Nolan K., Waller, Terry N., Abraham, David D., Brown, Bobby J., Johnson Billy E., Thomas, William A., Hubbard, Lisa C., Watson, Chester C., Abt, Steven R., and Thorne, Colin R. (1993). "Demonstration Erosion Control Project Monitoring," Fiscal Year 1992 Report, Volume 1: Main Text, Technical Report HL-93-3, U.S. Army Engineer Waterways Experiment Station, Vicksburg, MS.

Little, W.C., and Murphy, J.B., (1982). "Model study of low drop grade control structures," *Journal of Hydraulics Division*, ASCE, 108(HY10), 1132-1146.

U.S. Geological Survey. (1991 (Dec)). *Instrument News*, Hydrologic Instrumentation Facility, 20.

A Computerized Open Channel Flow Measurement Device

[1] Stan E. Malinky

Abstract

This is a technical information paper with the intention of introducing a new product. It is not the intent of the author to use this as a marketing tool. However, it is the intent to introduce the AquaCalc to the public. The author will discuss the history of discharge measurements, describe how the AquaCalc fits into the existing surface water records management system, discuss the savings associated with using the AquaCalc over conventional methods and present formatted output from the AquaCalc.

Introduction

The AquaCalc is a microprocessor based, hand held, computer, dedicated to open channel flow measurement. The unit readily adapts to all USGS accepted measurement equipment, wading and suspension, and has built in rating curves for standard and non-standard current meters. The unit is capable of storing transect information that describes characteristics of the cross section to be measured such as location and depths. The user may select from a number of modes of measurement ranging from a completely manual mode, requiring the user to input all of the transect information to a fully automatic mode requiring minimum user input. Once completed with a discharge measurement the operator can press a key and calculate total discharge and mean velocity.

The AquaCalc has storage capabilities for up to nine separate stream cross sections. This storage capability will allow a hydrographer to do a full days work without having to upload the AquaCalc to a personal computer. Back

---

[1] AquaCalc 5000 Project Engineer and Hydrologists, JBS Instruments, 311 D Street, West Sacramento, California 95605

in the office the operator can upload all transect information from the AquaCalc to a personal computer in the form of an ASCII file with the output being very similar to the USGS Measurement Notes Back Sheet. DOS based computer programs are available to convert the informational output from the AquaCalc to a format similar to that of the USGS Measurement Field Notes Front Sheet and to display graphs of Velocity Vs. Depth and Discharge Vs. Depth. Discharge measurement information will now be available in an electronic format for faster quality control of the field notes and to assist the hydrographer with his job as well as providing a source of information for engineers and designers for other types of studies.

Early measurement

Early man started measuring the flow of water prior to 1400 AD when the need for water required diverting steams to farms and cities. Through the centuries as man centralized the population and began farming, water became a necessity to everyday life. From this point on man begun studying the hydrological cycle and the impact's water had on the development of civilization. With ever increasing demands on this resource it became important that the limitations of this resource be understood so that it may be used wisely. Today this understanding is based an a network of stage recording stations and hydrographer's with the capabilities of collecting data with a high degree of proficiency. With the invention of the computer, sophisticated tools have been developed to increase the efficiency of data collection and use of surface water records. Even though computers were implemented into every facet of the surface water industry, discharge measurements were still made using manual methods. The hydrographer still used headsets to count the clicks, used a stop watch to time the rotations and recorded all of this information in the field notes while standing in a flowing river holding a wading rod and other equipment in a rain storm.

The first improvement in discharge measurements occurred in 1978 when the United States Geological survey developed the Current Meter Digitizer (CMD). The CMD replaced the headset and stop watch but the hydrographer was still writing his notes and doing his calculations by hand. In 1988 JBS introduced the AquaCalc, the last missing link in the evolution of computerized surface water records.

First records of stream measurement in the United States

The first records of stream discharge measurements in the United States were associated with gaging stations installed in Brooks County, New York in 1835. The first organized efforts of discharge measurements occurred on the lower Mississippi River in 1838. During this period the discharge measurements were made by sounding the depth with a weight and determining the mean velocity by observing a floating object traveling a known distance. This technique was very inaccurate but other methods were not yet developed.

Early equipment used for stream measurement

The first method used for early discharge measurements was the measurement of stage. Water managers in early civilizations controlled the water in irrigation ditches by observing the stage. Although this method did not yield a numeric value, it did allow the operator to control the quantity of water distributed to a system.

In 1790 a German engineer developed a vane type current meter. This meter was fitted with a counter that totaled the revolutions. At the completion of every measurement the meter was pulled to the surface the revolutions recorded and then reset back to zero. The development of this meter was the forerunner of the systems used today.

In 1867 Daniel Farrand Henry constructed a current meter with telegraphic capabilities. When a current meter is outfitted with this telegraphic electric circuit the signal could then be transmitted to the operator where he could count and record the revolutions. This development eliminated pulling the current meter from the water and resetting it to zero after each measurement.

In 1870 T. G. Ellis developed a current meter with cups revolving around a vertical axis similar to the anemometer. Then in 1882 Price modified Ellis' design and refined it into the present day current meter. The Price meter is the forerunner of the standard current meter used today by the United States Geological Survey (USGS).

From 1870 until 1978 the same method was used to make discharge measurements manually counting and timing the rotations. In 1978 the USGS developed the Current Meter Digitizer a data logger that timed and counted the clicks and when completed displayed the calculated velocity. Even with this improvement the hydrographer still had to write all of the information into his field notes and calculate the discharge by hand. This manual entry and calculation introduced the possibility of error into the system. JBS Instruments, in 1988, developed the AquaCalc 5000, a hand held computer, data logger dedicated to open channel discharge measurement. The AquaCalc replaces the headset, stopwatch, pencil, and notepad and when it is attached to a top set wading rod, stream measurement becomes a one handed operation. At the completion of the measurement the hydrographer can have an instantaneous display of the discharge and mean velocity.

Hand methods for stream measurement

Prior to the USGS' Current Meter Digitizer and the AquaCalc there were many shortcuts developed to reduce errors and simplify the hydrologist's job of calculating the depths, distances and velocities to determine total discharge. One major improvement was the introduction of the USGS' current meter digitizer which eliminated the current meter rating table and increased the efficiency and productivity of discharge measurements made by hydrographers.

For example, when making a measurement the operator counts and times the clicks. The operator must then look up the time and count in a rating table to obtain the velocity. To keep the table readable and still small enough for field use the time in seconds is given from 40 to 70 in one second increments but the velocity is given in the following values 3, 5, 7, 10, 15, 20, 25, 30, 40, 50, 60, 80, 100,150, 200, 250, 300, and 350. When using 40 seconds per measurement and 11 clicks are counted at 40 seconds the operator would have to wait until the 15th rotation to look up the corresponding velocity. This meant waiting 15 seconds and 4 more counts to obtain the velocity value from the table. On an average this can be as much as 20 seconds per measurement and when using 30 verticals this amounts to 20 minutes of increased total time. The normal measurement time for most streams is approximately 45 minutes and an additional 20 minutes is 44 percent more time in the stream. This amount of time saved is based on using the current meter digitizer, when using the AquaCal these savings and more can be realized.

The AquaCalc 5000

The AquaCalc 5000 was developed to eliminate the pencil, headset, stopwatch, calculator, and notepad as well as errors from transcription and simple calculations. The AquaCalc and a laptop computer make a perfect team in the field. The laptop has already been implemented in the field to setup and download continuous recording instruments such as stage recorders. With this type of equipment now readily available and inexpensive, the paper version of field notes made up of the front and back page can be eliminated altogether.

Field notes can now be stored on electronic media saving valuable storage space, time and paper. In the field where the hydrographer fills out his field notes he will now pull up a data entry screen on a lap top computer simulating the front page of his field notes and input all of the information concerning the station. The operator will then, with the same laptop down load the stage information from the continuous recording data logger. After this is all done he will then prepare his AquaCalc by inputting the information pertaining to the discharge measurement. The user will enter general information into the AquaCalc such as MANUAL OR AUTOMATIC DISTANCE & DEPTH, AUTO OR MANUAL POWER CONTROL, ENGLISH OR METRIC MEASUREMENT STANDARD, and the UPLOAD BAUD RATE, as well as more transect specific information such as USER ID, MEASUREMENT TYPE, SOUNDING WEIGHT, HANGER LENGTH, GAGE ID, GAGE HT, STAFF GAGE HT, ESTIMATED DISCHARGE, CURRENT METER ID, CURRENT METER TYPE, and MEASUREMENT TIME. The hydrographer is now ready to start the discharge measurement. At this point the steps for measuring the stream using the AquaCalc are the same as doing it by hand.

Faster more efficient method

Using an AquaCalc can increase the speed and efficiency of discharge measurements in the field and the review of measurement notes in the office. Calculating the velocity instead of using a table can save as much as 10 to 20 minutes per cross section. The average time to make a normal discharge measurement will vary from 45 to 60 minutes and sometimes longer if a sounding is necessary. Savings of just 15 minutes per measurement can increase a hydrographer's productivity by an additional transect per day in many cases. This increase in productivity alone can save a tremendous amount of money in a year's time.

The calculations done by the AquaCalc will also save time and eventually money, no more addition and subtraction mistakes. Once the discharge measurement is completed the hydrographer has total confidence in error free Discharge and Mean Velocity calculations. When calculating the discharge the AquaCalc compares the estimated discharge value with the measured discharge. This percentage indicates to the hydrographer the quality of his measurement. All of this is done when the last station is entered and the Calculate Discharge key is pressed. The hydrographer will no longer spend countless hours sitting on the stream bank or in the truck checking calculations for math errors.

Increased speed and efficiency in the office

Once the measurement is completed it can be uploaded to the laptop in the field or at the office. Quality control of the discharge measurement can now be done on the computer screen. The hydrographer's field notes no longer have to be checked for math errors. If changes are necessary they can be made right on the computer and instantaneously recalculated. The time necessary to review the field notes and check the calculations can be reduced dramatically, as much as one hour of office time can be saved per set of field notes.

This electronic method of handling the field notes can save approximately 1.25 hours per gaging station or measurement. Assuming a hydrographer makes $40,500 with overhead and an office manager $54,000 with overhead, with the office making 500 measurements a year the office can then realize a savings of approximately $15,000 per year. An office with two field personnel and 3 AquaCalcs would see a five year savings of $75,000.

Electronic format of data increases the usefulness of the data

Studies on streams and rivers often require cross section information and velocities. With the data available on computer it can now be readily available to anybody with an interest. With the data in electronic format plotting of the data becomes a very simple task. Cross section area curves showing Depth Vs Velocity can be overlaid on one another showing historic changes to the cross section of the stream or river. Information such as this can be useful to the hydrographer when the stream stage is suspected of shifting. Inexperienced users can also benefit from visualizing the information graphed.

Assists inexperienced users in making solid measurements

New software modules currently being developed for the AquaCalc will assist the beginner in making solid measurements by using a method of prompting. One example of this prompting is if the user exceeds 5% of the estimated Discharge in a subsection a warning will appear in the display. Another flag will warn the user that the wrong meter is being used, either because of depth or velocity. All of this will assist the inexperienced user in learning how to take a solid discharge measurement.

Another well-known problem but not talked about very much is making discharge measurements without leaving the truck. Security measures are being built in to eliminate "Dry Logging" a term that is used when a discharge measurement is made in the warmth of a motel room or truck instead of in the stream. The AquaCal not only saves time and money, it assures good solid measurements.

Personal computers and the AquaCalc

The AquaCalc comes with a simple Upload package for transferring the data from the AquaCalc to either a personal computer in the field for data integrity or to your work station back at the office. The AquaCalc can be uploaded to a PC using the supplied Upload software at baud rates of 300, 1200, and 2400. If the faster baud rate of 4800 is desired then standard commercial communications packages can be used. The file that is sent to the PC by the AquaCalc is a standard ASCII file. Once this file is uploaded onto your computer it can then be loaded into spreadsheet software for further checking and formatting.

JBS has developed a surface water package for in house use. This package uses Microsoft's Excel to check and process the transect records. The AquaCalc ASCII file is opened by Excel and then parsed into the cells on the Data Sheet. The Data Sheet is linked to the Analysis Sheet that has all of the formulas to recalculate and format the record. Two more graphical sheets are linked to the Analysis Sheet that graphically display the data. The graphs display Velocity Vs Depth and Discharge Vs Depth. These graphs provide a lot of information to inexperienced field personnel in evaluating their measurement techniques as well as engineers studying the resource.

TRANSECT	1
TYPE OF CROSSECTION	MAIN CHANNEL
USER ID#	3903
MEAS. SYSTEM	S.A.E. (English)
GAGE ID#	1
GAGE HEIGHT	0.69
STAFF HEIGHT	0.69
METER ID#	1
MEAS. TYPE	SOUNDING
SOUNDING WT.	100
HANGER LENGTH	1.5
METER TYPE	Price AA 1:1
START MEAS AT	LEW
METER CONST. C1	2.180
METER CONST. C2	0.020
METER CONST. C3	2.170
METER CONST. C4	0.030
MEASUREMENT TIME	40
DATE	11/23/93
TOTAL STATIONS	18
TOTAL WIDTH	18.0
TOTAL AREA	12.96
TOTAL DISCHARGE	6.09
EST. DISCHARGE	6.15
PERCENT DIFFERENCE	.97%
MEAN VELOCITY	0.47

Figure 1. Sample Transect Header Information

ST.	DIST	DEPTH	REVS	TIME	COS:VF	LOC	CLOCK	VEL	AREA	Q
1	0.5	0.00	0	0.0	1.00	.6	0:00	0.00	0.00	0.00
2	2.0	0.85	2	65.9	.00	.6	12:24	0.09	1.06	0.09
3	3.0	1.30	0	7.8	1.00	.6	12:27	0.00	0.00	0.00
4	4.0	1.80	13	42.5	1.00	.6	12:29	0.69	1.80	1.24
5	5.0	1.70	11	42.5	1.00	.6	12:30	0.58	1.70	0.98
6	6.0	1.20	12	40.3	1.00	.6	12:31	0.67	1.20	0.80
7	7.0	1.05	11	40.0	1.00	.6	12:33	0.62	1.05	0.65
8	8.0	0.90	12	43.7	1.00	.6	12:34	0.62	0.90	0.55
9	9.0	0.75	9	41.8	1.00	.6	12:36	0.49	0.75	0.36
10	10.0	0.75	7	42.1	1.00	.6	12:37	0.38	0.75	0.28
11	11.0	0.65	8	41.0	1.00	.6	12:38	0.44	0.65	0.28
12	12.0	0.65	6	47.2	1.00	.6	12:39	0.30	0.65	0.19
13	13.0	0.65	7	44.0	1.00	.6	12:41	0.37	0.65	0.24
14	14.0	0.50	8	42.3	1.00	.6	12:42	0.43	0.50	0.21
15	15.0	0.40	5	41.4	1.00	.6	12:43	0.28	0.40	0.11
16	16.0	0.45	2	59.7	1.00	.6	12:45	0.09	0.45	0.04
17	17.0	0.35	3	44.2	1.00	.6	12:47	0.17	0.43	0.07
18	18.5	0.00	0	0.0	1.00	.6	0:0	0.00	0.00	0.00

Figure 2. Sample Transect Back Page Information

JBS Discharge Calculation for use with "AA & Pygmy Current Meters"

Date of Measurement:	11/23/93			Processed:	14-Mar-94	
Stage Gage Height:	0.69					
Estimated Discharge	6.15 Difference					
Total Discharge:	6.09 0.99%					
Mean Velocity:	0.47					
Stream Name:	Griswold Creek					
Transect Number:	1		Current Meter ID#	1		
Measurement Units:	S.A.E. (English)		User ID#	3903		
Staff Gage Height:	0.69		Stage Gage ID#	1		
Total Stations:	18		Current Meter Type: Price AA 1:1			

Station	Distance (ft)	Depth (ft)	Revolutions	Time	Cosine Error	Velocity (ft/sec)	Area Sq-Ft	Flow (cu ft/sec)
EOW	0.5	0.00	0	0.00	1.00	0.00		
2	2.0	0.85	2	65.90	1.00	0.09	1.06	0.09
3	3.0	1.30	0	7.80	1.00	0.00	1.30	0.00
4	4.0	1.80	13	42.50	1.00	0.69	1.80	1.24
5	5.0	1.70	11	42.50	1.00	0.58	1.70	0.99
6	6.0	1.20	12	40.30	1.00	0.67	1.20	0.80
7	7.0	1.05	11	40.00	1.00	0.62	1.05	0.65
8	8.0	0.90	12	43.70	1.00	0.62	0.90	0.56
9	9.0	0.75	9	41.80	1.00	0.49	0.75	0.37
10	10.0	0.75	7	42.10	1.00	0.38	0.75	0.29
11	11.0	0.65	8	41.00	1.00	0.45	0.65	0.29
12	12.0	0.65	6	47.20	1.00	0.30	0.65	0.19
13	13.0	0.65	7	44.00	1.00	0.37	0.65	0.24
14	14.0	0.50	8	42.30	1.00	0.43	0.50	0.22
15	15.0	0.40	5	41.40	1.00	0.28	0.40	0.11
16	16.0	0.45	2	59.70	1.00	0.09	0.45	0.04
17	17.0	0.35	3	44.20	1.00	0.17	0.44	0.07
18	18.5	0.00	0	0.00	1.00	0.00	0.00	0.00
0	0.0	0.00		0.00	0.00	0.00	0.00	0.00

Figure 3. Spreadsheet Output

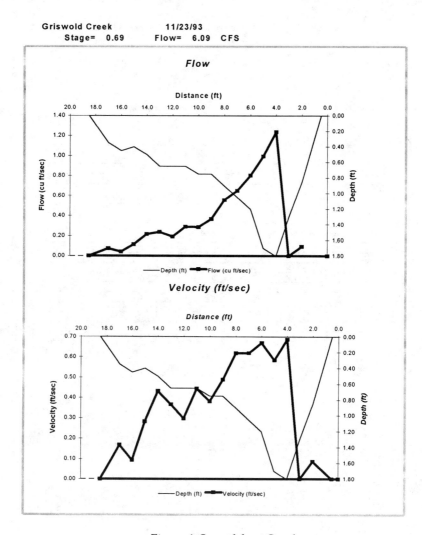

Figure 4. Spreadsheet Graphs

Reference

Chow, Ven Te, 1964 Handbook of Applied Hydrology, McGraw Hill Book Company, New York, New York

DEMONSTRATION PROJECT FOR SCOUR INSTRUMENTATION

J.D. Schall M. ASCE[1], W.R. Ivarson, M. ASCE[2], and T. Krylowski[3]

Abstract

A Federal Highway Administration (FHWA) Demonstration Project was developed to facilitate technology transfer of instrumentation related research to the highway industry. The objective of the Demonstration Project is to promote at the state and local level the use of new and innovative equipment to measure and monitor scour at bridges. This paper summarizes the instrumentation that will be demonstrated in both workshop and field settings.

Introduction

Bridge scour has become a topic of nationwide interest in recent years, and significant progress has been made in all aspects of the bridge scour problem. One area where concentrated research has occurred is scour instrumentation and development of techniques suitable for bridge scour monitoring and measurement. Historically, little field data has been collected on bridge scour, in part due to lack of suitable instrumentation. This lack of real world scour data has subsequently been a limitation in developing better analytical tools for scour prediction.

---

[1] Vice Pres., Resource Consultants & Engineers, Inc., P.O. Box 270460, Ft. Collins, CO 80527

[2] Vice Pres., Collins Engineers, Inc., 165 North Canal Street, Chicago, IL 60606

[3] Structural Engineer, Federal Highway Administration, 400 Seventh Street, SW, Washington, D.C. 20590

Furthermore, it is now recognized that accurate, reliable scour instrumentation can be an integral part of a scour monitoring program and may be a suitable countermeasure at a scour critical bridge. Therefore, the research and progress that has been made in bridge scour instrumentation will be a tremendous asset once the technology that has been developed is integrated into the highway community.

Objective of The Demonstration Project

To facilitate the technology transfer of instrumentation related research to the highway industry, particularly those responsible for inspection and maintenance operations, the Federal Highway Administration (FHWA) developed a demonstration project on scour monitoring and instrumentation. Unlike traditional training or many short courses, FHWA Demonstration Projects are specifically designed to incorporate physical demonstrations and/or hands-on experience to support and enhance classroom or lecture presentations. The objective of this Demonstration Project is to promote at the state and local level the use of new and innovative equipment to measure and monitor scour.

Project Organization

The Demonstration Project presents a two day workshop that will include instructional modules on fixed instrumentation, portable instrumentation, geophysical instrumentation and positioning systems. For each type of instrumentation an introductory lecture will be provided followed by a laboratory based equipment demonstration. An optional day of instruction may be developed that provides field demonstration. An Executive Overview at the beginning of each workshop will preview the entire Demonstration Project for both participants and others who may want a broad overview without having to attend the entire two days. The following paragraphs summarize the instrumentation that will be presented in the workshop.

Fixed Instrumentation

Fixed instrumentation refers to scour measuring or monitoring devices that are attached to the bridge. Fixed instrumentation is utilized when frequent measurements or regular monitoring are desired. The primary limitation of fixed instrumentation is that the maximum scour may not occur at the location where the instrument is installed, as a result of scour or stream instability conditions. Other limitations include maintenance of equipment exposed to adverse environmental factors,

SCOUR INSTRUMENTATION

and loss due to vandalism, floods, debris, etc. Fixed instrumentation readings should be independently confirmed by ground truthing with some form of observation or independent measurement technique.

Fixed instrumentation can be categorized as sonar based devices, sounding rods/devices, buried/driven rods and other buried devices. Sonar based devices that will be demonstrated include relatively low-cost instruments based on "fish-finder" type devices and survey grade devices that provide greater accuracy, multiple transducer operation and other enhancements. In both cases, the instrument can be connected to a data logger to provide a continuous record of scour conditions. Sonar transducers can be mounted directly to the bridge pier, or alternately a "bridge deck serviceable" installation will be demonstrated that permits transducer removal for cleaning and maintenance without boat or diver support.

Sounding rods in a fixed instrument application typically involve a sliding rod fastened to the bridge with a foot pad resting on the channel bottom. As scour develops, the rod drops downward. Both manually and automatically operated rods will be demonstrated. Automated sounding rods typically have some type of linear displacement measuring device to track rod location, which may or may not be attached to a data logger.

Buried/driven rods include all sensors and instruments supported by a vertical member such as a pipe, rail or column which could be driven or placed vertically in the bed. Scour measurement along the buried/driven rod can be by mechanical, electrical or electromechanical means. Buried/driven rods to be demonstrated include a magnetic sliding collar device, a conductance device and a sensor-based device.

Other buried devices consist of sensors that are buried in the bed at various elevations that become exposed as scour progresses. The simplest of these devices include chain or painted rocks buried in the bed, and the most sophisticated is probably a device based on hydrophone techniques developed from seismic exploration.

Portable Instrumentation

Portable instrumentation can be readily moved from one bridge to another, which facilitates real time flood and post flood measurements. Portable instruments also allow tracking and mapping thalweg movement and/or locations of maximum scour in the bridge cross section. The limitations of portable instrumentation include only monitoring discrete points in time and space, and labor intensive operation that often include the need for special work platforms such as boats, trucks or truck mounted articulated booms. Portable instrumentation can be classified as mechanical or physical probes and sonar based devices.

Mechanical or physical probes that will be demonstrated include sounding poles, lead lines, and diver probing. Sounding poles or lead lines can be used from the bridge, from a boat or sometimes by wading. The effectiveness of these devices can be limited by water depth and velocity.

As with fixed sonar-based instruments, portable sonar instruments can be low-cost "fish-finder" type devices or survey grade instruments. A low-cost sonar setup powered by a small battery, with the transducer suspended by a rod or from a small float, will be demonstrated. This instrument has potential widespread use by inspectors and maintenance personnel. Survey grade instruments, such as those normally deployed in conventional hydrographic surveying, will also be demonstrated.

Geophysical Techniques

Geophysical techniques provide not only information about the water-bed interface, but also subbottom information. Geophysical devices are typically based on instruments that measure reflected or transmitted acoustic or seismic and radar energy. Geophysical devices can provide high precision, but accuracy in interpretation and inferences drawn from measurements depend on the experience and expertise of the interpreter. Devices to be demonstrated include low frequency sonar instruments, a CHIRP acoustic profiler and a ground penetrating radar instrument.

Seismic or acoustic systems represent a well established method for the acquisition of subsurface data in water covered environments. They are based upon the generation and transmission of soundwaves through the water column and into bottom sediments as well as the reflection of those sound waves from boundaries between sediment layers. Physical parameters of bottom sediments which affect the transmission of sound waves are the densities and seismic P-wave velocities of the materials. Changes in sediment type, i.e., clay to sand, sand to rock, loose sand to dense sand, etc., can be detected and measurements of the thickness, structure, and distribution of lithologic formations can be accomplished if these different strata provide sufficient impedance (the product of density and seismic velocity) contrast.

Most acoustic systems operate in a frequency range of between 1 and 200 kHz. Generally the lower the frequency, the greater the penetration. However, the resolution of the method is limited to one-half the wavelength of the transmitted energy. Traditionally equipment that operates at approximately 3 kHz has been very popular for oceanographic survey applications. This equipment has also been used for the detection of infilled scour holes and will be demonstrated in the workshop.

A more recent development in acoustic equipment has been the application of swept band frequency modulated energy pulse instead of the relatively constant frequency pulse. The purported benefits to such a system are great penetration because of the low frequency part of the pulse and higher resolution of surface features because of the high frequency part of the pulse. This equipment will also be demonstrated in the workshop.

Ground penetrating radar (GPR) has been used for geophysical exploration in low on non-conductive environments for many years. It is also well known for detection of voids and other features beneath and within pavements and other structures. In clean water situations with granular bottom sediments, GPR has proven to be effective in detecting density differences in sediment strata and therefore infilled scour holes. It will be demonstrated as another useful tool in the detection and monitoring of bridge scour phenomenon.

Positioning Systems

In order to monitor scour at bridge sites it is necessary that the scour phenomenon be accurately located in time and space. For local scour it may be adequate to reference the scour measurement to the bridge structure. For contraction scour and overall channel stability in the vicinity of the bridge more precise positioning systems are required. The positioning systems typically used are those developed for hydrographic surveying, which can be categorized as visual/manual techniques and electronic systems. The current state-of-the-art of electronic positioning is changing rapidly because of rapid advances and falling prices in global positioning systems (GPS). The specific systems to be demonstrated will include an automated range-azimuth system and GPS.

Automated horizontal range-azimuth systems are presently the systems of choice for project condition type surveys. A condition survey is generally what is needed for bridge scour monitoring and evaluation (horizontal accuracy of ±6 m and vertical accuracy of ±0.3 m). The typical system consists of a ship mounted electronic echo sounder and data logger which is tracked by a shore station of a theodolite and electronic distance measuring devise (laser). The shore and ship stations are linked by radio so that all position and depth information are recorded on the ship board data logger. Depth sounding density of 1.5 m on center each way is within the capability of these systems.

Dynamic positioning (moving survey vessel) with GPS within the required accuracy is presently under development. Expensive systems are expected within the time frame of this demonstration project. Affordability of these systems remains an open question.

Conclusions

The Demonstration Project will provide an overview of a complete spectrum of equipment for measuring and monitoring scour at highway bridges. In addition to lecture presentations on the basic theory, operation and/or installation of specific instruments, participants will get hands-on experience with the equipment in a laboratory setting. At some locations, field installations will be available to provide additional experience based on prototype installations. As a preface to the equipment oriented lectures and demonstrations, a brief overview of scour and stream stability concepts will be provided, and the components of a monitoring program will be discussed as a concluding session.

Laboratory and Field Evaluation of Acoustic Velocity Meters

Tracy Vermeyen[1], A.M. ASCE

Abstract

A project is currently underway to evaluate the performance of 27 acoustic flowmeters used at Hoover, Davis, and Parker Dams on the lower Colorado River. Field surveys and laboratory testing are being used to evaluate and enhance the performance of the chordal-path acoustic velocity meters. A hydraulic model and a laser doppler anemometer are being used to determine velocity distributions for the nonstandard flowmeter installations.

Introduction

The material for this paper is part of a study requested by Reclamation's (U.S. Bureau of Reclamation) Lower Colorado Regional Office. The purpose of the study is to improve the flow measurement at the major dams along the lower Colorado River, namely Hoover, Davis, and Parker Dams. This study is only one of many being conducted in support of the LCRAS (Lower Colorado River Accounting System) program. LCRAS is a water management computer program which will allow Reclamation to better utilize water resources in the Lower Colorado River basin. LCRAS will be used to estimate water consumption by tracking consumptive use by: crops and phreatophytes, reservoir evaporation, municipal and industrial users, and groundwater recharge.

In an effort to improve the accuracy of flow measurement at Hoover, Davis, and Parker Dams, a two stage study was initiated. The first stage was to evaluate the existing flow measurement system which consists of AVMs (acoustic velocity meters) with four or eight acoustic paths. A field survey was conducted to determine if all 27 AVM installations conformed to ANSI/ASME Standards and ASME's

---

[1]Hydraulic Engineer, U.S. Bureau of Reclamation, D-3751, PO Box 25007, Denver, CO 80225.

Performance Test Code 18. The second stage was to determine if non-standard AVM installations were performing to manufacturer's specified accuracies of ±0.5 percent of the true discharge. To verify the flowmeters integration techniques when applied to an asymmetrical velocity distribution, a physical model was used to determine the penstock velocity distributions at the AVM measurement section. Model study results will be used to establish error bounds on discharge measurements and modifications which will reduce the discharge errors.

Field Surveys

To determine the accuracy of flow measurement at Hoover, Davis, and Parker Dams field surveys were conducted in September, 1992 to document and review: AVM equipment, AVM system parameters, as-built drawings, and perceived system performance. Each of the 27 AVM sites and installations was evaluated using ANSI/ASME Standard MFC-5M-1985, entitled *Measurement of Liquid Flow in Closed Conduits using Transit-Time Ultrasonic Flowmeters*. Likewise, ASME's Performance Test Code for Hydraulic Turbines (ASME PTC 18-1992) was used in evaluations because it is more stringent than the ANSI/ASME standard. Surveys at Hoover, Davis, and Parker Dams resulted in a large amount of site specific data and personal opinions as to how the AVM systems were performing. Survey information is summarized as follows:

Hoover Dam - Eighteen AVMs at Hoover Dam were installed over the period of 1989 to 1991. A review of AVM equipment, system parameters, and as-built drawings at Hoover Dam revealed that all AVM installations were according to standards and were configured properly. However, there were a few installations which had numbers transposed in the path length and/or angle entries. These types of setup errors can result in significant discharge errors, but are also easily corrected.

Davis Dam - Five AVMs were installed in 1989. A review of AVM equipment, system parameters, and as-built drawings for Davis Dam revealed that all five AVM installations were nonstandard because of inadequate length of straight pipe upstream of the meter section - 10 pipe diameters is the recommended minimum length in the ANSI standard. The amount of straight pipe upstream from the meter section ranged from ½ to 1-½ diameters, for each of the five, 6.7-m-diameter penstocks. However, these locations could not be avoided because of short penstocks. All AVMs were installed just upstream of the turbine scroll cases to maximize the length of straight pipe upstream. Because of short penstock lengths and bends upstream, cross flows (flows with nonaxial velocity components) were anticipated. Crossed path AVMs are used in difficult installations to eliminate cross flow errors. The shortest of the five penstocks was fitted with a crossed path AVM system. It should be noted that ASME's PTC 18 requires installation of two, four-path measurement planes, and that the intersection of the two planes shall be in the plane of the upstream bend. The crossed path AVM installations at Davis Dam do not meet the above criteria.

Parker Dam - Four AVMs were installed in 1989. A review of AVM equipment, system parameters, and as-built drawings at Parker Dam revealed that all four AVM installations were nonstandard because of an inadequate length of straight pipe upstream of the meter section. The length of straight pipe upstream from the meter section ranged from ½ to 6 pipe diameters, for each of the four, 6.7-m-diameter penstocks. However, these lengths could not be avoided because of short penstocks. Like Davis, all AVMs were installed just upstream of the turbine scroll cases to maximize the length of straight pipe upstream of the meter section. Therefore, two of the four penstocks were fitted with crossed path AVM systems, including the shortest penstock. The crossed path AVM installation at Parker does not meet ASME's PTC 18 requirement on acoustic path orientation with respect to the upstream bend.

In general, AVM system operators felt their systems were operating satisfactorily. However, our interviews indicated that there was a disparity in knowledge levels among system operators. There were varying degrees of expertise in system testing and troubleshooting depending on maintenance history. To alleviate this problem it was recommended that a training course be given to all system operators. It was also apparent that an experienced electronics technician is necessary to effectively operate and maintain an AVM system. We also recommended developing a database to log maintenance and repair data, as well as system parameters and error logs.

Some interesting equipment problems were identified during the surveys. At Hoover and Davis Dams, when acoustic transducers were removed for cleaning or when the penstock was dewatered, there was a large number of transducers which failed. Transducer failures have been prevented by keeping transducers submerged during maintenance operations. Another common concern was with field survey accuracies of path angles and lengths, and cross-sectional areas of the penstocks. These parameters are very difficult to measure accurately, and must be determined to a high degree of accuracy. Therefore, operators should be comfortable with the survey accuracy prior to going on-line with an AVM system.

AVM Data Analysis - Individual path velocities and discharge values were collected for the crossed path AVMs at Parker and Davis Dams to determine the errors associated with cross flows. Figure 1a contains a typical sample (~ 110 measurements taken over 2 minutes) of data collected from Parker penstock number 3 for a 50 percent gate opening. This penstock is equipped with a crossed path AVM, so a total of eight path velocities and two discharges were measured. Paths 1 and 4 are the upper and lowermost acoustic paths, respectively. Paths 2 and 3 are located in between paths 1 and 4. Analysis of the path velocities indicated that there is very little cross flow component, because velocities measured on similar acoustic paths agreed very well (e.g, path 1, planes 1 and 2 in fig. 1a). Likewise, if there was a strong cross flow component, flow-1 and flow-2 would be substantially different, but figure 1b shows there was good agreement between flow-1 and flow-2. However, the four path velocities indicated that the profile is distorted toward the penstock's invert. This is evident from the difference between the velocities (0.7 m/s) measured along paths 1 and 4, as shown in figure 1c. This distorted profile is caused by a 30° vertical bend two pipe diameters upstream of the AVM measurement section.

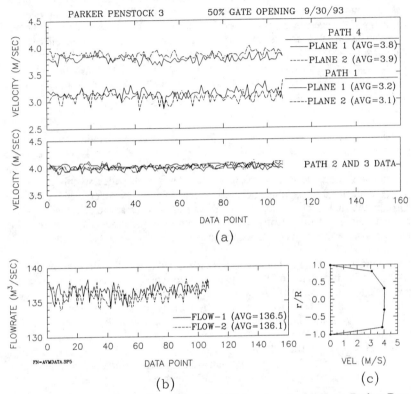

Figure 1. Raw data from the cross path AVM system installed on Parker Dam penstock no. 3.

A published error analysis by the AVM manufacturer (Lowell and Hirschfeld, 1979) does not adequately address the error related to the integration of an asymmetrical velocity distribution. The question remains, Can the integration method employed in the AVM discharge calculations accurately integrate an asymmetrical velocity distribution? To quantify the magnitude of the integration errors and to determine the velocity distributions a study was initiated which included hydraulic model studies of penstocks at Parker and Davis Dams. Likewise, a field demonstration comparing strap-on acoustic flowmeters to four-path AVMs was performed at Davis Dam.

Strap-on Acoustic Flowmeters - At Davis Dam, AVM discharges were compared to discharges measured using a portable, strap-on acoustic flowmeter. The strap-on flowmeter installation consisted of one diametral path. This comparison was performed as a demonstration of a strap-on meter, it was not intended to be an evaluation of the four-path AVMs. Accuracies for this uncalibrated discharge measurement application are normally expected to be within ±3 percent. However, discharges were measured for a full range of wicket gate openings and were consistently 6 percent lower than the four-path AVM measurements. This difference was likely attributed to the penstock's unknown coal-tar lining thickness, which affects the traveltime measurements, and the asymmetrical velocity distribution. While strap-on acoustic flowmeters cannot claim the installed accuracy of a four-path AVM, their cost is significantly less, and they are suitable for many discharge measurement applications. In addition, Taylor (1987) has suggested using strap-on flowmeters to determine the severity of asymmetric velocity distributions. Taylor's procedure may be useful in determining whether a crossed path AVM installation is warranted.

Laboratory Studies

In order to establish error bounds on the AVMs discharge measurement, it was necessary to define the velocity distribution for penstocks with nonstandard AVM installations. A physical model was constructed in Reclamation's Hydraulic Laboratory and is being used to study representative penstocks at Parker and Davis Dams.

The Model - A 1:22.9 scale hydraulic model is being used to determine the velocity distributions in penstocks at Davis and Parker Dams. The model includes features from the trashrack and inlet transition down to, but not including, the turbine scroll case. Measured velocity distributions will be analyzed to determine the deviation of the actual velocity distribution from a fully developed, turbulent velocity field. Model data will also be used to establish alternate measurement planes which minimize cross flow and/or integration errors. This model study is not intended to determine a calibration factor because of Reynolds number limitations. Its purpose is to determine the errors related to AVM integration techniques applied to an asymmetrical velocity distribution.

Velocity Measurements - Point velocities were measured using a fiber-optic LDA (laser doppler anemometer) system mounted to an automated one-dimensional traversing system (fig. 2). An LDA measures fluid velocity by determining the oscillation frequency of light pulses reflected from particles in the fluid as they pass through the probe volume. The probe volume is created where the two laser beams cross. Velocity data were collected at 12 locations along a radial path, for 24 equally spaced radii on the pipe section. This resulted in 288 point velocity measurements. Normally, each LDA reading was taken as the mean of 500 or more instantaneous velocity measurements. Strict signal validation criteria are used to assure data quality.

Velocity measurement locations were determined by dividing the pipe area into a center circle and 11 annuli, all of equal area. Velocities were measured at the midpoint of each annulus. These velocity measurements were later used in a velocity-area integration method used to calculate discharge. This integration technique is commonly referred to as the tangential method.

Figure 2. Photograph of the fiber-optic LDA probe, saddle mount, and single axis positioning table with stepper motor.

Refraction Through Wall of Penstock - Refraction at optical interfaces changes the laser beam paths, thereby changing the intersection point and the angle between the beams. The refractive properties for a cylindrical surface varies depending on whether the velocity component is being measured in the axial, tangential, or radial direction. For this study, we intended to simultaneously measure the axial and tangential velocity components (recognizing they would be at different radial locations) using a two-dimensional LDA system. Unfortunately, two-dimensional LDA processing software requires that the probe volumes for both laser beam pairs be coincident, which is not possible because of the different refractive properties through a curved surface. As a result, only axial velocities were measured. For measurement of axial velocity the optical system is oriented so that both laser beams are in a plane which passes through the axis of the cylinder, and the bisector between the beams is at a right angle to the axis. For this case, there is only refraction in the axial direction and the refracting surface will be perpendicular to the beam angle bisector, as with a flat window. For this orientation, only the beam intersection location is affected by refraction. However, because of small irregularities in the pipe wall thickness, a calibration was performed to determine the actual beam crossing location for several radii. The calibrated position varied only slightly from the theoretical position. Calibration data were used to develop a relationship between traverse system movement and beam intersection location in the penstock. A complete discussion on refraction and determining the intersection location is beyond the scope of this paper, but can be found in a text by Durst, et al., 1976.

Laser Mounting System - To efficiently collect velocity data on 24 different radii the LDA probe had to be easily rotated while keeping the laser beams in a plane perpendicular to the pipe's axis. This was done by machining a saddle-type mount with a slightly larger outside diameter than the model penstock. A plate and positioning table are attached at a 90° angle to the face of the saddle mount. A single axis positioning table was used to accurately position the LDA probe at the 12 different measurement radii. The positioning table consisted of a stepper motor system to move the LDA probe (resolution is 0.1 mm per step). The stepper motor was controlled by a personal computer and manufacturer supplied software. For this application, a heavy duty stepper motor with adequate holding torque (1 N·m) was necessary to maintain position under the probes weight. The LDA's software combined with a positioning system is capable of automatically collecting velocity profile data. However, our data collection was done manually because the LDA system parameters had to be adjusted as the sampling position was changed.

Discharge Measurements - Flows entering the model were measured using the laboratory's permanent bank of venturi meters. The venturi's are calibrated annually using a weigh tank; the 1993 calibrations for the 15-cm and 20-cm venturi meters used in this study were accurate to within ± 0.35 and ± 0.27 percent, respectively.

Integration of the measured velocity distribution was used to verify the quality of the LDA velocity measurements. Two velocity-area methods were used to calculate the discharge, the tangential method and the log-linear method. A thorough discussion of both methods is presented in a paper by Winternitz and Fischl, 1957. In general, velocity-area integration of the measured velocity distributions were within ± 1 percent of the discharges measured with the venturi meters.

Evaluation of the Acoustic Velocity Meters

Originally, the actual velocity distributions in the model penstock were to be used in evaluating the AVM's integration technique. This required using a commercially available, PC-based software package to create a gridded model of the velocity distribution. Model data were used in an interpolation scheme to assign velocity values to each of the grid nodes. Once the velocity distribution model was generated, a utilities package was used to determine the velocity profiles along the four acoustic

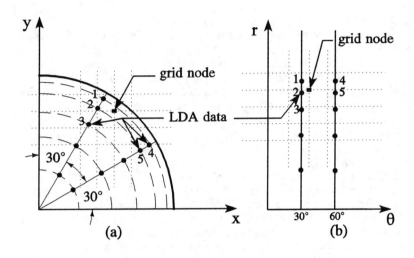

Figure 3. (a) search method for (x,y) coordinates, (b) search method for (r,θ) coordinates.

paths (projected onto the pipe cross section, $V = V_{path} * \cos\phi$, where ϕ is the angle between the acoustic path and pipe axis). We could then perform computer simulations of the AVM integration technique for acoustic paths as a function of path rotation. It was believed that this technique could be used to determine the optimum path position to minimize errors associated with cross flow and asymmetric velocity distributions. However, errors in the velocity distribution model were introduced by the software's interpolation search method.

When a gridded velocity distribution was numerically integrated to determine the discharge, errors of 2 to 4 percent were discovered. During discussions with the software developer it was decided that interpolation errors were caused by two factors: (1) limitations in the search method employed when selecting data points used to interpolate values for the grid nodes, and (2) large velocity gradients near the pipe wall. The problem with the search method is that it is based on a rectilinear coordinate (x,y) system. While our velocity data were collected in polar coordinates (r,θ). Problems were primarily confined to interpolation near the pipe wall. For example, the search method uses the four closest data points (points 1, 2, 3, and 5 in fig. 3a) to estimate a grid node value located between two radii. Consequently, the grid nodes located between two diameters are skewed to larger values with the inclusion of velocities measured at point 3. Ideally, the grid node value should be estimated using the points 1, 2, 4, and 5. This interpolation problem results in an

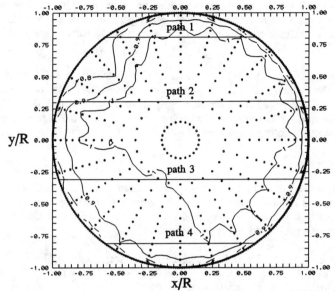

Figure 4. Isovel plot for the gridded velocity distribution. Notice how the interpolation creates a scalloped shape for isovel equal to 1.0.

overestimation of the discharge. An example is the distorted shape of the isovel equal to 1 in figure 4.

Another approach taken was to leave data in polar coordinate form (fig. 3b), but this resulted in the same search method problems because velocity data are spaced closer together near the pipe wall. Data were collected at irregular spacing as required for using velocity-area methods to calculate discharge.

Discussions with other gridding software vendors have not resulted in a product that will work with our velocity data set. However, there is an algorithm which is available that does not use a rectilinear search/interpolation method to estimate grid nodes. This gridding technique is called convergent gridding and is described in an article written by M. A. Haecker, 1992. Because of budget constraints this algorithm has not been tested.

To date, our method of analysis is focused on using mathematical functions which describe asymmetrical velocity distributions, as described by Salami, 1972. Mathematical velocity distributions which are similar to those measured in the hydraulic models will be studied. The true discharge can be determined by an exact integration of the velocity function. Likewise, the velocity profile along each acoustic path can be calculated and numerically integrated to determine the average path velocity. The average path velocities are then used in the flowmeter's integration equation to calculate discharge. A comparison of the true and calculated discharges will be used to establish the error bounds associated with the asymmetric velocity distributions.

Results from these studies will be used to determine if the current AVM installations can meet the discharge accuracy requirements of the LCRAS. If not, a potential solution would be to add more acoustic paths to the AVM system. Taylor (1987) showed that integration errors can be reduced to ± 0.2 and ± 0.1 percent using six and eight acoustic paths, respectively.

Conclusions

Field evaluations of AVM installations were valuable in identifying nonstandard installations and system parameter errors. A method for estimating integration errors associated with asymmetrical velocity distributions is needed to estimate the total uncertainty of an AVM discharge measurement.

Acknowledgements - Albert Marquez, Civil Engineer, Lower Colorado Region, U.S. Bureau of Reclamation, assisted in the field data collection and analysis and is the project manager for this study.

References

Durst, F., Melling, A., and Whitelaw, J.H., "Principles and Practice of Laser-Doppler Anemometry," *Academic Press*, 1976.
Fisher, S.G. and Spink, P.G., "Ultrasonics as a Standard for Volumetric Flow Measurement," Modern Developments in Flow Measurement, C.G. Clayton, Proceedings of International Conference, September 1971 at Harwell U.K., London, *Peregrinus*, 1972.
Haecker, M.A., "Convergent Gridding: A new approach to surface reconstruction," *Geobyte*, June 1992.
Lowell, F.C., Jr. and Hirschfeld, F., "Acoustic Flowmeters for Pipelines," *Mechanical Engineering*, October 1979, pp 28-35.
Salami, L.A., "Errors in Velocity-Area Methods of Measuring Asymmetrical Flows in Circular Pipes," Modern Developments in Flow Measurement, C.G. Clayton, Proceedings of International Conference, September 1971 at Harwell U.K., London, *Peregrinus*, 1972.
Taylor, J.W., "Prototype Experience with Acoustic Flowmeters," American Society of Civil Engineers, *Proceedings of WATERPOWER '87*, 1987.
Winternitz, F.A.L., and Fischl, C.F., "A Simplified Integration Technique for Pipe Flow Measurement," *Water Power*, June 1957, P. 225.

Mapping 2-D and 3-D Velocity Components in Circular Conduits Using an Electromagnetic Current Meter and a 5-Hole Pitot Probe

Joseph J. Orlins, E.I.T.[1]
Lawrence J. Swenson, P.E.[2]

ABSTRACT

The techniques presented in this paper facilitate the visualization of the two-dimensional and three-dimensional flow fields at arbitrary sections in a conduit. They have been used to evaluate both model and full-scale fish screening designs.

Velocity measurements were made in laboratory-scale and full-scale hydroelectric penstocks using a two-component electromagnetic current meter and a three-component five-hole pitot probe. The axial and tangential velocity components at a circular section normal to a pipe axis, and the three-dimensional velocity components at an elliptical section inclined with respect to the pipe centerline were measured. Data processing was accomplished using a combination of custom personal-computer based data acquisition programs and spreadsheet templates. Isovel and vector plots were created using commercially available software packages.

INTRODUCTION

Hydroelectric utilities are considering the use of screens in penstocks and turbine intakes to divert and bypass downstream migrating fish. The screens may be placed normal to the axis of the conduit or at an oblique angle. The criteria for evaluating screen designs include the uniformity of the flow field approaching the screen, as

---

[1] Associate Member, ASCE, Laboratory Engineer, ENSR Consulting and Engineering, 14715 NE 95th Street, Redmond, WA 98052.

[2] Associate Member, ASCE, Senior Technical Specialist, ENSR Consulting and Engineering, 14715 NE 95th Street, Redmond, WA 98052.

well as the velocity components normal to, sweeping along, and laterally across the face of the screen. Therefore, the axial and tangential velocity components in a plane normal to the conduit axis upstream of the screen, and the three-dimensional velocity components near the surface of the inclined screen are key to the design of the screen.

The penstock screen is constructed of wedge-wire screen mounted on a steel frame in the penstock and inclined at an angle to the flow. The screen is called an Eicher screen, named for George Eicher, developer of that screening concept (Swenson, 1993). Figure 1 shows a schematic of an Eicher Screen installation.

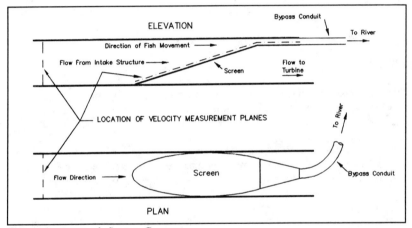

Figure 1: Penstock Screen Concept

This paper presents the methods used for measuring and mapping the axial and tangential flow field at a circular section normal to a pipe axis, and the three-dimensional flow field at an elliptical section inclined with respect to the pipe centerline. A two-component electromagnetic current meter and three-component five-hole pitot probe were used in laboratory-scale and full-scale penstocks.

MEASUREMENTS IN A PLANE NORMAL TO CONDUIT AXIS

The uniformity of the flow field approaching the model screen was measured with a custom-manufactured Marsh-McBirney Model 523 two-dimensional electromagnetic (EM) current meter.[3] Two velocity vector components were required to accurately assess approach velocities in the penstock upstream of the

---

[3] Marsh-McBirney, Inc., Frederick, MD, USA.

screen. The axial velocity component was measured parallel to the centerline of the penstock, and the tangential component was measured in the plane of the cross section, perpendicular to the centerline. These measurements were used to generate plots of isovels (contours of constant velocity), as well as plots indicating the tangential "swirl" or "twist" in the flow at the cross section.

The sensor-end of the EM meter was inserted into the penstock through valved fittings installed radially around a spool piece, as shown in Figure 2. The sensor shaft and clamping area were manufactured to allow traverses to be made across the 300-mm diameter pipe. The spool piece could be installed at any cross section upstream or downstream from the inclined screen. The fittings were installed so that velocity traverses across the diameter of the penstock could be made every 45° at any given cross section. For each radial traverse, measurements were made at six locations, starting with the centerline of the penstock and moving out. The axial and tangential velocity components (V_{axial} and $V_{tangential}$) were measured at a total of 41 (r, Θ) locations for each flow condition tested.

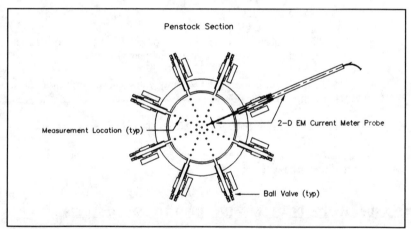

Figure 2: Velocity Measurement Locations in Normal Plane

The output from the EM meter was recorded using an IBM-compatible personal computer with a plug-in data acquisition card. The values of V_{axial} and $V_{tangential}$ were measured for 30 seconds, and the averages computed. Results were entered into a spreadsheet for each (r, Θ) measurement location. The spreadsheet converted the measured model velocities into normalized velocities, and converted the polar coordinate (r, Θ) measurement locations to normalized rectangular (x,y) coordinates. The velocities were normalized by dividing each value by the average of all the axial velocities measured. The rectangular coordinates were normalized so that (x,y) locations of $(1,0)$, $(0,1)$, $(-1,0)$, and $(0,-1)$ would be at the inside wall of the penstock. This normalization procedure was done to facilitate comparison of data collected at different flow rates and in different sized conduits. The spreadsheet then created an ASCII output file containing the normalized locations and velocities.

The normalized data were then passed to a commercial graphics package for data presentation. Batch programs were written to automate the task of data presentation. The program SURFER[4] was used for gridding, contouring, and creating vector plots. The procedure included

- creating a grid of the normalized axial velocity data using the Minimum Curvature method with a cell size of 0.04 times the pipe diameter (25 grid rows & columns);

- smoothing the grid to a cell size of .01 times the pipe diameter (100 grid rows and columns);

- blanking the grid outside of the conduit area;

- contouring the grid;

- posting scaled vector symbols at the proper location and orientation to represent the tangential velocity component;

- posting the normalized axial velocity value above the measurement location; and

- overlaying a reference grid on the completed isovel/vector plot.

A sample isovel/vector plot for measurements made in a plane normal to the penstock centerline in shown in Figure 3.

MEASUREMENTS IN A PLANE INCLINED TO CONDUIT AXIS

The three-dimensional velocity field near the inclined screen was mapped using a five-hole pitot probe connected to differential pressure transducer. Five-hole pitot probes measure total and static pressures, as well as yaw and pitch angles (α, β) of a moving fluid. With these measurements, the two velocity vector components parallel to the screen and one normal to the screen were computed. The detailed theory of operation of these types of probes is documented elsewhere (see for example, Treaster and Yocum).

---

[4] SURFER is a trademark of Golden Software, Golden, CO.

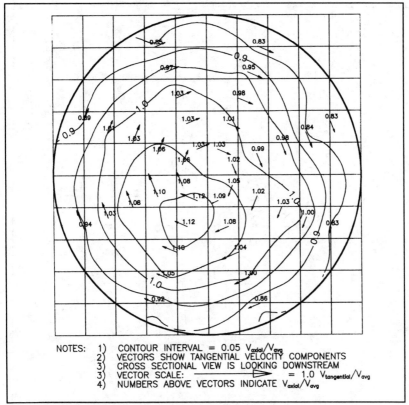

Figure 3: Normalized Isovels and Swirl Components

The five-hole probe used in the laboratory studies was a type DA-250 probe manufactured by United Sensors.[5] The probe was 610 mm long and 9.5 mm in diameter. The sensing tip of the probe was 51 mm long and 6.4 mm in diameter. The probe used in the prototype measurements was similar. It was 1.8 meters long and had a diameter of 64 mm, with a sensing head 76 mm long and 9.5 mm in diameter (Swenson, 1991).

Each of the five pressure sensing holes, designated P1 through P5 in Figure 4, is connected by stainless steel tubing to a fitting block at the opposite end of the probe. P1, P2, and P3 are used to sense the yaw angle, α. P4 and P5 are used to sense the pitch angle, β. Each of these pressure taps was connected to a manifold by plastic

---

[5] United Sensor Division, United Electric Controls, Watertown, MA.

58   HYDRAULIC MEASUREMENTS AND EXPERIMENTATION

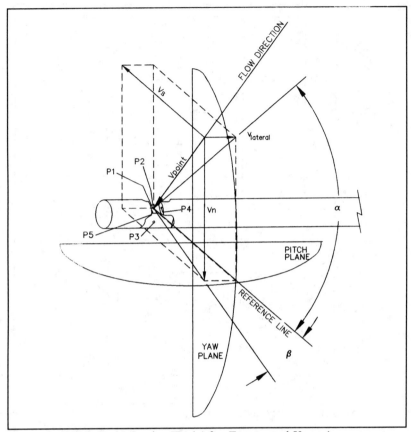

Figure 4: 5-Hole Probe Sensing Head (after Treaster and Yocum)

tubing. The manifold was connected to a differential pressure transducer. By opening and closing the appropriate valves on the manifold, the pressure differences between the probe taps were calculated by measuring the voltage output of the transducer.

The probes were inserted into the penstock of both the model and the prototype through valved fittings installed on the sides of the penstocks. The ports were spaced longitudinally along the screen to provide representative measurement coverage of the entire screen plane, as shown in Figure 5. For each test, the probe was inserted into the most upstream port and positioned at the first measurement location. The pressure lines in the probe were bled of all air bubbles, and the yaw angle, pitch angle, and total pressure were recorded. The probe was then moved to the next measurement location, and the process repeated.

The normal velocity component is denoted V_n. The velocity component in the plane of the screen and in the direction of the mean flow is called the sweeping component, and is denoted V_s. The component in the plane of the screen and perpendicular to the mean flow direction is called the lateral component, and is denoted by V_l. These velocity components are calculated values based upon the measurements of V_{point}, α, and β. Referring to Figures 4 and 5, the components are defined as follows:

$$V_{point} = \text{the total velocity vector}$$
$$V_n = V_{point}(\sin \alpha)(\cos \beta)$$
$$V_s = V_{point}(\cos \alpha)(\cos \beta)$$
$$V_l = V_{point}(\cos \alpha)(\sin \beta)$$

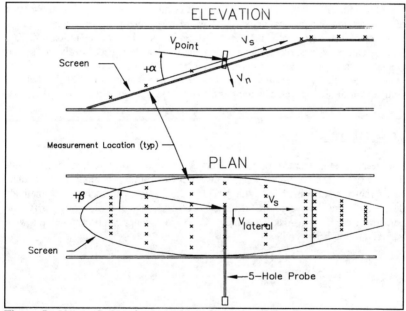

Figure 5: Vector Conventions and Measurement Locations

The measurements made at each location were entered into a spreadsheet template. For each measurement, the spreadsheet converted the measured model pressure differentials and yaw angle into the total velocity and pitch angle, and calculated the three velocity components. The velocities were normalized by dividing each value by V_{avg} (the average of all the point velocities measured). In addition, the ratios of V_n/V_{avg}, V_s/V_{avg}, and V_n/V_s were computed to allow comparison of measured values with established acceptance criteria. The measurement locations were normalized by dividing the distance from the upstream end of the screen by the total length of the screen (X/L) and the distance from the centerline of the screen by the total length of the screen (Y/L). The spreadsheet then created an ASCII output file containing the normalized locations and velocities.

The normalized data were then passed to a combination of custom batch programs and commercial graphics packages for data presentation. The data processing and presentation procedure was similar to those used for creating isovel and vector plots of the flow field in the section normal to the penstock axis. The three-dimensional screen velocity data were presented in two formats: (a) as profile velocity plots, as shown in Figure 6-a, and (b) as isovel plots, as shown in Figure 6-b. The profile velocity plots represent an elevation, or profile, view of the side of the screen. Velocity vectors are shown as arrows approaching the screen at the measured yaw angle, α. The length of each arrow is proportional to the average velocity magnitude for all points measured at that port location. The isovel plots show lines of constant non-dimensional velocity values. Three isovel plots were generated for each set of screen velocity measurements, corresponding to the ratios V_n/V_{avg}, V_s/V_{avg}, and V_n/V_s described above. Contour plots were chosen to display screen velocity data because areas of excessively high or low values of V_n or V_s could be readily identified. Presenting the data in dimensionless form facilitated the comparison of similar values among different test conditions.

CONCLUSIONS

The techniques presented in this paper facilitate the visualization of the two-dimensional and three-dimensional flow fields at arbitrary sections in a conduit. Measurements were made on both model and prototype penstocks. The methods described have been successfully used to assess, optimize, and verify the performance of in-penstock fish screens. These techniques can be applied to a wide variety of pipe-flow problems.

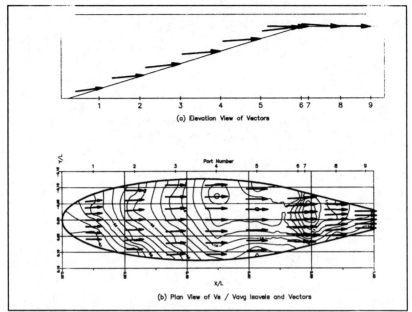

Figure 6: Inclined Plane Vectors and Isovels

REFERENCES

Swenson, Lawrence J., "Final Report: Prototype Velocity Measurements in the Eicher Screen at the Elwha Hydroelectric Project," Project Report No. 6400-004-520, prepared for Stone & Webster Engineering Corporation, Boston MA, by Engineering Hydraulics, Inc, Redmond, WA, July 1991.

Swenson, Lawrence J., "Final Report: Puntledge River Diversion Dam Permanent Fish Screen Project Hydraulic Model Studies," Project Report No. 0813-001-510, prepared for British Columbia Hydro and Power Authority, Vancouver, BC, by ENSR Consulting and Engineering, Redmond, WA, December 1993.

Treaster, A.L., and A.M. Yocum, "The Calibration and Application of Five-Hole Probes," *ISA Transactions*, Volume 18, No. 3, pp. 23-34, 1979.

Velocity and Turbulence Measurement from
the Illinois and Mississippi Rivers

by

Ta Wei Soong[1] and Nani G. Bhowmik[2]

Illinois State Water Survey
2204 Griffith Drive
Champaign, IL 61820-7495

Abstract

This paper presents a data measurement program that was used for collecting the velocity data associated with barge-tow traffic on the Illinois and Mississippi Rivers and includes a discussion of instrumentation, physical set-up, and deployment plans. Velocity changes dramatically in both temporal and spatial dimensions during barge-tow passage. Satisfactory results have been achieved. Data from two sites, one collected from the Kampsville site on the Illinois River and one from the Clarks Ferry site on the Mississippi River, are presented.

Introduction

The Illinois State Water Survey (ISWS) has collected physical data on the hydrodynamic changes associated with the movement of tow and barge traffic from the Upper Mississippi River System (UMRS). One of the goals of this project was to determine changes in velocity due to barge traffic at different reaches of the UMRS. Velocity changes induced by barge-tow have both temporal and spatial distributions. The passing time on the UMRS for a commonly configured barge convoy (3 columns by 5 rows; 15 barges) is between 2 and 5 minutes. The influences in velocity field, however, may appear before and after actual barge passage. In order to describe the disturbances,

---

[1]Director, Office of Hydraulics and River Mechanics, Illinois State Water Survey, 2204 Griffith Drive, Champaign, IL 61820-7495.

[2]Division Head, Hydrology Division, Illinois State Water Survey, 2204 Griffith Drive, Champaign, IL 61820-7495.

both average and instantaneous changes have to be investigated. And sufficient data before and after an event are necessary to address the impacted velocity. Electromagnetic two-dimensional current meters and an automated datalogging system are used for these purposes. Once a meter is in place in the field, it keeps collecting data for several days in an attempt to collect sufficient variations in barge types under different maneuvering and loading conditions. The appearance of barge-tows on the UMRS is also random. Therefore, the data measurement scheme has to cover a wide range of time scales and field conditions.

During barge-tow passage, the lateral variation of the impacted velocity is also subject to investigation. The changes in velocity field are impacts to juvenile fish and other aquatic animals and plants on the floodplain and in channel border areas. Geomorphologic features such as wide or narrow main-channels, bends or straight reaches, floodplains, side channels, backwater lakes, and sloughs can affect the lateral distribution patterns and are abundant in the UMRS. The ISWS data collection plan involves use of multiple electromagnetic current meters which are distributed in a transect or arranged in arrays. A total of 8 to 13 electromagnetic two-dimensional current meters have been used at different sites along the Illinois and Mississippi Rivers. The overall data collection program has been discussed by Soong et al. (1990).

Background

Many diverse physical changes are associated with navigation traffic. Within the environment of a large river, such as the Illinois or Mississippi, these changes are further complicated by the confined channel boundary and geometry. Other factors can also affect the barge-induced physical changes. These factors include barge maneuvering, barge configuration and loading, hydraulic conditions of the reach, and reach geomorphology. Some of the physical changes can be quantified and measured with available instrumentation. These changes could be related to basic hydraulic and physical factors. Many other factors can not be quantified and are difficult to measure in the field. To develop management alternatives for the entire UMRS, it is necessary to have a clear understanding about the induced changes and interactions between the barge-tow traffic and the hydraulic conditions, and the physical character of the specified river reaches.

Instrumentation

Velocity data were measured by utilizing Marsh McBirney meters, Model MMB527 and MMB511, and InterOcean current meters, Model S4. All these meters use the Faraday's electromagnetic principle and measure two velocity components in horizontal directions. The S4 meter is spherical in shape with a diameter of 25 cm (10 inches). It also has a compass to record the deviation of principal axis from magnetic north, which makes it easy to correlate the recorded x and y components of the velocities with the longitudinal axis of the river. All electronics and power units are contained within the S4.

The MMB527 has a 10.3-cm (4-inch) spherical sensor with four probes. The signal processing unit is housed separately and connects with the meter through coaxial cables that are laid on the bottom of the river. The signal processing unit provides a visual display of the x and y components of the velocities and a compass reading.

The MMB511 has a 3.9-cm (1.5-inch) spherical sensor with four probes. These meters function like the MMB527 except that they do not contain any compass. Therefore, these meters are generally installed closer to a survey station than other meters so that the direction of the principal axis of each meter can be identified externally. Table 1 gives an overview of the specifications for these current meters.

Table 1. Specifications of the Current Meters Utilized in the Field

Instrument	Manufacturer	Parameters	Range	Accuracy
S4	Inter Ocean	Vx, Vy	0-350 cm/sec	0.2 cm/sec
		direction	0°-360°	0.5°
MMB527	Marsh McBirney	Vx, Vy	±300 cm/sec	±2% of reading
		direction	0-360°	10°
MMB511	Marsh McBirney	Vx, Vy	±300 cm/sec	±2% of reading

A total of five S4s, six MMB511s, and two MMB527s were utilized at various sites on the Illinois and Mississippi Rivers. Modifications were made to the Marsh McBirney meters to increase the cable lengths. The cable lengths for the MMB527s were 500 feet (152.4 m) and 300 feet (91.46 m) respectively. Two of the MMB511s also had 500 feet (91.46 m) cables and four had cable lengths of 100 feet (30.49 m) each. During field investigation we also had the compass of MMB527 reversed to achieve better stability in mounting system.

Sampling Frequency and Duration: Velocity data were collected throughout the field working days (3-7 days). Data recording frequency was normally one sample per second for most meters. Several S4s were switched to record one sample per 2 or 5 seconds because of their capacity for data storage.

Data Storage: Digital data were converted to physical units of velocity and direction and stored in data storage units. The S4s have built-in storage. For the Marsh McBirney meters data were stored in CR10 data loggers. Programming time for all meters and field recording time were synchronized so that meaningful comparisons could be made later. Once the data loggers were full, these data were downloaded to a portable PC in the field. Diskettes were the final storage for all the field data. The ISWS has also developed several computer programs to graphically display velocity data once they were retrieved from the data loggers. Detailed field notes were also kept to indicate field conditions and any apparent change in the operation of the meters.

Mounting System

Current meters utilized for this project were all fixed-mounted. They were mounted either on a post or with a vertical support system (see Figs 1 and 2). Factors considered in all the mounting systems were their stability against the normal flow and ease of deployment and retrieval. These support systems were made of either stainless steel, plastic, or aluminum for connectors, bases, and housings. Each lead weight was 20 pound

(9.1 kg) and multiple weights are added to the bases. All the manufacturing systems were designed, manufactured, tested, improved and maintained by the personnel of the Water Survey.

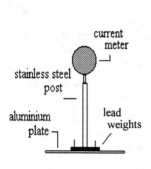

Figure 1. Mounting system for S4s and MMB527s

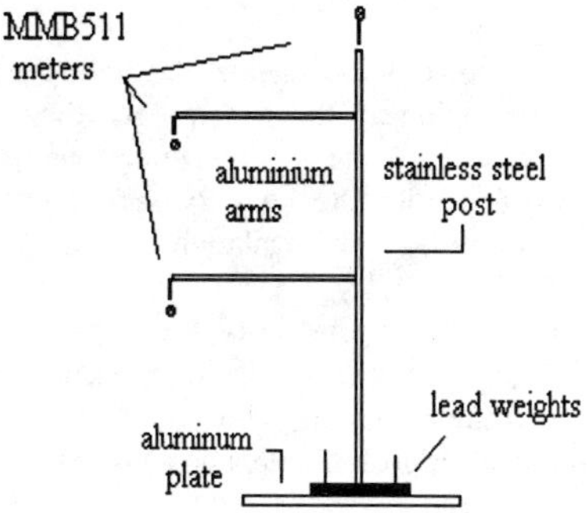

Figure 2. Mounting system for MMB511s

The S4s and MMB527s were supported on a post system (Fig. 1), where the height of posts ranged from 0.3 to 1 m above the bed. Similarly, the MMB511s used this type of support system and the heights could be closer to the bed. The closest height was about 0.22 m from the bed.

Another mounting system for the MMB511 was a "vertical array," where a series of meters were installed at various elevations above the bed (Fig. 2). The MMB511 meters were installed on a horizontal arm about 0.76 m away from the main post to avoid any disturbance from the main post to the surrounding velocity field. Three MMB511s were installed at various heights above the bed. Several versions of the vertical support systems were manufactured. The maximum depth of water for a vertical array installation was about 3.5 m.

Field Deployment

Two deployment plans were used in the field: the transect approach, i.e. meters were arranged along a selected transect, and 2) the array approach, that is meters were arranged to cover an area (generally in triangular or rectangular shapes) with main emphasis on the main (middle) transect. Most meters were distributed on one side of the river between a shore station and the navigation channel. In terms of placing the meters, the S4s were installed at deeper water and close to the navigation channel or at the farthest side from the shore station since they had self-contained data storage. The MMB527s were installed at intermediate distance from shore between S4s and MMB511s. Neither the MMB527 nor the S4 required the determination of their deviation from magnetic north since they contain a compass.

The MMB511s were used to measure changes in channel borders or in the vertical direction. Their angles of rotation were measured with a compass. Once installed, all the meters were periodically checked by the divers to make sure that they were functioning properly, i.e. installed correctly or had not tipped over because of debris.

Positioning the Meters A Micro-Fix System, manufactured by Racal Survey, was used to position all the meters once they were deployed. The system utilized in the field consisted of a set of three transponders, two installed on the shore and the third one on a boat. With the boat located at each meter's site, the position of each meter was determined by triangulation.

Field Data Collections
A total of 8 trips were made to 5 sites on the UMRS. Several sites were visited twice under different flow conditions. One event from the Kampsville, Illinois River, and one event from Clarks Ferry, Mississippi River, are presented here.

Kampsville Site The Kampsville site is located at River Mile 35.2 in Alton Pool, Illinois, 44.9 miles (72.26 km) downstream of the Lock and Dam at La Grange, and 3.2 miles (5.15 km) upstream of the town of Kampsville. This river reach represents a fairly straight channel. Two trips were made to Kampsville site. During trip 1 at the primary measurement transect, the river was 360 m wide with an average depth of 3.64 m. The average water surface slope on this reach was 0.096 ft/mile (0.018 m/km).

Location of Sensors: For Trip 1, a total of two S4s, two MMB527s, four MMB511s were utilized. All meters were deployed along a transect from the west side of the river; i.e., the right-hand side (RHS) of the river looking downstream. Fig. 3 is a cross-sectional view of the instrumentation at the site during Trip 1. Current meters were coded with manufacturer's name and serial number of the meter. For example, the MMB511 with serial number 998 was coded with an identification of MMB511/998, and so on. For Trip 1, MMB meters were grouped in four stations and deployed at distances of 12.9, 33.5, 47.2, and 65.5 m from the shore on the RHS of the river. Two additional stations were set up at dis-tances of 85.7 and 131.4 m from the left-hand side (LHS) of the river. The current meters (MMB527 and S4) were installed at heights of 0.70 and 0.92 m, respectively, above the river bed. Three MMB511s were mounted at three vertical heights of 0.31, 1.22, and 2.44 m above the river bottom, at the vertical array station (Fig. 3).

Fig. 4 illustrates measured velocity for barge Frank H. Peavey at three stations. These velocity data were rotated so that the x component was parallel to the longitudinal direction of the river, positive towards downstream, and y component was in the lateral direction of the main axis of the river, positive towards the left when facing downstream. Meter MMB511/1001 was the one closest to shore, MMB527/332 was the one farthest from the RHS and S4/040 was the one closest to navigation channel. This event was an upbound tow (Frank H. Peavey) pushing 4 rows 6 columns of cargo barge with one void at the end (23 barges) at a speed of 6.3 km/hr. The barges were mixed in loading conditions with an equivalent draft of 1.17 m, the distance from shoreline was 150 m.

VELOCITY AND TURBULENCE 67

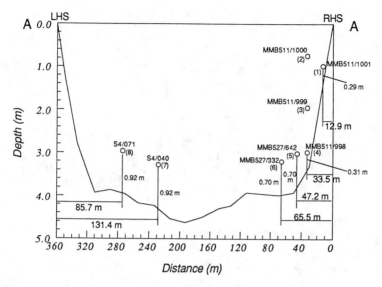

Figure 3. Instrumentation at Kampsville site for Trip 1.

The passing time (from the passage of bow to the passage of stern) was from 920.5 to 924.33 minutes (hours and seconds are converted to the unit of minutes). In order to demonstrate changes in the velocity field, the graphical representations covered approximately a period of 30 minutes to include three periods: pre-passage, actual passage, and post-passage. Magnitudes of changes at three different locations and intensity of fluctuations in time and space could be observed clearly.

Clarks Ferry Site

The Clarks Ferry site is located at River Mile 468.2 in Pool 16 at Muscatine, Iowa; about 14.7 miles (23.66km) downstream of Lock and Dam 15 at Rock Island, Illinois. With Andalusia Island located across the channel, this reach also represents a fairly straight channel on the Mississippi River. Two trips were made to Clarks Ferry site. During trip 1 at the primary measurement transect, the river was 633 m wide with an average depth of 4.18 m. The average water surface slope on this reach was 0.33 ft/mile (0.062 m/km) during Trip 1.

Location of Sensors: For Trip 1, a total of five S4s, two MMB527s, six MMB511s were used. The meters were deployed in an array fashion (Fig. 5). For Trip 1, five velocity stations were located on the right hand side (RHS) of the river at distances of 15.2, 28.0, 43.0, 73.5, and 104.0 m from the shore. Three additional stations were set up at distances of 67.1, 121.9, and 304.8 m from shore on the LHS. The current meters (MMB527 and

HYDRAULIC MEASUREMENTS AND EXPERIMENTATION

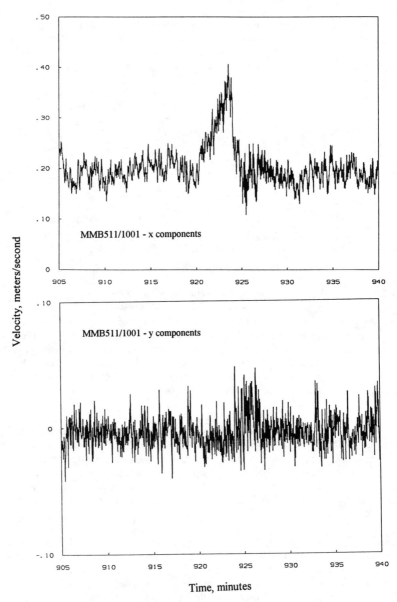

Figure 4a. Measured velocity for barge tow Frank H. Peavey (MMB511/1001)

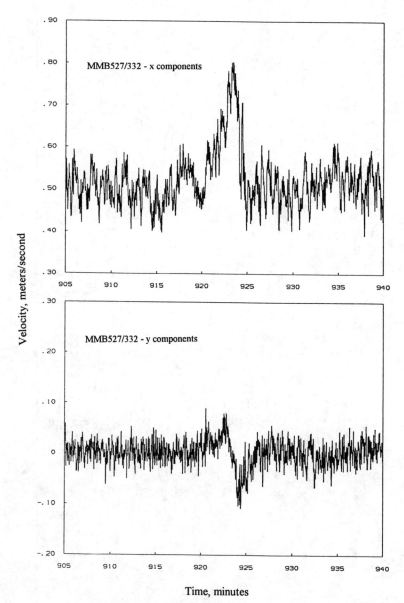

Figure 4b. Measured velocity for barge tow Frank H. Peavey (MMB527/332)

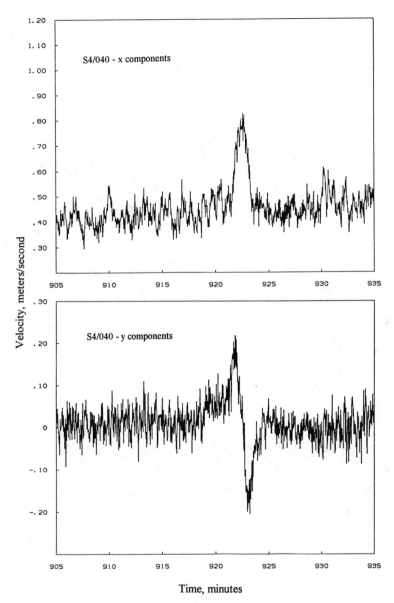

Figure 4c. Measured velocity for barge tow Frank H. Peavey (S4/040)

S4) were installed at heights of 0.70 and 1.0 m, respectively, above the river bed. Two MMB511 current meters were mounted at two vertical heights of 0.33 and 1.52 m above the river bottom, at station "b" from the RHS, whereas station "c" had three MMB511 meters installed, at vertical heights of 0.36, 1.62, and 2.53 m above the river bed.

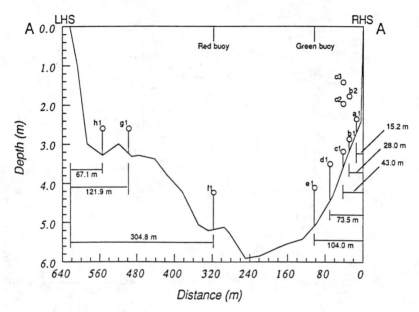

Figure 5. Instrumentation at Clarks Ferry site for Trip 1.

Fig. 6 illustrates measured velocity for barge T.S. Kunsman at four stations, which are locations c1, d1, e1, and i1 in Fig. 5. Meters S4/151 (d1), S4/834 (e1), and S4/832 (i1) form a triangular array on the studying site of the navigation channel; and MMB511/998 is the bottom meter on a vertical array c closer to shore. Note that an 11-point moving average has been applied and these plots are the moving average plots. This event was a downbound tow (T.S. Kunsman) pushing 3 rows and 4 columns of cargo barge (12 barges) at a speed of 12.3 km/hr. The barges were fully loaded with a draft of 2.74 m, the distance from the shoreline was 370 m, and the passing time was from 991. to 993.37 minutes.

Results and Discussion

Measured velocity can be used to develop plots such as those in figures 4 and 6 for visual inspection, or for data analyses from which changes in the velocity field can be assessed and then related to physical factors and hydraulic properties.

It could be noted that both components of velocities were impacted by the movement of barge tows (Mazumder et al., 1993). The x component normally

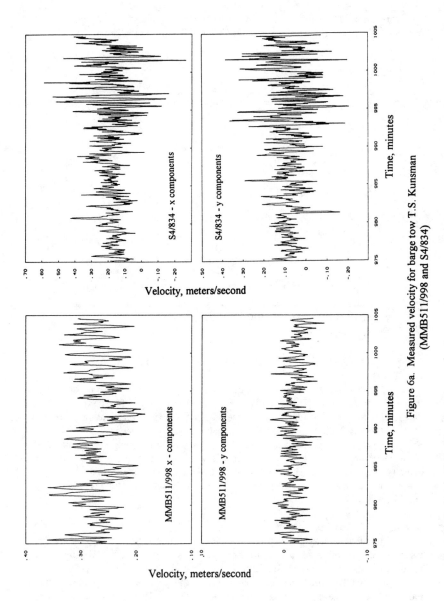

Figure 6a. Measured velocity for barge tow T.S. Kunsman (MMB511/998 and S4/834)

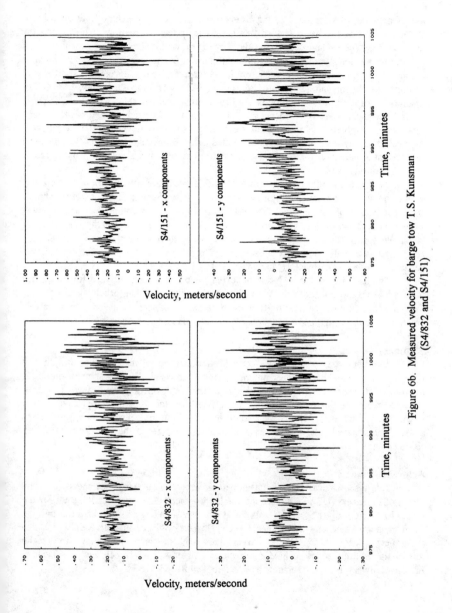

Figure 6b. Measured velocity for barge tow T.S. Kunsman (S4/832 and S4/151)

experiences the most changes during an event. These changes include flow opposite to the movement of the barge (return flow), thus temporary increasing or decreasing the mean velocity (depending on whether the barge is moving upstream or downstream) and altering turbulence intensity through different phases of barge passage. Many researchers assume the cross sectional distribution is uniform and have developed approximate solutions for the magnitude of return flows (e.g., Blaauw and van der Knapp, 1983). The assumption of uniform distribution is valid for narrow artificial ship channels. For wider natural channels, however, Hochstein and Adams (1989) used an exponential decay function to evaluate the distribution from barge to the shoreline. The validity of exponential decay is compared by the collected data. Bhowmik et al. (1992) proposed a different approximation for the lateral distribution pattern and a solution for the return flows in natural rivers.

The changes in the velocity are not uni-directional but rotate somewhat especially within the channel border areas during an event. These rotating flows obviously change the y component of the velocity. Generally, two peaks in the y components are observed for narrower channels (Fig. 4). This indicates that as traffic approaches the site the water is initially pushed toward the shore and then is sucked toward the main channel when the barge creates drawdown as it passes the site. When the channel is narrower, the rotation pattern becomes more obvious (Mazumder et al., 1993).

The measured velocity during nonevent periods has been analyzed for turbulence fluctuations (Bhowmik and Xia, 1993). This analysis indicates that turbulence is strong in the main channel near the river bottom and becomes weak near the channel border areas, and the velocity fluctuations also follow the normal (Gaussian) distribution.

Summary

This paper has presented a data collection scheme for investigating the physical changes induced by barge tows on the Upper Mississippi River System, including a discussion of instrumentation, physical set-up, and deployment plans. Two data sets, collected from the Illinois River at the Kampsville site and the Mississippi River at the Clarks Ferry site, respectively, were also presented. From the collected data, researchers at the Illinois State Water Survey are able to assess changes in the velocity field during barge passage.

Acknowledgments

This research was carried out as part of a project supported by the U.S. Army Corps of Engineers (USACOE) through the Environmental Management Program for the Upper Mississippi River System, Onalaska, Wisconsin. Funding was provided through the Environmental Management Technical Center, Onalaska, Wisconsin, with Ken Lubinski as the project manager. Many Water Survey staff members who worked very hard on the formulation and collection of field data include Rodger Adams, Bill Bogner, Ed Delisio, Jim Slowikowski, Bijoy Mazumder, Il Won Seo, and Renjie Xia.

References

Blaauw, H.G. and van der Knapp F.M.C. 1983. *Prediction of Squat of Ships Sailing in Restricted Water.* Presented at the 8th International Harbour Congress, Antwerp, Belgium, June 13-17, 1983. Delft Hydraulics Laboratory. Publication no. 302. 13p.

Hochstein, A.B. and Adams, C.E. 1989. *Influence of Vessel Movements on Stability of Restricted Channels.* Journal of Waterway, Port, Coastal, and Ocean Engineering, ASCE, 115(5), PP 444-465.

Mazumder, B.S., N.G. Bhowmik, and T.W. Soong. 1993. *Turbulence in Rivers due to Navigation Traffic.* ASCE Journal of Hydraulic Engineering, Vol. 119, No. 5, May 1993, pp. 581-597.

Soong, T.-W., W.C. Bogner, and W.F. Reichelt, 1990. *Field Data Acquisition for Determining the Physical Impacts of Navigation,* Proceedings of the National Conference on Hydraulic Engineering, San Diego, CA, July 30 to August 3, 1990.

Bhowmik, N.G. and R. Xia. 1993. *Turbulent Velocity Fluctuations in Natural Rivers.* ASCE 1993 National Conference on Hydraulic Engineering and International Symposium on Engineering Hydrology, San Francisco, CA, July 25-30, pp. 1677-1682.

Bhowmik, N.G., B.S. Mazumder, and T. W. Soong. 1992. *Return Flows in Large Rivers Associated with Navigation Traffic.* ASCE Proceedings of the 1992 National Conference on Hydraulic Engineering: Saving a Threatened Resource - - In Search of Solutions, Ed. Marshall Jennings and Nani G. Bhowmik, Baltimore, MD, August 2-6, pp. 760-765.

Velocity Measurements by the "One-Orange Method"

B. A. Christensen[1], M.ASCE

Abstract

Application of the most sophisticated instrumentation - so called state-of-the-art equipment - is not always necessary nor even desirable for measurement of mean velocities in verticals, discharges, bed shear stresses and other hydraulic parameters needed in sound engineering management of open channel flow. In many cases order of magnitude (or even more accurate) data may be generated by extremely simple and straight-forward means. For preliminary designs approximate values may very well suffice. Such a simple procedure based on the rising buoyant body method is introduced. The only tools needed are a slightly buoyant orange, a watch and measuring tape. A boat preferably with a fathometer is also desirable if the measurements cannot be accomplished by wading.

Introduction

Field measurements of velocity, discharge and quantities such as Manning's n, bed shear stresses, rates of sediment transport etc., that may be derived from these, are most often accomplished by introduction of more or less sophisticated equipment in the field, even when the needed accuracy does not justify the use of such equipment, but may call for faster, albeit less accurate, procedures. The old standbys, cup meters with a vertical shaft (Gurley meters) and propeller meters with a horizontal shaft (Ott meters) of various sizes and support equipment are most common. The Savonius meter, a similar rotary meter with vertical shaft, is used mostly in the oceanic environment. Non-intrusive electronic instrumentation, for instance hot wire - and film equipment, laser

---

[1] Professor of Hydraulics, Department of Civil Engineering, University of Florida, Gainesville, FL 32611

Doppler flow meters (LDV), and sonic Doppler flowmeters (SDV) (Kraus et al., 1994) is usually reserved for laboratory use but has been used in the field, in most cases with questionable success. A simple procedure based on the integrating buoyant body method requiring only the simplest and readily available equipment is therefore developed for use in natural and man-made water courses. Since this method was developed in Florida where oranges are plentiful, an orange has been used as the buoyant body. Its specific gravity is so close to unity that its rising velocity in water is well suited for the method. Hence the title of the paper.

A similar procedure for velocity measurements in sluggish stratified flows using a cloud of small air bubbles instead of a buoyant body was proposed earlier by the author (1985). In contrast to the method described in the present paper that method requires the introduction of a compressed air source and well defined air nozzles in the field.

Measuring Mean and Maximum Velocity in a Vertical

A slightly buoyant orange is released from the bottom at time $t = 0$ at the considered vertical. At time $t = T_1$ the orange surfaces the distance $\ell = L_1$ downstream from the vertical where it was released. It is then allowed to float on the surface until it reaches the distance $\ell = L_2$ from the vertical at time T_2 as shown in Fig. 1.

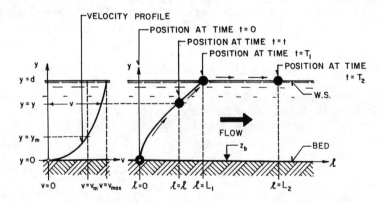

Figure 1. Rise of Buoyant Body in an Open Channel Flow. Notation.

From the observations of L_1, L_2, T_1 and T_2 it is obvious that the maximum velocity in the vertical may be found from

$$v_{max} = \frac{L_2 - L_1}{T_2 - T_1} \quad (1)$$

neglecting the influence of wind and assuming that the surface velocity is close to the maximum velocity.

The corresponding mean velocity in the same vertical, v_m, may be written

$$v_m = \frac{1}{d}\int_o^d v\,dy = \frac{1}{d}\int_o^{T_1} v v_r\,dt \quad (2)$$

in which d = depth, $v = d\ell/dt$ = local time-mean velocity at distance y from the bed and $v_r = d/T_1$ = constant rising velocity of the orange neglecting the short vertical acceleration distance near the bed.

Hence

$$v_m = \frac{1}{d}\int_o^{T_1} \frac{d\ell}{dt} \cdot \frac{d}{T_1}\,dt = \frac{1}{T_1}\int_o^{L_1} d\ell = \frac{L_1}{T_1} \quad (3)$$

regardless of shape of velocity profile and independent of the rising velocity v_r and depth d. The local discharge per unit width is $v_m d$ which integrated over the width B of the water course will yield the discharge.

Measurement of Roughness, Friction Velocity, and Bed Shear Stress

If the vertical is chosen in a section away from obstructions and bends it is reasonable to assume that the velocity profile is logarithmic corresponding to flow in the turbulent rough range, i.e., that

$$\frac{v}{v_f} = 8.48 + 2.5\,\ell n\,\frac{y}{k} = 2.5\,\ell n\,\frac{29.7y}{k} \quad (4)$$

in which v = velocity at distance y from the bed, v_f = friction velocity and k = equivalent sand roughness.

Applying this equation at y = d where $v \cong v_{max}$ and at y = d/e = 0.368d where $v = v_m$ yields

$$\frac{v_{max}}{v_f} = 2.5 \ln \frac{29.7d}{k} \tag{5}$$

$$\frac{v_m}{v_f} = 2.5 \ln \frac{29.7d}{ek} \tag{6}$$

In Eqns. (5) and (6) the two unknown quantities are v_f and k; e is the base of natural logarithms. Solving Eqns. (5) and (6) for v_f and k gives

$$\boxed{v_f = 0.4 (v_{max} - v_m)} \tag{7}$$

and

$$\boxed{k = \frac{29.7d}{e^{\frac{v_{max}}{v_{max} - v_m}}}} \tag{8}$$

in which v_{max} and v_m are found from Eqns. (1) and (3) after direct observation of L_1, L_2, T_1 and T_2. Manning's n may be found by a simple Strickler type formula once k is found from Eqn. (8).

Since the friction velocity is directly related to the bed shear τ_o by

$$v_f = \sqrt{\tau_o/\rho} \tag{9}$$

the latter may be found directly from Eqns. (9) and (7) giving

$$[\tau_o = \rho v_f^2 = 0.16 (v_{max} - v_m)^2] \tag{10}$$

To find the path of the rising orange consider the simple relationships

$$dy = v_r dt \tag{11}$$

and

$$d\ell = vdt \tag{12}$$

or by elimination of dt and assuming logarithmic velocity distribution in the vertical

$$\frac{dy}{d\ell} = \frac{v_r}{v} = \frac{v_r}{2.5 v_f \ln \frac{29.7y}{k}} \tag{13}$$

Solving this differential equation, introducing $v_r = d/T_1 = v_m d/L_1$, v_f from Eqn. (7) and using the boundary condition that $\ell = 0$ at $y = 0$ yields the equation of the orange's rising path

$$\frac{\ell}{L_1} = \left[\frac{\ln \frac{29.7d}{k}}{\ln \frac{10.93d}{k}} - 1\right] \cdot \frac{y}{d} \cdot \ln\left(\frac{10.93d}{k} \cdot \frac{y}{d}\right) \tag{14}$$

This dimensionless equation is plotted in Fig. 2.

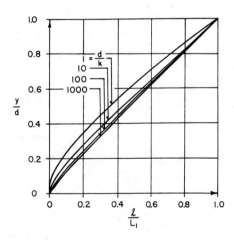

Figure 2. Path of Rising Orange. Turbulent Rough Flow Range.

Note that the inverse relative roughness, d/k, has only a minor influence on the path of the rising orange and that this path is nearly linear.

Measurement of Discharge

Measuring a chosen cross section's area, shape and maximum depth from soundings of the section and finding the roughness k from Eqn. (8), or simply estimating it, make it possible to calculate the total discharge Q.

Consider the fairly symmetrical section shown in Fig. 3.

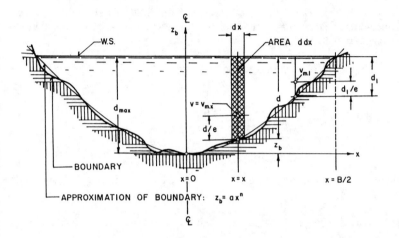

Figure 3. Cross-Sectional Geometry.

The wetted perimeter of this section may be approximated by the power expression

$$z_b = Cx^n \tag{15}$$

in which z_b = vertical distance from the section's deepest point to a point on the wetted perimeter located the distance x from the section's centerline as shown. C is a constant and the exponent n may be related to the cross-section's areal aspect ratio ω defined by

$$\omega = \frac{A}{Bd_{max}} \tag{16}$$

in which A = cross sectional area and B = width of the water surface, by expressing the area as

$$A = \frac{n}{1+n} B d_{max} \qquad (17)$$

Hence

$$\omega = \frac{n}{1+n} \quad \text{or} \quad n = \frac{\omega}{1-\omega} \qquad (18)$$

The areal aspect ration ω serves as a indicator of the sections general shape. For instance $\omega = 1$ indicates that the section must be rectangular. In a triangular section $\omega = 1/2$.

Eqn. (15) corresponds to the local depth d at $x = x$ being

$$d = d_{max}\left[1 - \left(\frac{2x}{B}\right)^n\right] \qquad (19)$$

where d = local depth at distance x from the centerline and d_{max} = maximum depth.

The discharge dQ passing through the narrow vertical area element dA = ddx at $x = x$ is

$$dQ = v_{m.x} d\,dx \qquad (20)$$

where $v_{m.x}$ = mean velocity in the vertical located at $x = x$. Assuming turbulent flow and a logarithmic velocity profile in the vertical it is easily seen by integration of the velocity profile that $v_{m.x}$ will occur at the distance $d/e \cong d/2.718$ from the bed. Hence, in the turbulent rough flow range,

$$v_{m.x} = v_{f.x}\left[8.48 + 2.5\ell n\left(\frac{d}{ek}\right)\right] = 2.5 v_{f.x} \ell n\left(\frac{29.7 d}{ek}\right) \qquad (21)$$

where the local friction velocity $v_{f.x}$ may be written

$$v_{f.x} = \sqrt{g d S_{EGL}} \qquad (22)$$

In Eqn. (22) S_{EGL} = slope of the energy grade line in the considered section and g = acceleration due to gravity.

Hence the total discharge may be written

$$Q = 2\int_{x=0}^{x=B/2} 2.5\sqrt{gdS_{EGL}}\,\ell n\left[\frac{29.7d}{ek}\right] d\,dx \qquad (23)$$

or by introduction of x found as a function of d from Eqn. (19), and the auxiliary variable

$$\zeta = \frac{d}{d_{max}} \qquad (24)$$

$$Q = -2\int_{x=0}^{x=B/2} 2.5\sqrt{gS_{EGL}}\,d_{max}^{\frac{3}{2}}\zeta^{3/2}\ell n\left[\frac{29.7d_{max}}{ek}\cdot\zeta\right]\frac{B}{2n}(1-\zeta)^{\frac{1}{n}-1}d\zeta \qquad (25)$$

in which n may be replaced by $\omega/(1-\omega)$, B by $A/(\omega d_{max})$, and $\sqrt{gS_{EGL}}$ related to the maximum velocity at $x = 0$, $v_{max.o}$, by

$$v_{max.o} = 2.5\sqrt{gd_{max}S_{EGL}}\cdot\ell n\left[\frac{29.7d_{max}}{k}\right] \qquad (26)$$

or

$$\sqrt{gS_{EGL}} = \frac{v_{max.o}}{2.5\sqrt{d_{max}}\,\ell n\left[\frac{29.7d_{max}}{k}\right]} \qquad (27)$$

giving

$$\boxed{Q = v_{max.o}A\cdot\eta\left(\omega,\frac{d_{max}}{k}\right)} \qquad (28)$$

where the η - function may be defined by

$$\eta\left(\omega, \frac{d_{max}}{k}\right) = \int_0^1 \frac{1-\omega}{\omega^2} \cdot \frac{\ell n\left[10.93\frac{d_{max}}{k} \cdot \zeta\right]}{\ell n\left[29.7\frac{d_{max}}{k}\right]} \zeta^{3/2}(1-\zeta)^{\frac{1}{\omega}-2} d\zeta \qquad (29)$$

since the lower limit of the integral in Eqn. (25), $x = 0$, corresponds to $\zeta = 1$ and the upper limit, $x = B/2$, to $\zeta = 0$. This function is evaluated by numerical integration and plotted in Fig. 4.

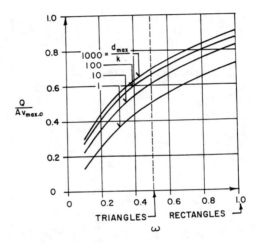

Figure 4. $Q/(v_{max.o}A)$ as a Function of Areal Aspect Ratio ω and Inverse Relative Roughness d_{max}/k.

For ω-values in the range $0.5 < \omega < 1$ the upper limit of the integral in Eqn. (29) has been replaced by a value slightly less than unity to facilitate the evaluation of the integral.

Once area A, maximum depth d_{max}, and areal aspect ratio ω are found from a survey of the section, and v_{max} and k determined by the one-orange-method, the discharge may be estimated directly from Fig. 5. The method is fast and therefore especially suited for rate of flow measurements in tidal rivers and in other water courses where the discharge may be changing substantially during the time needed for a conventional discharge measurement.

Estimation of Bed Load Capacity

The potential rate of bed load transport in a water course may be estimated by the one-orange-method that will provide values and distribution of the local bed shear stress. The corresponding local rates of bed load may be estimated by bed load formulas such as DuBoys' deterministic formula or Einstein's well-known stochastic bed load formula as reported for instance by Raudkivi (1990). Integration over the width of the water course will yield its total capacity for bed load transport. It must be emphasized that this procedure will not give the actual rate of sediment transport in the water course but only its potential capacity.

References

Christensen, B.A., "Measurement of flow, bed shear and roughness characteristics in open channels by an air bubble device," Proceedings of FLOMEKO 85, International Conference on Flow Measurements, pp. 35-42, Melbourne, Australia, 1985. Published by H.S. Stevens and Associates, Bedford, England.

Kraus, N.C., Lohrmann, A. and Cabrera, R., "New Acoustic Meter for Measuring 3D Laboratory Flows," Journal of Hydraulic Engineering, Vol. 120, No. 3 pp. 406-412, ASCE March 1994, New York.

Raudkivi, A.J., "Loose Boundary Hydraulics," 538 p., Pergamon Press, New York, NY, 1990.

EXPERIMENTAL DESIGN AND MEASUREMENT TECHNIQUES FOR INVESTIGATION OF TWO-PHASE FLOW

Andreas J. Kuck[1]

ABSTRACT

Discussing the relevant hydraulical and inter-phasial processes of two-phase flow in open tanks the necessary requirements for set-up and operation of physical model studies are worked out. Comparative investigations enable to identify the influence of the overflow rate, the *Froude* number and the detention time in prototype and model. Measurement techniques which are applicable under field and laboratory conditions are presented and applied to detailed studies of the internal flow patterns as well as for describing the overall performance of two-phase flow in hydraulic structures.

INTRODUCTION

Investigation of two-phase flow has become an important field in hydraulic research. Many technical structures which are applied in environmental engineering depend on the transport of dispersed particles in water rsp. the extraction of those substances from the fluid. Optimisation of these structures requires detailed knowledge of the related flow processes which often can only be obtained from studies of scaled-down physical models.

Despite a long history of scientific investigations on separation processes, the internal flow structures and interacting processes of different phases in open tanks still need further consideration. In order to obtain a deeper understanding of the hydraulics in the flow and the dynamics between the fluid phases, theoretical investigations have to be accomplished by experimental model studies carried out with appropriate measurement techniques.

---

[1] Research Engineer, Institute for Hydraulic Engineering and Water Resources Management, Aachen University of Technology, Germany

INTERACTION OF PHASES

Two-phase flow in hydraulic structures mainly depends on the physical processes of flow stratification, coagulation and sedimentation rsp. buoyancy. A basic problem in modelling this flow is that the two independent phases, the so-called outer or continuous phase and the dispersed phase usually have different properties and therefore often require separate experimental treatment. Regarding the dispersed phase special attention has to be given to the usually extremely complex and time-depending processes of aggregation and coalescence.

In a system of two different liquids, the actual size of the dispersed liquid particles is a result of a dynamic balance between dispersion on the one hand and aggregation and coalescence on the other hand which is generally not in equilibrium. RUMPF & SCHUBERT (1973) have shown that dispersion, i. e. the formation of small particles, is caused by high pressure and velocity gradients occurring at local disturbances in the flow.

Turbulence and thermal oscillation enable the accumulation of dispersed particles into a swarm of still individual elements. Hydrodynamically this aggregate shows a reduced flow resistance thus leading to an increased ascending velocity. However, the aggregates are not stable; eventually the liquid film between the particles will rupture and smaller elements join to a larger one. Therefore aggregation can be regarded as a step leading to coalescence which occurs in three forms (fig. 1): (A) fusion of individual elements within the continuum; (B) fusion of single elements with a free fluid surface; (C) assimilation of dispersed elements on a solid surface (SMITH, 1991).

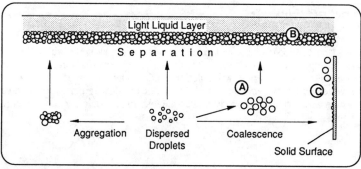

Figure 1. Principles of Particle Aggregation and Coalescence

Because the contrary processes of dispersion and aggregation rsp. coalescence are taking place simultaneously it is practically impossible to clearly separate both effects. Nevertheless a thorough acquisition of the dispersion characteristics is a basic requirement for investigations on phase-interaction in two-phase flow. Therefore it is appropriate to apply chemical or physical measurement techniques which do not focus on single particles but investigate the characteristics of the dispersed phase in a sample as a whole.

HYDRAULICAL ASPECTS

The flow patterns and hydraulical behaviour of two-phase flow is determined by the interaction of the flow structures of the continuous phase on the one hand and the dispersed phase on the other hand which generally is influenced by gravity effects like bouyancy and sedimentation. The relative movement of dispersed particles in the continuous phase follows *Stokes*'s law, giving a terminal velocity of an idealised spherical particle

$$v = \frac{1}{18}\left(1 - \frac{\rho_d}{\rho_c}\right)\frac{g\,d^2}{v} \qquad (1)$$

where ρ_d density of dispersed phase [kg/m³]
ρ_c density of continuous phase [kg/m³]
d diameter of dispersed particle [m]
g gravitational acceleration [m/s²]
v kinematic viscosity of continuous phase [m²/s]

This relation is valid for a *Reynolds* number of the dispersed phase

$$Re_d = \frac{v\,d}{v} \qquad (2)$$

in the range $10^{-4} < Re_d < 1$.

The main influences on the flow patterns of two-phase flow are according to FISCHERSTRÖM (1955) density and kinetic currents. In open tanks where the inflow has a higher specific density than the tank content due to temperature or density differences the inflowing liquid will reach the outlet in a bottom-near flow, while the surface water will turn backwards and dead zones with stagnant water bodies do occur. These zones can significantly reduce the hydraulic efficiency τ of the particular structure

$$\tau = \frac{t}{t_{th}} \qquad (3)$$

where t actual detention time [s]
t_{th} theoretical detention time [s]

A similar negative effect occurs if the kinetic energy of the inflow water is not equally distributed over the cross-sectional area of the tank. Then the jet from the inlet can reach far into the tank thus evoking a hydraulic short circuit. A suitable parameter for the stability of flow, i. e. its ability to withstand disturbing factors like density and kinetic currents is the *Froude* number Fr; this parameter describes the ratio of gravitational to inertial forces acting on a volume of liquid and should be high in order to guarantee a stable flow:

$$Fr = \frac{v_c^2}{g\, r_{hy}} \qquad (4)$$

where v_c velocity of continuous phase [m/s]
  r_{hy} hydraulic radius [m]

VERIFICATION OF MODEL INVESTIGATIONS

An inevitable tool for the verification of model tests are comparable investigations carried out in the prototype itself. Though this in some cases may require enormous technical and economical efforts, it is indispensable to guarantee a reliable prediction of the flow structures in the prototype. Therefore special interest lies in the efficient design of the required prototype tests to minimise the economical costs.

The discussion of internal tank hydraulics has shown the importance of the *Froude* number on the separation process. Other common approaches to design separation tanks either take the *Hazen* number

$$Ha = \frac{v}{v_o} \qquad (5)$$

where v_o overflow rate [m/s]

as a decisive parameter for the separator performance or they provide a particular detention time t to enable time-depending coagulation effects like aggregation and coalescence. Therefore physical modelling of the hydro-dynamic processes in separation tanks basically has to consider the following three hydraulic parameters:

  Fr   *Froude* number
  Ha   *Hazen* number
  t   detention time

Because in model studies only one parameter can be kept identical to the prototype at a time it becomes necessary to determine the dominant parameter in comparative investigations of prototype and model.

In an applied study the prototype of a rectangular tank designed for separation of dispersed light-liquid droplets in water was investigated. The separator (fig. 5) was operated with an overflow rate of 1.1 mm/s, a concentration of 0.5 % dispersed phase and a rate of dispersion of $3.95 \cdot 10^{-3}$. The respective model tests were performed with an undistored physical model of scale 1 : 5. In three distinctive test runs always one of the mentioned hydraulic parameters was kept identical to the prototype at a time; the other operating parameters like discharge and test duration were scaled accordingly. The concentration of the dispersed phase and the rate of dispersion in the inflow were identical to the prototype for all model tests.

The comparison between prototype and model (fig. 2) shows significant divergences for the *Hazen* model. The reason for the exaggerated values lies in *Hazen*'s simplified approach which assumes turbulence free flow and equally distributed velocity profiles, which obviously do not correspond with reality. On the contrary the detention time model shows better performance than the prototype due to over-proportional provision of ascending time for the rising particles. The best agreement between prototype and model results is achieved for the *Froude* model (KUCK & ROUVÉ, 1993). The maximum divergence of about 20 % above the prototype values is caused by partly neglecting time-depending effects. In practical applications this is tolerable as it leads to hydraulic structures slightly over-designed.

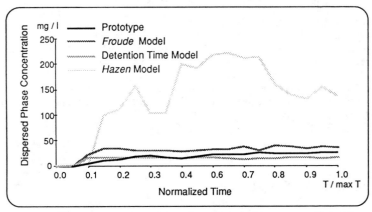

Figure 2. Comparison of Outlet Concentrations in Prototype and Models

MEASUREMENT TECHNIQUES

The basic inflow parameters for a physical model like discharge and hydrograph of the continuous phase can be easily calculated according to the chosen scaling law. Generally it becomes more difficult to model the inflow condition of the dispersed phase adequately, because a precise knowledge of the flow properties in the prototype, i. e. the size spectrum rsp. settling or ascending velocity of the dispersed particles, is required.

The characteristics of the dispersed phase can be analysed with an integrative measurement technique which exposes a sample of 1 liter to a separation process in a standardised flask (fig. 3). After 15 minutes the sample is divided into two fractions:

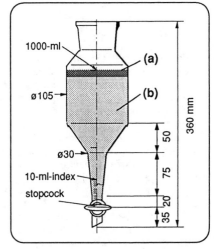

Figure 3. Standardised Separation Flask

(a) the upper layer with the floating light liquid volume, and (b) the remaining sample volume. The concentration of light liquid is chemically analysed in both fractions and interpreted in fraction (a) as directly separable and in fraction (b) as non-directly separable dispersed phase, the sum of both contents giving the total concentration of the sample. As a characteristic parameter the rate of dispersion D is defined as

$$D = \frac{c_n}{c_t} \tag{6}$$

where c_n concentration of non-directly separable dispersed phase [%]
       c_t total concentration of dispersed phase [%]

This technique forms an integrative access to the droplet size distribution of the dispersed phase and has proved as a robust and reliable method for application in the field as well as in the laboratory. With the obtained data it is possible to operate the model with a calibrated particle dispersion which guarantees inflow properties identical to the prototype.

In the prototype investigation the described approach has been applied to study the development of the rate of dispersion along the flow through a pipe. From the results it can be assumed that the process of dispersion is influenced by a number of physical parameters like discharge, pipe material, length, slope, disturbances in the pipe etc. An example for this investigation method is presented in fig. 4 for different discharge situations.

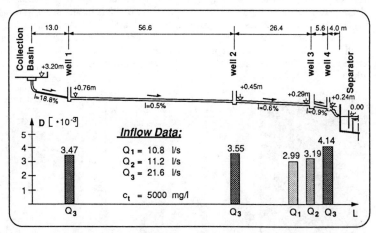

Figure 4. Development of Rate of Dispersion D in the Pipe Flow

The investigation of internal flow structures like density or kinetic currents and dead zones in scaled-down models is carried out with a technique based on Digital Image Processing (DIP) which has proved as a powerful and accurate tool. Activating the fluorescence property of the dispersed phase with UV-light, the interface between both phases is recorded with a video-camera through transparent tank walls. The images are digitised by a micro-computer and the position of the high concentration of the dispersed phase is determined at characteristic instants (fig. 5).

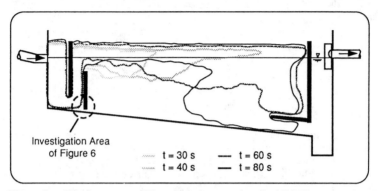

Figure 5. Displacement of Phase Interfaces analysed with DIP

In areas of particular interest, the technique of Particle Image Velocimetry (PIV) is applied for the optimisation of hydraulic structuresi. Illuminated with a pulsed light sheet the particles in a vertical plain are recorded at regular intervals; digitising and superimposing the particle position of consecutive instances provides a 'multi-exposed' digital image which is statistically analysed by 2-dimensional Fourier transformation to obtain velocity fields (Fig. 6). This procedure is automatically performed and enables to investigate critical tank sections where unequal velocity distribution or short circuits do occur. The combination of both techniques, DIP and PIV, allows the direct control of macro and micro flow structures and therefore proves as an excellent tool for the stepwise optimisation of two-phase flow structures.

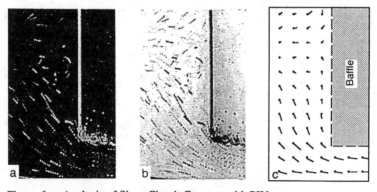

Figure 6. Analysis of Short Circuit Currents with PIV
a: Original Image   b: Inverted Image   c: Velocity Field

CONCLUSIONS

The application of scaled-down physical models allows to investigate the complex hydro-dynamic process of two-phase flow in open tanks. As a basic requirement to achieve comparable results the model has to be operated according to *Froude*'s scaling law with identical dispersion characteristics of the inflow in prototype and model tests. Methods of Digital Image Processing rsp. Particle Image Velocimetry prove as suitable tools to study large scale flow structures as well as to investigate the movement of single dispersed fluid particles. The application of these measurement techniques provides a better and deeper understanding of the separator hydraulics and enables a technically and economically effective optimisation of hydraulic structures.

REFERENCES

FISCHERSTRÖM, C. N. H. (1955), "Sedimentation in Rectangular Basins", *Proceedings ASCE*, Vol. 81, pp. 1 - 29

KUCK, A. J.; ROUVÈ, G. (1993), "Physical Modelling and Prototype Verification of Separation Processes in Large Oil / Water Separators", in: WANG, S. S. Y. (ed.): *Advances in Hydroscience and - Engineering*, Vol. 1, pp. 1952 - 1957

RUMPF, H.; SCHUBERT, H. (1973), "Verfahrenstechnische Gesichtspunkte bei der Leichtflüssigkeitsabscheidung", *Gesundheits-Ingenieur,* Vol. 94, pp. 225 - 256

SMITH, J. M. (1991), "Coalescence Phenomena", in: HEWITT, G. F.: *Phase-Interphase Phenomena*, Hemisphere Publishing Corporation, New York

Sediment Measurement Instrumentation
A Personal Perspective

E. V. Richardson[1] F. ASCE

Abstract

The development of the suspended sediment samplers in the 1940s was a major contribution to the science of sedimentation and provided for the proper development of dams, reservoirs, and river-control structures. The development of the samplers provided a means for measuring the total sediment discharge of rivers and development of a reliable method of determining the total sediment discharge, the Modified Einstein procedure. Many people were involved in developing the samplers and Modified Einstein procedure, but Paul Benedict and Bruce Colby deserve major credit. Benedict and Colby envisioned the need for and developed the sampler, and provided the resources for developing a method to determine total sediment discharge, and Colby developed the Modified Einstein procedure.

Introduction

In the 1940s, the first scientific instruments were developed to accurately measure the velocity weighted concentration of the sediment moving in the vertical or at a point in the vertical of a stream or river. The production models, which were the instruments used extensively in the field, were the D-43, DH-48, D-49, P-46, and P-50. The DH-48 was a depth-integrating hand sampler, the D-43 and D-49 were depth-integrating, cable-suspended samplers, and the P-46 and P-50 were point samplers. The DH-48, D-49, P-50, and P-63 samplers are illustrated in **Figure 1**. (The number in the designation refers to the year a standard design was adapted.) These

---

[1]Senior Associate, Resource Consultants & Engineers, Inc., P.O. Box 270460, Ft. Collins, CO, 80527 and Professor Emeritus, Civil Engineering Dept., Colorado State Univ.

suspended sediment-measuring devices provided the first reliable method of measuring the suspended sediment discharge of the streams and rivers of the world. Also, these instruments permitted the measurement of the total bed material discharge on streams using specially constructed or naturally occurring sections with highly turbulent flows. From these measurements, the Modified Einstein sediment discharge equations were developed and other equations checked.

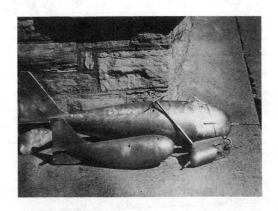

Figure 1. Sediment samplers: top: the DH-48, D-49, and P-50 samplers; bottom: the P-63 sampler.

This paper describes the development of the suspended sediment samplers and their use in measuring the total sediment discharge and the development of the Modified Einstein total sediment discharge method.

Measurement of Suspended Sediment Discharge

There were many prominent engineers involved in the development of the suspended sediment samplers and their use in measuring the total sediment discharge of sand-bed streams. Lane and Benedict were primarily responsible for the development. Lane recognized the importance of instruments for the measurement of sediment discharge in alluvial sand-bed streams and conceived the idea of an extensive study to develop such an instrument (pers. comm., Koelzer, 1993). Benedict was responsible for the design, laboratory, and field testing the initial experimental suspended sediment samplers (Inter-Agency Report No. 6, 1952) and their use to measure total sediment discharge.

AT THE TIME (1946 to 1952), SEDIMENT DISCHARGE OF A STREAM WAS CALLED LOAD, SUSPENDED SEDIMENT LOAD, TOTAL SEDIMENT LOAD, ETC., PROBABLY BECAUSE THE DISCHARGE WAS REPORTED IN TONS PER DAY. LOAD IS A MISNOMER, IN THAT IT IS THE TIME RATE OF THE QUANTITY OF SEDIMENT MOVING PAST A SECTION, WHICH IS A DISCHARGE. ANOTHER TERM, WASH LOAD, DESIGNATES THAT PART OF THE SEDIMENT DISCHARGE THAT IS NOT FOUND IN APPRECIABLE QUANTITIES IN THE BED (EINSTEIN, 1950). BENEDICT OBJECTED TREMENDOUSLY AND MADE THOSE OF US WHO WERE WRITING ON THE SUBJECT USE THE TERM, FINE SEDIMENT LOAD, INSTEAD OF WASH LOAD. FORTUNATELY, WASH LOAD HAS SURVIVED, BUT HOPEFULLY LOAD WILL NOT.

Lane recognized the need for a reliable and accurate sampler to measure the sediment discharge in a stream or river and convinced Hathaway of the Corps of Engineers of this need (pers. comm., Koelzer, 1993). Hathaway helped form an Inter-Agency Advisory group consisting of Fry (Tennessee Valley Authority), Paulsen and Love (U.S. Geological Survey), Brown (Department of Agriculture), and Hathaway (Corps of Engineers) (Rouse, 1980). The USBR and U.S. Indian Service joined and the name became the Interdepartmental Committee representing the above agencies. In 1946, this committee was superseded by the Federal Inter-Agency River Basin Subcommittee on Sedimentation (Inter-Agency Reports Nos. 1-6, 1940-1952).

Early work on developing a sampler and determining methods of analyzing sediment samples was performed by a Research Team headed by Lane at the Iowa Institute of Hydraulic Research. Other members of the Team were Koelzer (USGS), Horne (Corps of Engineers), Palmer (SCS), and Boyll (TVA). Other participants in various aspects of the program were Dubrow and Christensen (Corps of Engineers), Byrnon Colby and Benedict (USGS), Parker (USIS), Rhinehart (USFS), and Noble and Stanley (USBR).

With the advent of World War II in 1942, the leadership of the program was placed in the hands of Benedict and Nelson (Corps of Engineers). Benedict was mainly responsible for the development of the field instruments that were adopted for use. After August 1946, development of new models and laboratory and field research to determine the characteristics and limitations of the samplers were carried out by Byrnon Colby (USGS) and Christensen (Corps of Engineers) under the general supervision of Benedict and Nelson.

Six requirements were set early in the project for the developed depth-integrating and point samplers: (1) sampler must fill at a rate such that the velocity in the nozzle at the point of intake is equal to the local stream velocity, (2) intake nozzle should point into the stream parallel to the flow and project sufficiently far in front of the sampler body so that the flow streamlines at the nozzle are not affected by the sampler, (3) sampler should be smooth filling (not gulp water), (4) point samplers must not have an initial in-rush, (5) sample should be contained in a removable container, and (6) sampler should collect samples close to the streambed.

The developed samplers have, with very little modifications, met the test of time. The DH-48, D-49, and P-50 samplers are still in use. However, additional samplers have been developed by the Inter-Agency Subcommittee, such as pump samplers to sample the flow automatically at selected frequencies.

The Inter-Agency Subcommittee was not only a source of sediment-measuring equipment, but became a source of many fine engineers for the cooperating agencies. Many undergraduate and graduate students at the Universities of Iowa and Minnesota worked or became familiar with the project.

Measurement of Total Sediment Discharge

Suspended sediment discharge in a stream is only part of the sediment a stream transports. The suspended sediment discharge that was measured using these samplers in different streams could range from 30 to 95 percent of the total sediment discharge. Since the sediment samplers could only measure within 0.3 to 0.4 ft of the bed, there was an unmeasured zone that consisted of the contact sediment discharge and suspended sediment in the unmeasured zone. Streams transporting large concentrations of suspended sediment would have most of the sediment discharge measured (80 to 95 percent); whereas, streams transporting small quantities of suspended sediment would have a smaller portion of the sediment load measured (40 to 80 percent). A method was needed to determine the total sediment discharge of a stream if proper development of river-control structures was to be made on alluvial sand streams.

Benedict left the Inter-Agency project around 1946 to become Regional Engineer in charge of the quality of water studies in the Missouri River Basin. However, he was still involved with the research as the USGS representative on the Subcommittee. Sediment investigations in the Loop River basin in Nebraska indicated that the unmeasured load could be a significant, but unknown, quantity. The unmeasured discharge was also thought to be a major problem on other streams and rivers in the United States and World; for example, the sedimentation problems on the Lower Colorado River as a result of Hoover Dam, the erosion and sedimentation problems in the Wind River basin in Wyoming as a result of irrigation development, and the Indus River in Pakistan. Benedict, Lane, Borland, and others believed that if all the sediment discharge could be placed into suspension and if the suspended sediment sampler could sample the total depth, existing suspended samplers could be used to measure the total discharge of a stream.

One suggestion was to suspend all the sediment discharge using jets of air; another was the use of baffles. Benedict asked Albertson if he thought baffles could be used to suspend the total sediment discharge and Albertson replied in the affirmative. It was decided to build a turbulence flume under a bridge at Dunning, Nebraska. The flume used baffles to generate sufficient turbulence to suspend all the sediment and measure the suspended sediment with the DH-48 and P-46 samplers (Benedict et al., 1955). A sill was placed at the downstream end of the flume so that the depth-integrating samplers would traverse the total depth. The effort was sponsored by the USGS in cooperation with the USBR. Lane, Maddock, and Borland (USBR) assisted in the project. To design the flume, a model study was conducted at Colorado A & M College in Fort Collins, Colorado, under the supervision of Albertson. Matejka, a graduate student, played a major role in the model study and subsequently joined the USGS and collected data on the prototype. At this time, I entered the picture by working on the model study as an undergraduate student. Oltman and Vice (USGS) supervised the construction of the flume in the winter of 1948-49. In the model, 62 percent of the total concentration was transported in suspension, 22 percent was contact load, and 16 percent was unmeasured load. In the prototype, the percentages were

53, 29, and 18 percent, respectively (Benedict et al., 1955).

Modified Einstein Procedure for Total Sediment Discharge

The Dunning, Nebraska, flume studies were indicative of the problem of the unmeasured load, but did not give a general solution. Benedict decided, with the resources at his disposal, to research methods of determining the total sediment discharge. He assigned a portion of his staff at Lincoln, Nebraska, and the field offices in the Missouri River Basin to conduct these studies in addition to their ongoing duties of collecting quality of water (sediment and chemical) data in the basin. The majority of the theoretical work was carried out in the Lincoln, Nebraska, office by Bruce Colby, Hembree, Matejka, and Hubbell. To begin, they used the Dunning data to investigate existing equations (Hubbell and Matejka, 1959).

At the same time, it was decided to investigate the use of naturally occurring, highly turbulent stream sections to collect additional data on the total sediment discharge. The streams were the Niobrara River near Cody, Nebraska (Colby and Hembree, 1955), and Fivemile Creek near Riverton, Wyoming (Colby et al., 1956). These early studies of existing equations convinced Colby that the Einstein equation and procedure was the most theoretically sound (pers. comm., Colby, 1960). Using the above data, Colby, working with Hembree, Hubbell, and others modified the Einstein equation so that measured suspended sediment data could be used to determine the sediment discharge in the unmeasured zone. He called it the Modified Einstein method to give credit to Einstein's concepts and equation (Einstein, 1950). Einstein, for some reason, took exception to being "modified" and so expressed himself to Colby, which hurt his feelings. Before Colby's death, I was able to convince Einstein of Colby's intentions and reconcile the two.

Hubbell should write about the development of the Modified Einstein method of determining the total sediment discharge of a stream, as he worked directly with Colby and played a major role in its development. It was a major contribution to sedimentation knowledge, inasmuch as it is the most accurate method of determining the total sediment discharge of a stream (outside of measuring it in specially constructed or naturally occurring turbulent stream sections). I was in Riverton, Wyoming, at the time (1949 to 1953) and collected data on Fivemile Creek near Riverton and constructed the turbulence flume and collected data at Fivemile Creek Near Shoshoni, Wyoming.

Fivemile Creek is an ephemeral stream that, due to return flow from an irrigation project, was converted to a perennial stream. The result was a stream that had a low annual sediment discharge was changed into a stream that transported a large annual sediment discharge (up to 3 million tons per year of suspended sediment). Investigations were conducted into the source of the sediment discharge on Fivemile Creek and methods devised in the 1950s to decrease the discharge. Other streams in the area (Wind and Popo Agie Rivers, Beaver, Poison, Badwater, Muddy, Cottonwood Creeks, and others) as well as irrigation drains, were investigated. A major impetus to the studies was the construction of Boysen Reservoir on the Wind River at the head of Wind River Canyon and upstream of the Wedding of the Waters. In the studies, the daily suspended sediment discharge was determined by daily stream sampling. High school students were hired in the summer to collect samples and help in the studies. One of the students, Jim Goodman, became a Structural Engineering Professor at Colorado State University and is now Vice President of South Dakota School of Mines. I like to think that we influenced him to become a Civil Engineer, but we also, probably, discouraged him from hydraulics. The studies showed that Fivemile Creek was the largest contributor of sediment into Boysen Reservoir. The USBR constructed erosion control on Fivemile Creek that decreased total sediment discharge into Boysen Reservoir from Fivemile Creek by 90 percent.

To determine the total sediment discharge in Fivemile Creek, a narrow, steep section of the stream (8 to 10 ft wide) incised into bedrock was used. This section was approximately in the middle of the irrigation project (Fivemile Creek near Riverton). Up- and downstream of this bedrock section, the creek was wide and had a sand bed. Across this narrow section, a 2 by 12 ft plank was placed from which to sample using the DH-48 hand sampler. A study of the sediment size distribution in the vertical showed that all sizes that the sampler sampled were uniformly distributed in the vertical. A DH-48 sampler, fitted with a cap that could be opened or closed to get point samples, was used in the study. Depth of flow in the narrow section was under 3 ft and the initial in-rush could be neglected. The sampler, of course, could not sample material coarser than 3/16 or 1/4 inch. Albertson and I tried to determine the quantity of large material not sampled by sampling with a specially constructed wire basket on a long steel rod, but we lost the basket in the high velocity flow (12 to 16 ft/sec) and almost lost ourselves trying to hold it.

The magnitude of the suspended sediment discharge in Fivemile Creek was so large that it was decided to construct a narrow section at lower Fivemile (Fivemile Creek near Shoshoni) in order to determine its total sediment discharge into Boysen Reservoir. Albertson and I placed large rocks (approximately 3 by 4 in cross section by 4 to 7 ft long) to form a channel approximately 10 ft wide in the gaging reach at Lower Fivemile Creek. We guided the rocks into place as they were lowered by a crane furnished by the USBR. After the narrow channel was formed, the stream channel was backfilled behind the larger rock with smaller rock and soil. The gaging reach was sand bed over bedrock. The narrow reach was swept clean of sand which I confirmed by being knocked into the flume when placing the rock during construction. Sampling was conducted from a 2 by 12 ft plank as was done for Middle Fivemile Creek, and as at Middle Fivemile Creek, all sizes were uniformly distributed in the vertical. These measurements were used to determine how to decrease the sediment discharge into Boysen Reservoir and supported the development of the Modified Einstein equation.

In 1964, as part of the alluvial channel studies being conducted by Simons and me working for the USGS at Colorado State University, a weir was constructed across the Rio Grande Conveyance Channel near Bernardo, New Mexico (Harris and Richardson, 1964; Gonzales et al., 1969; Culbertson et al., 1972). This weir allowed for the measurement of the total sediment discharge as well as the total water discharge in the Conveyance Channel. In the winter, when the weather was bad in Washington D.C. and the Conveyance Channel was in full operation, Dawdy, Carter, and Benedict from Washington, D.C.; Culbertson, Nordine, Gonzales, and Curtus from Albuquerque, New Mexico; Ames from Denver, Colorado; and Simons and me from Fort Collins, Colorado, collected total sediment discharge, water discharge from the weir, and data from a normal reach of the Conveyance Channel. Among other studies the data were used by Burkham and Dawdy (1980) to eliminate some of the empirical adjustments contained in the Modified Einstein procedure.

Conclusions

The development of the suspended sediment samplers by the Federal Inter-Agency Subcommittee on Sedimentation (formerly the Interdepartmental Committee) was a major contribution to sediment transportation knowledge. The samplers accurately measured the suspended sediment discharge of rivers and streams, as well as the total sediment discharge at turbulence flumes and naturally occurring highly turbulent stream sections.

These samplers provided the information needed to develop the Modified Einstein procedure for determining the total sediment discharge of any stream. Because this method utilizes measured suspended sediment discharge, it is the most accurate and reliable method of obtaining the total sediment discharge of streams and rivers. It also greatly advanced our knowledge of erosion and sediment processes.

Many people were involved in the development of the sampler and the Modified Einstein procedure. However, credit largely goes to Paul Benedict and Bruce Colby. Benedict for the sampler and envisioning the need and providing the resources for developing a method to determine total sediment discharge and Colby for developing the Modified Einstein procedure.

References

Benedict, P.C., Albertson, M.L. Albertson, and Matejka, D.Q. (1955). "Total Sediment Load Measured in Turbulence Flume." Trans. ASCE, Vol 120

Burkham, D.E. and Dawdy, D.R. (1980). "General Study of the Modified Einstein Method of Computing Total Sediment Discharge." USGS Water Supply Paper 2066.

Colby, B.R. and Hembree, C.H. (1955). "Computation of Total Sediment Discharge Niobrara River near Cody, Nebraska." USGS Water Supply Paper 1357.

Colby, B.R., Hembree, C.H., and Rainwater, F.H. (1956). "Sediment and Chemical Quality of Surface Waters in the Wind River Basin." USGS Water Supply Paper 1373.

Culbertson, J.K., Scott, C.H., and Bennett, J.P. (1972). "Summary of Alluvial-Channel Data from Rio Grande Conveyance Channel, New Mexico, 1965-69." USGS Water Supply Paper 562J.

Einstein, H.A. (1950). "The Bed-Load Function for Sediment Transportation in Open Channel Flows." U.S. Dept. of Agriculture Tech. Bulletin 1026.

Gonzales, B.B., Scott, C.H., and Culbertson, J.K. (1969). "Stage-Discharge Characteristics of a Weir in a Sand-Channel." USGS Water Supply Paper 1898-A.

Harris, D.D. and Richardson, E.V. (1964). "Stream Gaging Control Structure for Rio Grande Conveyance Channel near Bernardo, New Mexico." USGS Open File Report.

Hubbell, D.W. And Matejka, D.Q. (1959). " Investigations of Sediment Transportation, Middle Loup River at Dunning, Nebr." USGS Water Supply Paper 1476.

Inter-Agency Subcommittee on Sedimentation, The District Engineer, Corps of Engineers, St. Paul, Minnesota:
Report No. 1, (1940). "Field Practice and Equipment Used in Sampling Suspended sediment."
Report No. 2, (1940). "Equipment Used for Sampling Bed-load and Bed Material."
Report No. 3, (1941). "Analytical Study of Methods of Sampling Suspended Sediment."
Report No. 4, (1941). "Methods of Analyzing Sediment Samples."
Report No. 5, (1941). "Laboratory Investigation of Suspended Sediment Samplers."
Report No. 6, (1952). "The Design of Improved Types of Suspended sediment Samplers."
Report No. 7, (1943). "A Study of New Methods for Size Analysis of Suspended Sediment Samples."
Report No. 8, (1948). "Measurement of the Sediment Discharge of Streams."

Rouse, H. (1980). "Early Years of the Federal Inter-Agency Sediment Project." Report in Files of E.V. Richardson, Ft. Collins Colorado.

Real-Time Data Collection of Scour at Bridges
David S. Mueller[1], M.ASCE and Mark N. Landers[2], M.ASCE

ABSTRACT

The record flood on the Mississippi River during the summer of 1993 provided a rare opportunity to collect data on scour of the streambed at bridges and to test data collection equipment under extreme hydraulic conditions. Detailed bathymetric and hydraulic information were collected at two bridges crossing the Mississippi River during the rising limb, near the peak, and during the recession of the flood. Bathymetric data were collected using a digital echo sounder. Three-dimensional velocities were collected using Broadband Acoustic Doppler Current Profilers (BB-ADCP) operating at 300 kilohertz (kHz), 600 kHz, and 1,200 kHz. Positioning of the data collected was measured using a range-azimuth tracking system and two global positioning systems (GPS). Although differential GPS was able to provide accurate positions and tracking information during approach- and exit-reach data collection, it was unable to maintain lock on a sufficient number of satellites when the survey vessel was under the bridge or near the piers. The range-azimuth tracking system was used to collect position and tracking information for detailed data collection near the bridge piers. These detailed data indicated local scour ranging from 3 to 8 meters and will permit a field-based evaluation of the ability of various numerical models to compute the hydraulics, depth, geometry, and time-dependent development of local scour.

INTRODUCTION

Scour of the streambed at bridge piers and abutments during floods has resulted in more bridge failures than all other causes in recent history (Murillo, 1987). Numerous studies to understand and model hydraulic and sediment transport

---

[1]Hydrologist, U.S. Geological Survey, Water Resources Division, Kentucky District, 2301 Bradley Ave., Louisville, KY 40217

[2]Hydrologist, U.S. Geological Survey, Water Resources Division, Office of Surface Water, 415 National Center, Reston, VA 22092

processes resulting in scour of the riverbed at bridges (bridge scour) have been conducted in laboratory settings. Design and evaluation methods based only on laboratory studies that do not adequately reproduce the effects of turbulence, flow instability, heterogeneous bed material, and other factors present in natural rivers are frequently limited in their application and reliability for real- world conditions. Historically, very little emphasis has been placed on the collection of field data to study the processes associated with scour of the riverbed at bridges. The catastrophic failures of the New York State Thruway bridge spanning Schoharie Creek and the U.S. 51 bridge over the Hatchie River in Tennessee highlight the need for improved techniques for design, evaluation, and monitoring of scour at bridges. At the heart of the research to improve current practice is the need for field data suitable for studying the processes causing scour.

The U.S. Geological Survey (USGS), in cooperation with the U.S. Federal Highway Administration, is researching, evaluating, and integrating instruments that can be used to make detailed scour measurements (Landers and others, 1993). Although this effort is not complete, the record flood on the Mississippi River during the summer of 1993 provided a rare opportunity to collect data on scour at bridges during extreme hydraulic conditions. These conditions provided a severe test of the equipment in its current state of development. This paper describes the instrumentation used and its limitations and capabilities for collecting data on scour at bridges.

DATA COLLECTION EQUIPMENT AND TECHNIQUES

Collecting hydraulic and sediment transport data during floods in sufficient detail to study scour processes and improve design and evaluation methods is a complex task. A complete data set should include three-dimensional velocity measurements, channel bathymetry, bed-material load, bed material samples, water-surface elevation, water-surface slope, water temperature, and discharge. These data should be collected during the rising limb, at the peak, and during the recession limb of the flood hydrograph in the reaches upstream, at, and downstream of the bridge with increased detail around bridge piers and abutments. Although the measurement of bed-material load is still lacking, recently introduced technology and the improvements in existing technology have made the collection of the other data feasible.

Deployment

The development of a remote control boat discussed by Landers and others (1993) was not completed at the time of the flood. The use of a manned boat as the survey vessel requires sufficient clearance beneath the bridge to avoid safety hazards and collect data under the bridge. This severely limited data collection on smaller rivers where clearance under bridges is often less than 1.5 meters (m) during floods; however, navigable rivers such as the Mississippi River had sufficient clearance under nearly all bridges to permit a manned boat to safely pass beneath the bridge. Therefore, detailed data collection efforts were concentrated on bridges over the Mississippi River.

Reliability, handling, and adequate launch facilities are also important considerations on use of a manned boat during floods. During extreme flood conditions, boat ramps are flooded and velocities can be high, even near the shore. Flooded local streets with sufficient slope and the river side of levees were used to launch the survey vessel during the 1993 flood. The support and cooperation of local citizens and government agencies were valuable in locating adequate launching facilities at all sites. The reliability and handling characteristics of the vessel are important for safety and are required to maneuver the vessel near and around the bridge piers.

Horizontal Positioning

Channel bathymetry and detailed three-dimensional velocity data were collected from a moving boat. The value of these data are dependent on knowing the horizontal position for each data point. Technology related to horizontal position determination has improved dramatically during the past several years. Hydrographic surveyors have used positioning systems for many years, but the accuracy attainable by most of these systems is less stringent than that required for the purpose of analyzing and modeling river processes. Previous technology could not typically provide real-time position measurements with an accuracy of less than 1 m and often required time-consuming pre-surveying of setup locations. Current technology can provide better than 1 m accuracy and fast setup with no pre-surveying requirements. A range-azimuth system and two Global Positioning Systems (GPS) were used to determine the horizontal position of the data collected.

Range-azimuth tracking systems are similar to total stations used for land surveying. Tracking systems have broad-beam lasers and do not have to be pointed directly at a target in order to acquire data; however, if the instrument is not pointed directly at the target the horizontal and vertical angles will not be correct and some error will result. These systems usually have a powerful, eye-safe laser, capable of reflecting from objects up to 300 m away and from prisms up to 10,000 m away. They have the capability to update readings every 0.5 seconds, to filter the data using gating and maximum and minimum features, and to automatically resume tracking even after the target is lost for a period of time. The system selected for this project has a distance accuracy of 0.1 m and a horizontal and vertical angle accuracy of 5 seconds. Although these numbers would indicate accuracy in the order of 0.2 m, in practice, it is very difficult to keep the instrument pointed directly at a moving target. During the data collection on the Mississippi River, it is believed that an accuracy of approximately 0.7 m was achieved when tracking the target mounted on the moving survey vessel.

The power of the laser allowed setup points to be referenced to the bridge quickly and often without the need of a prism. At many setups, the laser allowed the setup to be referenced to the centerlines of several piers by pointing the instrument at the centerline of each pier and reflecting the laser directly off the concrete pier. This capability allowed for fast and efficient setups and insured that the majority of the time on site was spent collecting data rather than setting up instruments and surveying control points. Plotting of the pier and setup point locations showed that an accuracy of about 0.3 m was achieved when using the instrument to survey setup points in the manner described.

During extreme events, such as the 1993 flood, locations for instrument setup that provide an adequate view of the bridge, approach, or exit sections are often difficult to find, especially at sites with very wide flood plains. The range-azimuth tracking system worked well for tracking the boat near the bridge. Tracking the boat during surveys of the approach and exit reaches was complicated by the presence of vegetation on the river bank which limited the view upstream downstream. GPS is well-suited for the approach- and exit-reach surveys.

GPS is a rapidly evolving technology. Differential GPS (DGPS) increases the accuracy of conventional GPS by making corrections based on GPS data collected at a known location. During initial evaluations of positioning systems, real-time, kinematic, DGPS with decimeter accuracy was being proposed but was not readily available. In addition, this accuracy required four satellites to be locked in at all times; if failure occurred, it could take several minutes for the system to reacquire lock and be ready to track again. The capabilities of DGPS have evolved rapidly over the last 18 months, since it was reviewed for applications to bridge scour data collection. Real-time, kinematic, DGPS can now achieve centimeter accuracy, and during a kinematic survey, the new technology can automatically begin position updates immediately upon reacquiring lock on a sufficient number of satellites. These enhancements increase the potential value of DGPS for collection of scour data at bridges.

During the flood, the U.S. Army Corps of Engineers (COE), St. Louis District operated a GPS base station in St. Louis, Mo. They broadcasted differential correction data from a radio tower in St. Louis, and, via phone links to St. Louis, from radio towers located south along the Mississippi River. The USGS rented a (roaming) GPS unit for use on the survey vessel and the COE provided a data radio to allow use of their transmitted real-time differential corrections. Because the COE had set their base station up rapidly to facilitate work on the flood, they had not determined the accuracy attainable. The accuracy of the position data collected is believed to be about 1 m or less.

Real-time, kinematic, DGPS used for navigation reference and positioning allowed rapid collection of velocity and bathymetric data in the approach and exit reaches of the river. Because DGPS requires no setups on shore and no personnel to track the boat, data was collected very rapidly and over a much longer reach of river than would have been feasible with the range-azimuth tracking system. However, data collection near the tree lines and bridges was hampered by loss of adequate satellite coverage caused by blockage of the sky by trees and bridge structure. At one site, which has a high and narrow-lane bridge, the GPS worked underneath the bridge because there were sufficient satellites low in the sky upstream and downstream of the bridge. However, coverage was lost when the boat was positioned next to a pier. On narrow tree-lined streams, the utility of DGPS may be limited, depending on the configuration of the satellites at the time of the survey. Additional investigations on the use of DGPS on small streams is needed.

The optimum positioning system for collecting real-time scour data at bridges allows both detailed positioning data to be collected under the bridge as well as in the approach and exit reaches. Using available technology, the optimum system is a combination of DGPS and range-azimuth tracking systems. The DGPS provides accurate positions in areas where adequate satellite coverage can be maintained. The range-azimuth system provides accurate positions under the bridge, around the piers, and on small streams where DGPS may not be usable because of lack of adequate satellite coverage.

Riverbed Elevation Measurement

The elevation of the riverbed is determined by measuring the distance from a known datum to the riverbed. Echo sounders are commonly used to measure the distance from a submerged transducer to the riverbed. The distance from the transducer to a known elevation must then be determined. Heave of the boat, change in vertical position caused by wave action, causes rapid vertical change in the distance from the transducer to a known datum. This vertical movement is too rapid for accurate measurement by the range-azimuth tracking system. In addition, pitch and roll of the boat moves the transducer out of its vertical orientation; therefore, the distance measured may not be a vertical distance. Several instruments that measure vessel attitude (heave, pitch, and roll) have been evaluated for use in collecting scour data, however, the instruments designed to measure vessel attitude accurately in dynamic conditions are very expensive (approximately $40,000). No vessel attitude instruments were used during the scour data collection. The water-surface elevation was recorded and assumed to be constant near the bridge. In the approach and exit reaches, the water-surface elevation was adjusted to the average water-surface slope in the area estimated from upstream and downstream gages and concurrent water-surface surveys. The depth of the transducer was measured, and the riverbed elevation was computed as the water-surface elevation less the depth of the transducer and the depth measurement. The accuracy of the riverbed elevation measurements was a function of the dynamic motion of the boat and slope of the water surface. The estimated accuracy of the riverbed elevation data is about 0.3 m.

The quality and performance of the echo sounder is also important for

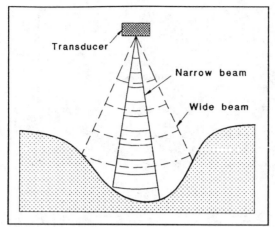

Figure 1. Effect of transducer cone angle on the acoustic footprint of an echo sounder.

accurate riverbed elevation measurements. For hydrographic surveying, it is important that the echo sounder provides signal processing with a digital output that can be recorded by a computer. Two types of signal processing schemes are commonly employed. The most common scheme is threshold detection, which measures the distance based on the time from acoustic release until the reflected acoustic energy exceeds a predetermined threshold. Peak value detection analyzes all reflected energy and computes the distance associated with the peak of the return signal. The acoustic footprint of the echo sounder is a function of the cone angle of the transducer. A wide cone angle results in a large footprint and less accurate measurements of steep slopes or rapidly changing bottom (Fig.1). The peak detection method is less sensitive to acoustic reflectors in the water column (sediment, fish, debris, etc.) and tends to measure the approximate center of the acoustic footprint rather than the edge of the footprint effectively reducing the acoustic footprint. Most survey-grade echo sounders also provide a hard copy record in the form of an analog strip chart; this is valuable for quality assurance of the digital data.

The echo sounder used for data collection employed peak detection with a 3° cone angle transducer and provided RS-232 compatible digital output as well as an analog paper chart. The value of the analog paper chart was not fully appreciated before these data-collection activities. The paper chart records the vertical location of all objects causing acoustic reflection, while the digital processing produces only a single value. Figure 2 shows a situation with the boat very near a bridge pier where the riverbed had scoured below the top of the footing. Portions of the acoustic signal reflected off the pedestal, footing, and seal of the pier before reflecting off the bed. The reflection off the pier was strong enough to cause the signal processor to digitize a depth shallower than the actual depth to the riverbed. Utilizing only digital data, the

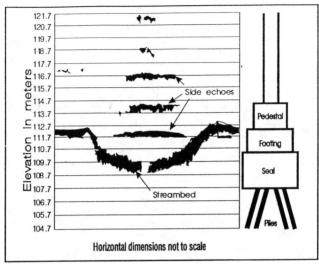

Figure 2. Example of side echoes from bridge pier.

actual depth and extent of the scour hole would have been missed; however, the depth and extent of the scour hole is clearly indicated on the paper chart. Therefore, it is important that the paper chart be used to verify the digital data and that any necessary corrections to the digital data be made.

Velocity Measurement

Traditional river velocity measurements made with horizontal or vertical axis meters contained only velocity magnitude. The inclusion of a flux gate compass in the weight deployed with the meters allows the horizontal direction of the velocity to be measured. However, these techniques only measure one point at a time and do not measure vertical velocities. The development of the Broadband Acoustic Doppler Current Profiler (BB-ADCP) allows a three-dimensional velocity profile to be collected from a moving survey vessel.

The BB-ADCP measures velocity magnitude and direction utilizing the Doppler shift associated with the reflection of acoustic waves off moving objects (acoustic reflectors). The BB-ADCP transmits an encoded pair of acoustic pulses from four transducers through the water column and records the backscattered acoustic energy from acoustic reflectors (sediment and organic matter) in the water. The reflected signal is then discretized by time differences into individual segments representing specific depth cells within the water column. Additional acoustic signal analysis provides the velocity of the acoustic reflectors along each of the beams. The geometric arrangement of the four transducers allows the three-dimensional velocity vectors to be computed for each depth cell. Although, theoretically, only three beams are needed, the fourth beam provides a quality check of the measurement (RD Instruments, 1993).

To utilize the BB-ADCP on a moving boat, the speed and direction of the boat must be measured and used to correct the water velocity measured by the BB-ADCP. Under most conditions, the BB-ADCP does this by tracking the bottom speed and direction relative to the instrument. The BB-ADCP sends bottom-track acoustic signals and analyzes the Doppler shift of the backscattered energy associated with the riverbed. If the riverbed is stationary, the velocity of the boat can be measured accurately based on this principle. However, if the riverbed is actively transporting sediment this technique may not accurately measure the speed and direction of the boat.

During the 1993 flood, four different BB-ADCP's were used. Two of the BB-ADCP's operated at 1,200 kilohertz (kHz), one at 600 kHz, and one at 300 kHz. The two 1,200 kHz BB-ADCP's differed in the geometric arrangement of the transducers, one had a convex arrangement and the other a concave arrangement. The concave arrangement is slightly smaller and provides more protection for the transducers but may trap air along the face of the transducers. The first units used for collecting hydraulic data associated with scour at bridges during the 1993 flood were the 1,200 kHz units. The position of velocity measurements near the bridge made with the 1,200 kHz BB-ADCP, were recorded using the range-azimuth tracking system. The tracking system is not directly compatible with the software used with the BB-ADCP, so the position of the boat was recorded separately and notes were taken to associate the velocity measurements with the recorded position data. These position data also provided a check on the bottom track accuracy of the 1,200 kHz BB-ADCP. The 1,200 kHz units generally do not have adequate acoustic energy to sound the water column for the conditions being measured during the flood. The measurement conditions included maximum depths greater than 25 m, high suspended sediment load, and bed-load transport characterized by 2 m dunes.

A 300 kHz unit was used in cooperation with the COE, New Orleans District. The lower frequency provided greater acoustic penetration of the water column and of the mobile layer of the channel bed. The 1,200 kHz and 300 kHz units were run while anchored to a fixed barge in the flow. The 1,200 kHz unit tracked the surficial mobile bed layer and measured a moving bed while the instrument was fixed; thus its bottom tracking could not have been used to measure the instrument position. The 300 kHz unit was able to penetrate the mobile bed layer and track off of the immobile channel bed; thus, it correctly indicated an immobile instrument position. Traverse distances measured by the 300 kHz BB-ADCP compared closely with measurements using DGPS. Velocity measurements made in the approach- and exit-reaches using the 300 kHz BB-ADCP were positioned by making a non-differential GPS measurement at the beginning and end of each cross section measured. The accuracy of this positioning method is approximately 30 m.

A 600 kHz instrument was used during the recession limb of the hydrograph. DGPS is compatible with the software used with the BB-ADCP and was utilized with the 600 kHz unit to provide accurate position information and a second source for measuring boat velocity. Although the 600 kHz unit occasionally reported bad data, comparisons of the bottom track data with the DGPS data showed the bottom track to be accurate.

The BB-ADCP's allowed very detailed velocity data to be collected in the approach and exit reaches and near the bridge. However, the BB-ADCP's could not be used to collect velocity information in the vortices at the bridge piers. The BB-ADCP's assume the water velocity along a horizontal plan passing through the four beams to be uniform. The size of the vortices were often smaller than the area bounded by the four beams so that flow measured by one beam was not continuous with flow measured by another beam. Although the instruments could compute the velocities parallel to the beams, they were unable to resolve these velocity vectors into horizontal and vertical components due to the lack of flow uniformity between the beams. Therefore, although BB-ADCP's were able to measure a three-dimensional velocity profile under normal river conditions, but were not able to measure the velocity profile in the vortices that result in scour around bridge piers.

DATA COLLECTION PROCEDURES

Two sites on the Mississippi River were selected for detailed data collection, Interstate 255 (I-255) bridge located just south of St. Louis, Mo. and Missouri State Route 51 (S.R. 51) bridge located at Chester, Illinois. Bathymetric data were collected along the upstream edge, centerline, and downstream edge of the bridges. On the basis of these data, piers with scour were surveyed in detail. Cross sections extending about 20 m each side of the pier were made from as close to the nose of the pier as the survey vessel could get, to about 50 m upstream, in 8 m increments. Longitudinal sections parallel to the long axis of the piers were collected. Longitudinal sections were surveyed at increasing distances from the pier until evidence of the scour hole no longer showed on the echo sounder chart. These longitudinal sections were run from about 50 m downstream of the bridge to 50 m upstream of the bridge. Additional cross sections and velocity data were collected upstream and downstream of the bridge for 1 to 2 channel widths in 150 m increments. Approach- and exit-reach bathymetric and velocity data were collected at cross sections 300 to 600 m apart with the BB-ADCP using GPS and DGPS for positioning.

River stages at the I-255 bridge were read and recorded every few hours from a staff gage located about 300 m downstream of the bridge. The S.R. 51 site is a USGS streamflow and sediment discharge station. At this site, the stage is automatically recorded, and was routinely checked using a wire weight mounted on the bridge. Suspended-sediment samples are routinely collected at this site and were collected during the flood. In addition, bed-material samples were collected using a

BM-54 sampler deployed from the bridge deck. No sediment samples were collected at I-255, although bed-material and suspended-sediment samples were collected at other sites in the St. Louis area during the flood and bed-material data are available for the I-255 site from work prior to the flood.

SUMMARY AND CONCLUSIONS

State-of-the-art instruments were used to collect very detailed data on scour at bridges during the 1993 flood on the Mississippi River. The instruments used performed well when used in conditions that were within the specified operating ranges. The use of these state-of-the-art instruments allowed concurrent bathymetric and hydraulic data to be collected with an accuracy and level of detail not attainable with prior technology. Continued development of instruments utilizing new technology is necessary to provide adequate velocity measurements that completely characterize the vortices around bridge piers and accurately measure bed load.

REFERENCES

Landers, M. N., Mueller, D. S., and Trent, R. E., 1993, Instrumentation for detailed bridge-scour measurements, *in* Shen, H.W., Su, S.T., and Wen, Feng, eds., Hydraulic Engineering '93: New York, American Society of Civil Engineers, p. 2063-2068.

Murillo, J. A., 1987, The scourge of scour: Civil Engineering, American Society of Civil Engineers, v. 57, no. 7, p. 66-69.

RD Instruments, 1993, Direct-reading broadband acoustic doppler current profiler technical manual: San Diego, Calif., 224 p.

LOW-COST BRIDGE SCOUR MEASUREMENTS

J.D. Schall, M. ASCE[1], J.R. Richardson, M. ASCE[2], and G.R. Price[3]

Abstract

This paper reports on the use of a low-cost, highly portable sonar system. The ease of use of the instrument and the short time required to make measurements should make this system valuable for use by bridge inspectors. Some of the limitations of low-cost sonar devices are also discussed.

Introduction

Sonic sounding has been widely utilized for hydrographic surveying, and more recently for bridge scour measurement. Survey grade sonar instruments are often used for these measurements; however, a limited number of researchers and highway department personnel have been experimenting with lower cost, recreational type sonar devices for bridge scour measurement and monitoring. This paper reports on a low-cost instrumentation system that was developed specifically for bridge scour measurement and has been used in a variety of flow environments. Based on this experience the advantages and limitations of this type of instrumentation for bridge scour monitoring are discussed.

---

[1] Vice Pres., Resource Consultants & Engineers, Inc., P.O. Box 270460, Ft. Collins, CO, 80527
[2] Asst. Prof. Civil Engineering Dept., Univ. of Missouri, Kansas City, MO, 64110
[3] President, ETI Instrument Systems, 1317 Webster Ave., Fort Collins, CO, 80524

Development of a Low-Cost Sonar System

Portable sonar systems were investigated as a method to ground truth fixed scour monitoring instrumentation developed under a National Cooperative Highway Research Program project (Project 21-3). The initial "portable sonar" unit developed for ground truthing activities was a Lowrance LMS-200 graphic screen sonar mounted in a toolbox with a small battery to allow portability. The battery could be recharged in a car by plugging into the cigarette lighter, or at the office by plugging in a built-in trickle charger.

Two methods of positioning the sonar transducer were utilized. A hand-held cable suspension using a small torpedo weight with the transducer incorporated as an integral part of the torpedo was effective, but only up to velocities of about 1 mps. The second method used an extendible painters pole (up to 5 m) with the transducer attached to the bottom of the pole. In high velocities (as much as 2.5 mps) the drag on the transducer was significant which limited how deep into the water the transducer could be placed. Results indicated that only a narrow cone transducer (8°) would function in the high sediment concentrations typical of the Rio Grande test site (where the fixed scour instrumentation was located), particularly at low-flow depths. This was attributed to the more concentrated energy of the narrow cone, and the smaller areal coverage limiting the scatter of the sound wave by suspended sediment. Note that the ability to operate in sediment laden water is also a function of the sonar output power. The LMS-200 is a relatively high power unit at 75 watts RMS, as well as having advanced signal processing capabilities to assist in discriminating noise.

While this instrument facilitated ground truthing during weekly site visits, its size and weight still made it cumbersome for one person to operate. Using a small dash-mount style sonar with digital readout only, a more portable instrument was developed. The instrument selected for use was an Interphase model DG-1 with a narrow cone transducer. The low current drain of this sonar (less than 0.3 amp) allowed powering it by a smaller battery source. To minimize the overall size of the instrument, conventional RC (remote control) battery packs were used with a 1500 mah rating. These ni-cad battery packs are readily available and can be quick recharged (15 minutes each) using an RC battery charger that plugs into a cigarette lighter or household current. Two 7.2 volt battery packs wired in series provided the proper voltage to operate the sonar. The sonar and battery packs were mounted in a PVC pipe not much bigger than a large flashlight (7.5 cm diameter, 25 cm long) and equipped with a neck strap to allow one person operation The transducer was again mounted to a painters pole, but instead of a 5 m pole, a 7.5 m pole was used that allowed measurement off higher bridges.

One limitation of this instrument was its low output power (35 watts RMS) and lack of more sophisticated signal processing capability. Consequently, it did not work well in sediment laden flows and it had a minimum depth capability of 1 m. However, testing of this instrument on tidal bridges in Florida found that the instrument functioned well in the relatively clear tidal water conditions, as compared to the highly sediment laden flows in the Rio Grande.

Work in Florida also lead to development of a pontoon type float built of PVC pipe to position the transducer rather than using an extendible pole. A short pole with a hook on the end was utilized to maneuver and position the float. This device performed well in tidal water conditions, up to moderate velocities (e.g., 1 mps), and was effective on bridges as high as 15 m off the water when a standard transducer extension cable was utilized. During scour inspections, detailed information on scour and cross section conditions were obtained by simply walking across the bridge, towing the float and reading the depth at every pier or bent. Although this device worked well in tidal flows and was simple to use even on large, long bridges, there might be problems in high velocity riverine flows if significant wave disturbance existed due to the relatively small size of the float.

To achieve portability in an instrument that would function well in sediment laden water, a higher powered graphics screen sonar was adopted for use. Unlike earlier graphics devices (e.g., the LMS-200 mounted in a toolbox), the newer generation graphics readout sonar devices are much more compact. A Lowrance model X25 powered by a small battery was selected for use. The instrument and battery were attached to a specially fabricated mounting bracket with a neck and waist strap that allowed one person operation. The higher power (75 watts rms) and advanced signal processing capability of this instrument provided accurate readings in depths as low as 25 cm. The battery can be recharged with a trickle charger overnight. This instrument has been utilized with an 8-degree transducer in both float and pole suspension in a variety of flow environments with good results.

Limitations of Low-Cost Sonar Devices

A common concern when using low-cost sonar devices for survey/inspection type work is their accuracy. Unlike survey grade instruments, recreational sonar devices cannot be field calibrated for density variations. The speed of sound in water varies with density, which is primarily a function of temperature and suspended or dissolved solids content. Recreational sonar devices typically use an average value of the speed of sound to estimate flow depth. For the Lowrance fishfinder type units used in much of this research, the speed of sound is assumed to be 1,463 mps, which is about the average value over a temperature range of 0 to 40°C, and is representative of freshwater at about 16°C.

An analysis of the error introduced over a range of flow depths when the actual speed of sound deviates from the assumed 1,463 mps was completed for NCHRP Project 21-3 (RCE, 1994). The results indicated that when the water temperature was near freezing, the maximum flow depth that can be measured without exceeding an error of +0.3 m is about 9 m. Similarly, when the water is warm (near 25°C) the maximum flow depth that can be measured without exceeding an error of -0.3 m is about 12 m. Careful interpretation of these results suggests that these error conditions are probably not a serious limitation in the practical application of low-cost sonar devices in bridge scour measurement/inspection work.

One observation is that flow depths of 9 to 12 m are typical of larger rivers and many rivers where low-cost instrumentation will be utilized will be shallower. Furthermore, even on large rivers flow depths in excess of 9 to 12 m would typically occur during flood season when the water temperature would probably be close to 16°C. Consequently, during flood season there will be little impact on the speed of sound and hence the accuracy of low-cost sonar devices. The overall conclusion is that temperature compensation or correction when using low-cost sonar devices will only be warranted or necessary under extreme environmental conditions (e.g., very deep water at near freezing conditions).

The effect of density on the speed of sound also raises concerns about the accuracy and use of low-cost sonar devices in at tidal bridge sites. The assumed speed of sound (1,463 mps) occurs at about 21°C in seawater, rather than the approximate 16°C in freshwater (RCE, 1994). At a tidal bridge site it would be unusual for the temperature to be greater than about 24°C or colder than about 10°C. These extremes would create errors of about -0.1 m in 24 m of water, to about +0.3 m in about 12 to 15 m of water, respectively (RCE, 1994). Consequently, as with low-cost sonar use in a freshwater environment, temperature compensation or correction should not be a serious limitation in the use of these devices for bridge scour measurement/inspection. If temperature correction is desirable, a correction based on theoretical considerations was developed during NCHRP Project 21-3 (RCE, 1994).

Conclusions

Low-cost, portable sonar devices discussed in this paper were originally developed for ground truthing of fixed scour monitoring instrumentation; however, the ease of use and short time required to make measurements should make these devices very valuable for bridge scour evaluation and regular inspection work. The instruments have been useful in scour inspections conducted in Florida as part of their Phase I scour evaluation program. Cross sections and scour measurements have been related to the bridge deck or guard rail, from which the sonar data can be converted to an absolute datum based on bridge plan information. On small bridges the measurements have taken from 20 to 30 minutes, and on very large bridges less than one hour.

References

Resource Consultants & Engineers, Inc. (1994). "Instrumentation for Measuring Scour at Bridge Piers and Abutments." Final Phase II Report for National Cooperative Highway Research Program Project 21-3.

Calibration of Movable Bed Model for Armant Area

F.S. El-Gamal* and A.F. Ahmed*

Abstract

The morphology of the Nile River, in Egypt, is undergoing continuous changes which causes navigation bottle-necks in some local reaches. About 200 km downstream Aswan, at Armant area, the shallow water depths in the river bend caused a navigation difficulties for the ships sailing on Luxor-Aswan route. Because of the importance of this reach and because of the limited time available for the study of this problem, a desk study was conducted and a reliable temporary solution for the problem was recommended. The recommended solution was a dredging of navigation channel crossing the river width to connect between the deepest parts in the problem area. The navigation channel has a width of 100 m and has a length of 450 m, with a bed level of 67 m, above MSL, which gives a water depth of 2.3 m in the winter closure period. In the mean time, a mobile bed model was constructed to test different alternative solutions including the recommended one and regularly monitored the morphological changes in the area, to be informed when dredging will be needed again. Moreover, another objective of the mobile bed model, with its ability of predicting the morphological time scale is to select an optimal solution concerning cost for a safe navigation in this river bend.

The main purpose of this paper is to introduce the technique used in both scaling of the processes involved and the calibration of the mobile bed model.

Introduction

Due to navigation difficulties caused by too shallow water depths in the river bend at Armant area, The River Transport Authority, RTA, has requested the

---

* Senior Researchers, The Hydraulics & Sediment Research Institute, Delta Barrage, Egypt

Hydraulics and Sediment Research Institute (HSRI) to study one of the navigation problems at Razaqat village (km 212 downstream Aswan) and propose a reliable economic solution.

An area, covered from km 209.5 to km 215, was surveyed as shown in Figure 1, and the data was analyzed. The bathymetric survey indicated that the problem is located at a bend where the cruises have to cross the river from one side to another to follow the navigation line. The water depth in the problem area, corresponding to a discharge of 1590 m^3 / s, vary from 3.1 m to 3.6 m, but during the winter closure period, where the water level decreases to about 69.3 m, the depths vary from 1.4 m to 1.65 m which are not enough for navigation. There is a secondary channel just upstream the problem area which carries about 15 % of the river discharge. Also, there is a small island, at the middle of the river, in the area just downstream the secondary channel, with a length of about 80 m and an average width of about 50 m. A sand bar is existed, just downstream the island, and extended for about 500 m. A series of groins are located on the left bank, outer curve, cause a difficult navigation condition near this bank.

Based on the results of a disk study, and because of the limited time available for the study of the problem, a reliable temporary solution was recommended. The recommended solution was a dredging of a navigation channel crossing from the right side at cross section 16 to the left side at cross section 22. The navigation channel has a width of 100 m and has a length of 450 m, with a bed level of 67 m above MSL.

A mobile bed model was constructed to test different alternative solutions including the recommended one and regularly monitored the morphological changes in the area, to be informed when dredging will be needed again.

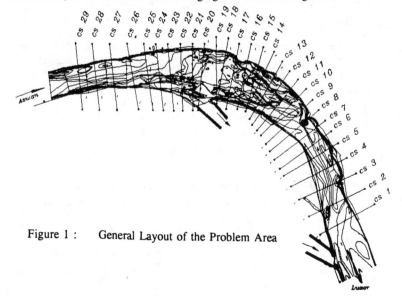

Figure 1 :   General Layout of the Problem Area

Dominant Parameters

Based on the recorded water levels, at El-Matana gauge station (km 174.7) and at Armant gauge station (km 213.85), the water surface slope was calculated and the corresponding water level at the problem area was determined. The water surface slope was found to be ranged from 4 cm/km to 4.2 cm/km with an average value of 4.1 cm/km.
The water level at the problem area ranged from 69.20 m to 73.20 m above MSL. The discharge of the Nile River at the problem area was calculated, based on the measured discharge downstream Esna Barrage. At the problem area, using least square method, the stage discharge relationship was found to be :

$$H = 67.66 + 0.029 \, Q^{\frac{2}{3}} \tag{1}$$

where H is the water level (m) and Q is the discharge (m^3 / s).

In order to design the movable bed model, the average parameters represented the problem area should be determined. On that base the dominant discharge, which defines as the discharge corresponds to the average sediment transport, all over the year, was calculated using (Engelund and Hansen 1967) sediment transport formula. The dominant hydraulic parameters corresponding to the dominant discharge were also calculated.

Model Design

In the design of a mobile bed models, the simulation of the water and the sediment movements are a must. Since the flow field for any alluvial channel can be characterized by Froude number (F_r), Reynolds number (R_e), and roughness distortion ratio (T_r). To provide a sufficient turbulence in the model, R_e must exceed a certain value. Generally this requirement can be easily fulfilled.
For rivers with shallow friction-controlled flow and dominant bed load transport, as in the case of the Nile River, (Struiksma, 1980 and Ahmed, 1990) showed that F_r is of secondary importance for the reproduction of the flow pattern since the flow field is mainly governed by the bed topography and the roughness distortion ratio T_r. This implies that this ratio has to be reproduced at full scale which leads to the so-called roughness condition :

$$n_C^2 = \frac{n_L}{n_h} \tag{2}$$

where the scale factor n_x of any parameter x is defined as the ratio of the value of x in the prototype to that in the model, C is the Chezy roughness coefficient, h is the average flow depth and L is a characteristic length, e.g. meander length.

In case of modelling alluvial streams, the satisfying of roughness condition leads in most cases to distorted model ($n_L > n_h$) because n_C is larger than one.

In the same time, correct reproduction of the bed topography is expected if the sediment transport scale is constant in space. According to (De Vries, 1973), this is achieved when the ideal velocity scale is fulfilled. This scale follows from the assumption that there is a unique function (sediment transport equation) between the transport parameter Ψ, and the flow or Shields parameter θ which is written as:

$$\psi = \frac{s}{\sqrt{D^3 \Delta g}} \quad (3)$$

$$\theta = \frac{h\, i}{\Delta\, D} \quad (4)$$

where $\Delta = [(\rho_s - \rho) / \rho]$ is the relative density of the sediment, ρ_s is the density of the sediment, ρ is the density of water, i is the water surface slope, D is the grain size, and s is the sediment transport in volume (including pores) per unit width.

Since similarity of the transport is aimed at, the ideal velocity scale can be obtained from the condition that ($n_\theta = 1$) which together with Chezy' relation and with using sand in the model leads to :

$$n_u = n_c \sqrt{n_D} \quad (5)$$

Equation (5) reveals that F_r in the model will be, in most cases, larger than that in the prototype because ($n_u < n_h^{1/2}$). This means that the condition of the ideal velocity scale leads generally to a deviation from Froude condition. Accordingly, model velocity will be exaggerated to provide sufficient bed material transport, and hence, the water surface slope is exaggerated in the model. In order to compensate such effect the model has to be constructed with the technique of tilted datum plane. Applying the above condition automatically leads to $n_\Psi = 1$, and therefore using the transport formula of (Engelund and Hansen, 1967), the sediment transport scale becomes:

$$n_s = n_D^{3/2}\, n_c^2 \quad (6)$$

To clarify the meaning of the tilting, one may assume rotating the whole longitudinal center-line of the modeled reach around an imaginary cross section. Considering the above definition the tilting angle i_t can be derived as follows:

$$i_t = i_m - i_p \frac{n_L}{n_h} \quad (7)$$

where i_m and i_p are the water surface slope in the model and prototype respectively.

The morphological time scale can be calculated using the formula :

$$n_t = (n_L \cdot n_h) / (n_D^{3/2} n_c^2) \qquad (8)$$

The above equation implies that the time scale is dependent on the vertical and horizontal scales, as well as the sediment transport scale. If the sediment transport formula does not give an acceptable accuracy, it is worth to mention that it is not sufficient to estimate the time scale in a simple way.

Therefore, the only method to give a proper value to the time scale can be obtained by reproducing an event in each of the prototype and model such as a dredged trench. A second best method is to measure the sediment transport in the model and prototype to determine the scale of the transport directly or via selection of a transport formula for prototype and model.

Summary of the results of the previous design procedures applied for the case of dominant discharge, using $n_L = 150$, are presented in the following table:

Parameter	Prototype	Scale	Model
Discharge m³/s	1484.4	13059	0.1136
Width m	400.1	150	2.67
Depth m	4.81	47.06	0.1022
Velocity m/s	0.771	1.85	0.415
Water surface slope	4.2 x 10⁻⁵	0.02333	0.0018
Chezy coeff. m^{1/2}/s	54	1.785	30.2
Grain Size mm	0.289	1.078	0.268

The sediment transport scale (n_s) and the morphological time scale (n_t) were calculated, using equations 6 and 8, and found to be equal to 3.57 and 1979 respectively. All the previous calculation of the scales of the hydraulic parameters and the corresponding model values are based on the initial assumed value of C_m. The correct value of C_m can be determined theoretically or experimentally in a separate flume using the previous calculated hydraulic parameters.

Model Description

About 5 Km reach of the prototype located between Km. 210 and Km. 215 was modeled. Banks and islands were shaped according to the field survey. The bed consists of about 0.40-m thick layer of medium sand with mean particle size of 0.268 mm. The sand was placed above a plane concrete layer of 0.15 m thick, covered with an impervious material to minimize or eliminate the seepage of the water. The physical model contains the model entrance which consists of a brick / mortar basin to accommodate a horizontal 2.5 m long, 0.4 m diameter steel-pipe. This contains a number of about 20 holes 70 mm diameter each which are uniformly distributed along the lower side of the pipe. The purpose of these holes is to equalize the distribution of the water and sand mixture along the model entrance. In order to establish the right inflow condition with respect to that in the prototype, a group of wooden strips (Piano) was fixed in a wooden frame at 0.30 m downstream of the entrance. The wooden strips can be easily adjusted, manually, to simulate and reproduce the trend of the inflow velocity distribution at the model entrance. In order to adjust the water surface level a flap gate 4.0 m wide was installed at the lower end of the model. The width of the tail-gate was designed to accommodate for different discharges. A sump was constructed downstream the tail-gate with a dimension of 4m x 3m at the top surface, 0.5m x o.5 m at the bottom and 4 m deep. The recirculating system consists of an electric centrifugal pump, 8 inch diameter plastic pipe-line, and a suction pipe connected to the sump. Water and sand were pumped from the sump and delivered to the model inlet through the delivery pipe-line. The measurements of water surface levels were carried out using four pairs of side-well stations distributed along the model length. Each was firmly affixed in a vertical position and connected from the lower end with a static head meter and equipped with a point gauge with vernier scale which give accuracy of 0.1 mm. Measurement of the flow rate was obtained by means of an electrical flowmeter which was installed at half way on the delivery pipe-line. A total number of 29 pairs of supports, located at distances ranged between 1 m and 1.5 m apart, were firmly [erected] along both banks of the model. These supports were utilized to accommodate a portable bridge which was used to measure the bed profile of each cross-section along the model.

In order to calibrate the model, some tests were Preliminarily taken place to assess the simulation of the scale model with the condition in the prototype. Using the flow condition, that represents the dominant discharge, tests were conducted to adjust the inflow condition, adjust the flow distribution, calibrate the averaged parameters and calibrate the local parameters. The procedures applied and the obtained results are summarized as follows:

Inflow Condition

As the flow discharge and slope were adjusted in the model, tests were conducted to adjust the mid-depth velocity distribution at the model entrance in such a way as to simulate that measured in the prototype. As the prototype measurements were carried out during a discharge of $Q_p = 1590$ m^3 / s which is larger than the calculated dominant discharge, the corresponding values for the velocity in the prototype will be larger than that in the model, therefore, in order to compare the velocities in model and prototype, as well as to take into account the decreasing in flow depth, the following procedures can be used:

1. The water level during the discharge measurement, Q_p, was 71.39
2. From the stage discharge relationship, equation 1, the water level corresponding to the dominant discharge, Q_d = 1484.4 m³/s, was calculated and found to be equal to 71.43 m.
3. Considering steps 1 and 2, we conclude that there is no change in the water level and consequently the water depth and the river width are consider constant.
4. Using the continuity equation, the relation between the dominant and the measured discharges can be written in the form:

$$\frac{Q_d}{Q_A} = \frac{u_d}{u_A} = \frac{1484.4}{1590} = 0.93 \qquad (9)$$

From equation 9, it was concluded that, during the comparison between model and prototype velocity distribution, every point value for the velocity corresponding to prototype must be multiplied by factor equal to 0.93.

Many attempts were made to adjust the inflow velocity pattern and the final result of these trials, together with the measured velocity in the prototype, for cross section 26, is shown in Figure 2.

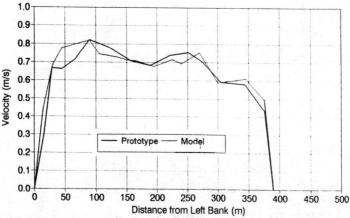

Figure 2 : Comparison of Inflow Velocity Profiles in Prototype and Model at Cross Section 26

Discharge Distribution

Attempts were made to adjust the discharge distribution through the second channel and the main branch of the river. Using the Electro-magnetic current-meter, mid-depth velocity profiles at a certain locations across the two branches were measured and the percentage of discharges were determined. As some difference from the condition in the prototype was resulted some adjustment was made at the upstream end of the secondary channel until the correct discharge distribution was achieved.

Calibration of the Average Parameters

The third phase of model calibration was conducted to adjust the average parameters, obtained from the design procedures. The model was continuously run, day and night, with the dominant discharge and the required slope. A sounding bridge was regularly applied twice a day along the flume length. As a result of the measurements the variation occurred in the average flow depth within the reach were recorded. During this stage the flow discharge and slope were regularly monitored and the necessary adjustments were made.

Results of average flow depth obtained from the tests showed stocastic variation which is due to the propagation of bed forms and some other influences, e.g. small variation in the discharge and water temperature. After a working time of about one month, during which a total number of 20 measurements were taken, equilibrium was seemed to be established. This stage was assigned when analysis of the measured data show no significant variation in the average values of the measured depth, as shown in Figure 3.

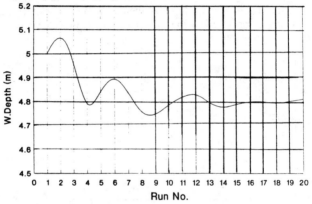

Figure 3 : Average Values of the Measured Depth in the Model

Calibration of the Local Parameters

Up to this stage, it can be concluded that the average flow depth and slope are adopted and the roughness condition is fulfilled. However, comparison of the bed topography in the model with that of the prototype, showed a local discrepancies in some cross sections. For this reason a fourth stage for the calibration took place during which the local parameters were refined.

To improve the condition in the model in such a way as to simulate the real flow pattern of the river reach, a fourth stage was conducted. To achieve such purpose some necessary modifications were locally introduced in the model. Those modifications are summarized as follows :

    a.   Due to the fact that, it is not possible to adjust both the flow and sediment at the model entrance using either the holes opening in the distribution pipe or the wooden strips (piano), a surface vanes were placed at the model entrance to adjust the distribution of the sediment transport.

b. The shape and side slope of the existing groins were reproduced in a more realistic way and in such a way as to minimize the scale effect.

c. The vertical boundaries for some locations at the banks and islands were tilted in such a way as to simulate the actual condition. This was done by considering the vertical and horizontal scales of the model as well as the real slope in the prototype.

Using the sounding bridge, measurements of bed profiles were regularly acquired then comparison with the corresponding prototype data shows a local discrepancies, specially along the outer bank where the groins are existed. Therefore, an attempt was made to reduce both the volumes and the depth of the scour holes at the groins site. This was done by replacing the solid groins with another type with the same dimensions. The new groins have holes on it to minimize the eddies at the front edge of the groins. Comparison of three longitudinal bed profiles, at the center of the river, at 50 m from the left bank and at 50 m from the right bank, for both model and prototype indicate that good results were obtained as shown in Figure 4.

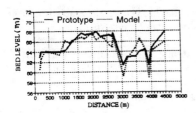

50 m From Left Bank   50 m From Right Bank

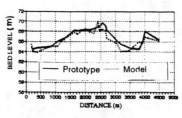

Axis of The River

Figure 4 :   Comparison of Longitudinal Profiles in Prototype and Model

Summary and Conclusions

Due to the navigation difficulties caused by too shallow water depths in the river bend at Armant area on the Nile River in Egypt, a dredging of navigation channel was recommended. The navigation channel has a width of 100 m and has a length of 450 m with a bed level of 67 m above MSL, which gives a water depth of 2.3 m in the winter closure period, during minimum discharges.

A mobile bed model was constructed to study the morphological time scale in the Nile river and to investigate the most economic solution for the problem. The scale rules of the model design were introduced and the different techniques used for the model calibration were presented.

The results of the model calibration including the inflow velocity distribution, the average water depth in the model and the bed profiles were in good agreement with the measured data in the prototype.

References

Ahmed, A. F. (1990). "River Scale Models with Movable Bed". Mission Report, The Hydraulics and Sediment Research Institute, Delta Barrages, Egypt.

De vries, M. (1973),"Application of Physical and Mathematical Models for River Problems", Delft Hydraulics Laboratory, Publication No.112.

Engelund, F. and E. Hansen (1967). " A Monograph on Sediment Transport in Alluvial Streams". Danish Technical press, Copenhagen, Denmark.

Struiksma, N. (1980)."Recent Developments in Design of River Scale Models with Mobile-Bed", IAHR Symposium on River Engineering and its Interaction with Hydrological and Hydraulic Research, Belgrad, May 26-28.

Using the SedBed Monitor to Measure Bed Load

Roger A. Kuhnle[1] and Robert W. Derrow II[2]

Abstract

An acoustic distance measuring device (SedBed Monitor) was developed to accurately measure bed surface transects in a sediment and water recirculating flume. From these transects the rate of bed load sediment transport was calculated as the product of the mean migration rate of the bed forms, the mean height of the bed forms, the density of the sediment, and a constant related to the shape of the bed forms. The sediment transport rates calculated from the bed surface transects were very close to mean transport rates measured using a density cell connected to the flume return pipe. Using bed surface transects to measure sediment transport rates has the advantage of a much shorter period of time necessary to obtain an accurate average transport rate when compared to conventional methods.

Introduction

Measurements of bed surface irregularities in an alluvial stream are important for determining channel resistance, flow depth, and for use to calculate sediment transport. An accurate knowledge of these parameters is necessary for developing design criteria for maintenance or restoration of alluvial stream channels. Yet measurement of these parameters is very difficult in most stream channels.

The bed forms that occur on a sand bed in an alluvial

---

[1]Research Hydraulic Engineer, National Sedimentation Laboratory, USDA - Agricultural Research Service, P.O. Box 1157, McElroy Drive, Oxford, MS 38655

[2]Research Scientist, National Center for Physical Acoustics, University of Mississippi, University, MS 38677

channel for different flows are well known (Simons and Richardson, 1961; Middleton and Southard, 1984). With increasing flow for a given flow depth over an initial plane bed of coarse sand the sequence of bed configurations are lower flat bed, dunes, upper flat bed, and antidunes. The dunes that form on sand beds tend to be nearly triangular in cross-section and migrate downstream. The experiments of this study were restricted to flows under which dunes are stable.

In order to facilitate the measurement of bed surface transects the National Center for Physical Acoustics (NCPA) has developed a high resolution PC-based underwater distance measuring device (SedBed Monitor). The use of the SedBed Monitor has made the collection of high quality bed surface transects routine. This data has simplified and improved the calculation of flow depths and has also been useful for the calculation of mean bed load transport rates for a given set of flow and sediment conditions.

Measurement of the mean rate of bed load transport for a given flow and sediment at a given channel cross-section generally requires that transport rates be averaged for periods on the order of hours (Willis and Kennedy, 1977; Gomez et al., 1990). The large spatial and temporal fluctuations that are characteristic of the transport of bed load by alluvial channels require long periods of sampling to calculate a representative mean rate.

One way to reduce the sampling time necessary to obtain an accurate mean bed load transport rate is to average the transport over a reach of channel in one measurement. This can be done by taking successive streamwise traverses of the bed surface, calculating mean size and migration speed for the bed forms and using this information with the density of the sediment to calculate the bed load transport rate. This idea is not new. Hubbell (1964) states that the idea of calculating the bed load from the speed and size of the bed forms was proposed in the 19th century. More recent studies concerning the calculation of bed load from bed form transects were conducted by Simons et al. (1965), Willis (1968), Willis and Kennedy (1977), and Engel and Lau (1980). These researchers either measured two transects in the streamwise direction over the same reach of channel or used two stationary probes and measured the time it took for the bed forms to migrate past them.

The unique aspect of this study is that only one pass of the instrument carriage was required for each bed load calculation. The SedBed Monitor made possible the collection of high resolution bed height data that was necessary to calculate bed load transport.

SedBed Monitor
    The SedBed Monitor (SBM) is a computer-controlled, high-resolution distance measuring device. The monitor

consists of a high frequency (1 MHz), acoustic pulse-echo system which graphically displays the bedform depth on a high resolution computer screen as well as storing the data to disk. The system is capable of temporal (time dependent) and spatial data collection with a distance measurement resolution of ±0.5 mm. The system is field configurable and can be equipped with up to 8 channels to allow bedform correlation calculations or multiple data collection sites. The sampling depth of the SBM ranges from 0.033 to 3.0 m. The user friendly software allows universal control over the system and unattended data collection. Data sets, which are in an ASCII format, can also be remotely downloaded via a serial communications port or on standard floppy disks.

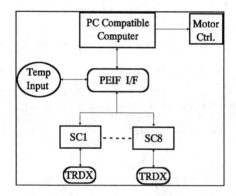

Fig. 1. SedBed Monitor system block diagram.

A block diagram of the major system components is shown in Figure 1. The PC compatible computer handles the display and system timing and control. The digital Pulse Echo Interface (PEIF) card performs the clocked data acquisition and control between the Sonar Cards and the computer. It also handles the temperature input data and control. The transmitter and receiver are built into the Sonar Cards, which send and receive the acoustic signal from the transducer (TRDX). There is one Sonar Card (SC) for each channel. The TRDX's can be used up to 300 ft. from their respective SC. Both the PEIF and the Sonar Card's were designed and built at NCPA and integrated into the overall system.

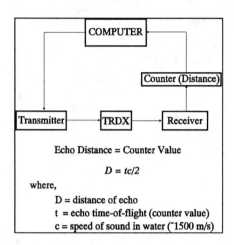

Fig. 2. Pulse-echo distance measurement method.

A simplified system diagram and distance measurement technique for the SedBed Monitor are shown in Figure 2. The distance from the TRDX face to the bed surface is calculated by multiplying the speed of sound in water (c, see Kinsler et al., 1982) by the echo time-of-flight (EToF). The EToF is simply the time it takes the transmit pulse to leave and the echo to return. The EToF x c product is then divided by 2, which gives the one-way travel distance. To achieve the high resolution in the distance measurements, the water temperature must be constantly monitored because of its effect on c, hence the SBM incorporates an automatic water-temperature measurement circuit with a resolution of ±0.5 degrees Celsius.

A diagram of a typical lab operational environment for the SBM is shown in Figure 3. The instrument carriage moves linearly along the flume rails with the aid of a stepper motor. The motor is controlled by the SBM computer via the Motor Driver Unit. The motor steps in sync with the software to move at a rate of 2 mm/step. Temporal measurements are basically the same as spatial measurements except the motor is turned off and sampling is done in a stationary position, unless the user provides an external motor drive system, as was done for the data collected for this paper. The SBM then provides a simple synch signal to allow the external motor to correlate movement with the SBM data acquisition loop.

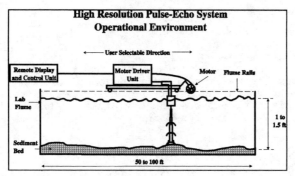

Fig. 3. Operating environment of SedBed Monitor.

Calculation of Bed Load Transport From Bed Surface Transects

Calculation of bed load transport rate using the migration rate of bed forms of given size and geometry is theoretically straightforward. In reality, however, complications arise as bed forms vary with distance along the channel and individual bed forms vary with time as the they migrate downstream. After Engel and Lau (1980), the volumetric bed load transport rate at a given point of a bed form is given by

$$q_{vp} = U_d (a-a_0) \tag{1}$$

where q_{vp} is the volumetric transport rate per unit width, U_d is the speed of the bed form migration, a is the local bed elevation, and a_0 is the elevation of zero transport. The mean transport rate over one bed form is given by

$$q_{vb} = U_d \frac{1}{S} \int_0^S (a-a_0) \, dx \tag{2}$$

where q_{vb} is the average volumetric transport rate over a bed form, S is the length of a bed form, and x is the streamwise direction. When a whole reach of channel with several bed forms is considered one obtains

$$q_{vr} = U_d \frac{1}{L} \int_0^L (a-a_0) \, dx = U_d a_e \qquad (3)$$

where q_{vr} is the average volumetric transport rate per unit width over the reach, L is the length of the reach, and a_e is the average thickness of the moving sediment layer. It is convenient to express the deviation of the bed profile relative to the mean bed plane ($\bar{a}$)

$$\bar{a}' = \frac{1}{L} \int_0^L |a-\bar{a}| \, dx \qquad (4)$$

where $\bar{a}'$ is the mean absolute deviation of the bed from the mean bed plane over the length of the transect. The mean thickness of the moving sediment layer becomes

$$a_e = \alpha \, \bar{a}' \qquad (5)$$

where α is a coefficient which depends on the bed form shape. The sediment load from migrating dunes can then be computed from

$$q_s = U_d \alpha \, \bar{a}' \rho_s (1-p) \qquad (6)$$

where q_s is the mean sediment transport rate from the dunes in dry mass per unit width, ρ_s is the density of the sediment, and p is the porosity of the sediment. Engel and Lau (1980) calculated dune load with a value of α equal to 1.32. They arrived at this value by assuming that the point of zero transport on a dune was slightly above the lowest point of the trough. Using the assumption that bed forms were composed partly of a triangle and partly of a sine wave, Willis (1968) derived a value of α that was about 1.6. The assumptions of Willis (1968) appear to apply more closely than those of Engel and Lau (1980) to the bed forms of this study. The value of α used in this study was 1.58.

Experiments

All experiments were conducted in the 30.48 m long water and sediment recirculating flume at the USDA National

Sedimentation Laboratory. The channel of the flume is 0.61 m deep by 1.22 m wide with the sand bed occupying approximately 0.15 m of the channel depth. The bed was composed of quartz sand with a median diameter of 0.795 mm and a standard deviation of 0.45 ϕ ($\phi = -\log_2 d$, where d is grain diameter in mm). The size and narrow distribution of the sand assured that transport would be predominantly as bed load, which simplified the comparison to total sediment load measurements. Slope of the flume is adjustable using gearmotor-driven jacks located upstream and downstream from the center pivot. Flow discharge to the channel is selected by changing the pump speed with a variable belt drive and is continuously variable between about 0.085 and 0.440 m^3 s^{-1}. Discharge is measured using a Venturi meter in the 0.406 m flow return line. The Venturi meter was connected to a pressure transducer and the voltage output was read at 1-s intervals and averaged during measurement of sediment transport.

Total sediment load was measured using a density cell to determine the density of a fraction of the sediment water mixture from the return pipe of the flume. The pressure difference across the pump was used to drive flow through a 1.27 cm diameter sampler nozzle through tubing to the density cell and back through tubing to a tap just upstream from the pump. The sampling system is similar to that used by Willis and Kennedy (1977). The density cell consisted of a vibrating U-tube of which vibrational amplitude/frequency changes associated with density changes in the U-tube generated proportional changes in voltage. The output voltage of the density cell was read by a personal computer. The difference of the output voltage and the voltage at zero sediment concentration of the density cell was proportional to the sediment concentration in the water.

The density cell was calibrated using the bed sediment over the full range of sediment concentrations encountered in the experiments before the investigation began. During data collection periods the voltage output of the density cell was read at 1-second intervals. At periods of approximately one hour, a valve in the density cell sampler line was closed and the voltage of the density cell at zero sediment concentration was recorded for several minutes. Sediment concentrations were calculated for every 1-s interval of sampling as the difference between the logged voltage and the linearly interpolated zero voltage taken just before and after the concentration data was taken. Mean transport rates were calculated from the approximately 7 hr of transport data collected in each of the experiments. These were used to compare to the transport rates calculated from the migrating bed forms.

Bed-surface and water-surface profiles were taken from a motorized carriage that rode on steel rods on top of the channel walls. Speed of the carriage was precisely controlled by a stepper motor and ranged from 0.0126 to 0.0379 m s^{-1}, depending on the migration speed of the dunes. The two sensors from the SedBed Monitor and a commercially available ultrasonic distance measuring device that operated in air were mounted on the carriage over the center of the channel. Flow parallel transects began about 5 m downstream from the headbox for approximately 15 m with distances to the bed and water surfaces collected by computer every 0.1 s and 0.4 s, respectively. Flow depths were calculated from the mean distances between the bed and water-surface profiles. Water-surface slope was calculated as the sum of the slopes of the flume rods and the slope of the water surface relative to the flume rods. Ten transects were taken in each experimental run.

Six experimental runs were completed (Table 1). Before the first run the bed was screeded flat using an adjustable-height blade mounted on a carriage. The flume was run for a period of 40 hr before data was collected for the first experiment to assure that a stable bed configuration had formed. In all of the other experiments the bed forms remaining from the previous experiment were used as the starting condition. Higher flow experimental

Table 1. **Experimental Conditions.**

Experiment number	Flow discharge (m^3 s^{-1})	Flow depth (m)	Water surface slope	Mean flow velocity (m s^{-1})	Water temperature (°C)
TNP-1	0.2029	0.362	0.00034	0.459	19.2
TNP-2	0.2565	0.368	0.00107	0.571	21.4
TNP-3	0.3023	0.366	0.00190	0.678	19.7
TNP-4	0.3412	0.365	0.00278	0.766	22.4
TNP-5	0.3869	0.363	0.00313	0.874	21.3
TNP-6	0.4258	0.367	0.00374	0.952	23.7

runs followed the lower flow experimental runs in all cases. The flume was run for 20 hours before data was collected for the other 5 experiments.

Application To Bed Transects

The SedBed Monitor was used to collect two bed surface transects simultaneously with the probes placed 3.1787 m apart. The distance between the probes used was a compromise between getting a measurable migration distance and being able to complete the transects in a reasonable amount of time. Care was also taken to assure that the bed forms had not significantly changed in shape from one record to the next.

A computer program was written in Fortran to analyze the bed surface transect data. The distance between the probes was subtracted from the beginning of the record of the upstream sensor and from the end of the downstream sensor. The mean bed plane for each transect was calculated and subtracted from each transect. Next, cross-correlations were calculated for the two bed height records for lags from zero up to 1/2 the total length of the transects. The mean absolute deviation from the mean bed plane ($\bar{a}'$) and the mean migration speed of the bed forms (U_d) were also calculated by the program. Values of $\bar{a}'$ and U_d corresponding to the maximum correlation coefficient were read from the output file of the program and used in equation (6) to calculate the bed load transport rate.

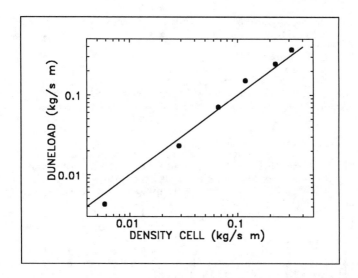

Fig. 4. Sediment transport calculated from migrating dunes vs. sediment transport measured from the density cell.

Bed Load From Bed Surface Transects

The comparison between total sediment transport measured using the density cell and that calculated from the migration of bed forms is shown in Figure 4. The transport rates from the density cell were all measured over periods of about 7 hours, while the transport rates calculated from the bed surface transects generally were an average of 10 transects. As shown in Figure 4 the transport rates from the two methods show a close correspondence.

The SedBed Monitor has been shown to be a practical instrument for collecting bed surface transect data in alluvial channels. Using the SedBed Monitor to collect two simultaneous bed transects allows mean bed load transport rates for a given flow and sediment to be measured in a fraction of the time required to measure bed load at a fixed location of a channel.

Acknowledgements
The experimental runs of this study could not have been made without the assistance of John Cox. The manuscript was reviewed by Douglas Shields and James Sabatier whose comments led to improvments in the manuscript.

Appendix I. - References

Engel, P., and Lau, Y. L. (1980). "Commputation of bed load using bathymetric data", ASCE J. Hydr. Div., 106(3), 369-380.
Gomez, B., Hubbell, D. W., and Stevens, H. H., Jr. (1990). "At-a-point bed load sampling in the presence of dunes", Water Resources Research, 26(11), 2717-2731.
Hubbell, D. W. (1964). "Apparatus and techniques for measuring bedload", USGS Water-Supply Paper 1748.
Kinsler, L. E., Frey, A. R., Coppens, A. B., and Sanders, J. V. (1982). Fundamentals of Acoustics, 3rd ed., John Wiley and Sons, Inc., New York.
Middleton, G. V., and Southard, J. B. (1984). Mechanics of Sediment Movement, 2nd ed., SEPM Short Course no. 3.
Simons, D. B., and Richardson, E. V. (1961). "Forms of bed roughness in alluvial channels", ASCE J. Hydr. Div., 87(3), 87-105.
Simons, D. B., Richardson, E. V., and Nordin, C. F. (1965). "Bedload equation for ripples and dunes", USGS Professional Paper 462-H.
Willis, J. C. (1968). "A lag-deviation method for analyzing channel bed forms", Water Resources Research, 4(6), 1329-1334.
Willis, J. C., and Kennedy, J. F. (1977). "Sediment discharge of alluvial streams calculated from bed-form statistics", Iowa Institute of Hydraulic Research, Report No. 202, Iowa City.

VERTICAL SORTING WITHIN DUNE STRUCTURE

Mohamed Abdel-Motaleb[*]

Abstract

Sediment transport occurs in natural stream beds at least in part as bed load. A striking feature of many of these natural stream beds is the distortion of these beds into trains of waves. For low Froude numbers, these bed forms (waves) are typically classified as ripples or dunes, depending on the bed morphology and flow conditions. Their equilibrium (fully developed) shapes are determined by interaction between the flow, the bed geometry, and sediment transport field. Thus the bed form is created by the flow and conversely the flow is acted on by the bed form.

Vertical sorting of sediment mixtures within dune structures was measured experimentally by conducting three types of experiments: running water experiments, still water experiments, and air experiments. Five different sediment mixtures with known initial gradations were used. The median grain diameters for the five sediment mixtures were between 0.35 mm and 0.86 mm, the geometric standard deviations of the same sediment mixtures were between 2.30 and 2.9.

In the running water experiments, each experiment was continued until the dunes were fully developed down the flume. Then each dune was sampled along several horizontal layers. In the still water experiments, a delta shape was deposited, foreset by foreset, following one another in a continuous way. In the air experiments, the sand mixture was deposited as in the still water experiments. These experiments were to study the effect of the gravitational force on the vertical sorting process.

The results of the running water experiments showed clearly demonstrated the vertical sorting process (vertical reduction in the sediment grain diameter) within the two-dimensional dunes. Also, the still water and air experiments showed the importance of the hydrodynamic force on the sorting process.

Senior Researcher, the Hydraulics and Sediment Research Institute, Delta Barrage (13621), Egypt.

Introduction

Sediment sorting is one of the bedform characteristics which is not physically fully understood despite the attention of earlier researchers of the mid-sixties, such as Brush (1965a), Jopling (1964,1965), McKee (1965) and others. In general, sorting is the fundamental process which leads to the formation of primary sedimentary structures and textures in an alluvial channel. According to Brush (1965, p. 25) "Fall velocity, turbulence, diffusion, gravitational sliding, and shear stress in proximity to the stream bed reflect the physical controls which lead to temporal and spatial segregation of sediment particles by size, and which interact with the population characteristics of the particle distributions available."

A physical understanding of how sorting contributes to the sediment deposited structure will enhance our understanding of the environment in which the sediment has been deposited. Another aspect of understanding this phenomenon is that it is a step further toward the relationship between the grain size destribution within the dune structure and the shape of the dune. But before going to this step, the phenomenon of vertical sorting needs to be documented and explained.

For this study attention is focused on sorting associated with sediment transport by water, specifically the sorting that occurs in individual dune structures. Also there will be a comparison between the roll of hydrodynamic forces and gravitational force on the sorting process.

Evidence of vertical sorting

Brush (1965b) studied the formation of primary sediment structures in and along an alluvial channel sand bed which occurs as result of interaction between gravity, the physical characteristics of the sediment and fluid as well as the hydraulic environment. The occurrence of many of these structures in channel beds resulted from the presence of ripples, dunes, bars, and antidunes on the bed. However, the actual process by which recognizable structures developed arises from sorting of sediment with respect to size, shape, and density along the bed and within the stream. The settling rates of the particles, turbulent diffusion, gravitational sliding, and boundary shear stress contribute to the sorting processes.
Brush (1965b) took local surface samples from the dune surface of several experimental runs in a large recirculating flume. These samples were taken at the crest, at the half of the dune back height, at the toe of the dune, and at a point in the lowest portion of the trough which occurs immediately downstream from a dune, (Figure 1). In addition, he took a number of core samples and surface samples for the entire dune surface. The sand used for this study had a mean diameter of 0.37 mm. A similar but much less extensive study had been made earlier by Brush in the flume at Colorado State University during several runs made by Simons and Richardson in 1962 for sand of 0.45 mm in diameter and with a standard deviation of about 1.6 (uniform sand).

According to Brush, finer sizes occur in the trough due to settling from the suspension of the material passing over the crest of the dune and subsequently caught in the stable eddy zone. Also, he reported that an equally important aspect of the sorting is that the larger particles tend to reach the bottom of the dune in greater concentrations because they would meet fewer obstructions in their downward movement (gravitational sorting).

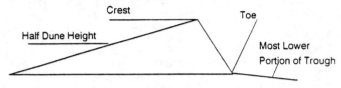

Figure 1 Position of samples (after Brush, 1965b)

Ribberink (1983) conducted a study in the framework of a research project concerning the development of a mathematical model for morphological computions of rivers in the case of nonuniform sediment. The study consisted of series of laboratory experiments in a straight flume, 40 m long, 0.5 m deep, under steady uniform equilibrium conditions with a restriction to bed load transport and dune regime. The flume was fed upstream by different mixtures of two very narrow sieve size fractions (0.78 mm, 1.29 mm). When equilibrium was reached, in addition to regular registrations of water and bed level, the dunes were sampled.

The sediment was siphoned out, layer by layer (thickness of each layer was about 0.5 cm) into a sieve tray. One of the main results of the experiments is as follows:

Vertical sorting of size fractions occurred in all experiments. At the steep lee side of the dunes, the coarse size fraction was more often deposited at the lower level than the fine size fraction. Differences in volume concentration per size fraction up to 30% occur between upper and lower layers.

Ribberink also pointed out that vertical sorting in bed forms not only occurs in laboratory conditions but also takes place in natural conditions. Ribberink reported that Zanke (1976) found similar vertical sorting of a wide range of grain size mixture in the Weser river in Germany.

Experimental setup
Flume

The flume used in these experiments is 10 m long, 0.2 m wide, and 0.2 m deep. It recirculates water and sediment (Abdel-Motaleb, 1992).

Bed sample

In order to measure the composition of the sediment mixture in the transport layer (dune) of the bed, the bed sampling technique by Ribberink (1983) was used after changing the procedure. At the end of each run, the water flow was stopped leaving approximately 10 cm of water on the top of the bed. A glass tube (outer diameter 1 cm) mounted on the end of the point gage was used to siphon the sediment (Figure 2). Connected to this glass tube was a rubber tube to carry the sediment to the sieve tray.

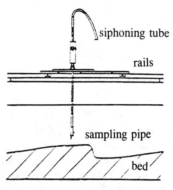

Figure 2 Schematic diagram for the bed sampler

On the rubber line there was a clamp to control the flow of water through the rubber line, so the sand structure would not be destroyed by the siphoning procedure. The siphoning tupe could be positioned anywhere above the dune by means of the instrument carriage which could be moved along the flume. The point gage was used to read the layer thickness before the siphoning process was started. In most cases the layer thickness was about 5 mm and extended across the flume and along the entire dune. The weight of the sample varied according to the layer length but the minimum weight was about 600 gm, and the number of layers depended on the dune height.

Sediment Mixture

Five different sediment mixtures were used in these experiments. The median diameters and geometric standard deviations of the five mixtures are as follows:

Mixture	d_{50}	σ_g
1	0.35	2.85
2	0.45	2.30
3	0.60	2.90
4	0.72	3.30
5	0.86	2.68

Procedure

There were three types of experiments, each with a different procedure. The first one involved running water, the second, still water, and the third air. The first procedure was as follows: Befor the experiment was begun, the d_{50} of the mixture was known and the slope set by using the Simons and Richardson (1966) graph so the depth and discharge could be approximately predicted. The slope of the flume bed was adjusted appropriately. When the sand was put in, it took the slope of the flume. Then clear water was pumped into the flume at a very low rate, the discharge gradually increased with continuous adjustment of the downstream gate to establish uniform flow conditions at each gradual increase in the discharge. In each run the bed configuration was assumed to have reached equilibrium when the bedforms had not only developed fully down the flume but also had ceased to show overall change in size or geometry with time. Runs continued for about 24 to 48 hours to ensure sufficient time for attainment of equilibrium. During this time, readings for bed level, water surface, and discharge were taken several times. Just before stopping the run, These readings were taken for a last time. After flow was stopped, then the samples were taken from three or four dunes in the middle section of the flume.

The second procedure (experiment with still water) was as follows: First, a wooden foreset of a dune was put on the flume bed. Then clear water was pumped into the flume. As the water level was rising, the downstream steel gate was adjusted until a proper depth of water was reached. After that the pump was stopped, the controlling valve was closed and the downstream steel gate was sealed to keep the water level stable during the experiment. The water was about 2 cm higher than the dune height needed. The sand mixture was poured into the flume using a wooden funnel with the same width as the flume width (Figure 3). The sand was poured into an inclined layer with an angle equal to the angle of repose of the sand mixture. The funnel was as low as possible to the water surface in order to remove the impact factor of the sand mixture grains on these already deposited. After reaching an appropriate dune lenght, the dune was sampled layer by layer.

The third procedure (experiment with air) was the same as the experiment with still water but with air as the media in which deposition occurred. To sample the dune layers using the same siphoning method as before, the dune was covered completely with water. After that the sampling of the dune layers was started.

Experimental Data

Tables 1 contains the experiments data of the flume measurements for the six experiments with running water. Five mixtures were used in these six experiments.

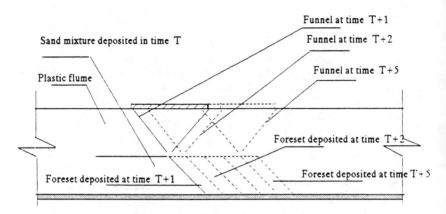

Figure 3 Diagram showing the deposition of each sand foreset using a funnel.

Table 1 : Summary of the measured data for running water experiments.

Run No.	d_{50} (mm)	σ_g	d_{ave} (cm)	Q (m/sec)	S_w	S_e
RW-1	0.35	2.85	18.5	0.021	0.0024	0.0022
RW-2	0.45	2.3	17.1	0.022	0.0028	0.002
RW-3	0.60	2.9	16.1	0.021	0.0022	0.002
RW-4	0.72	2.3	17.8	0.022	0.0024	0.0022
RW-4.1	0.72	2.3	17.8	0.022	0.0023	0.0022
RW-5	0.86	2.68	17.8	0.022	0.0027	0.0025

Experimental Results

Running water experiments

From the grain size distribution of each layer of the different dunes from different mixtures, there was clear evidence of the vertical sorting within the entire length of the dune structure. Figure 4 shows that the dune layers were distinctive one from another and also an increase of d_{50} of each layer downward.

For Mixture #1 the ratio between the median diameter (d_{50}) at the top of the dune (y=d) was 2.6, for Mixture #3 this ratio was 2.36, and for Mixture #5. The ratios

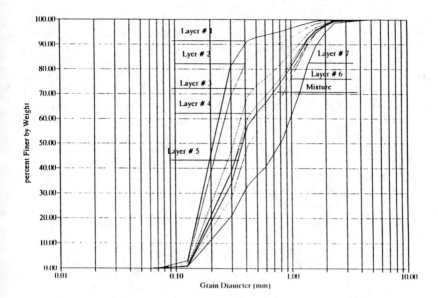

Figure 4 Size distribution curves for dune layers (Dune # 1, Mixture # 1)

for any of the five mixtures. For example, the geometric standard deviation (σ_{gm}) for the five mixtures was 2.85, 2.30, 2.90, 2.33, 2.70, and they had a ratio between the lower layer grain diameter and the upper layer grain diameter equal to 2.6, 2.38, 2.36, 1.45, 1.67 respectively.

Still water experiments

One of the five mixtures, # 5, $d_{50} = 0.86$ mm and $\sigma_{gm} = 2.67$, was used to study the vertical sorting process (vertical decrease in sand grain size within the dune) in still water. The main reason for this kind of experiment was to remove the effect of the hydrodynamic forces in order to study the effect of the gravitational force only on the vertical sorting process.

The height of the deposited sand for the first experiment was 0.08 m; this height was chosen to be as close as possible to the running water dune height and at the same time could be practically done. The results of the sieve analysis showed that it is clear that there was no sorting at all within the dune structure. All the layers were almost falling on top of each other. The reason for the lack of vertical sorting may be that the length of the lee slope, which is a function of the height of the dune, was not long enough to give the gravitational sliding a chance to develop the sorting of the sand particles. Based on this conclusion, the next stage was to do another experiment but with a deposited sand mixture height higher than this first experiment.

The height of the deposited sand for the second experiment was double that of the first experiment, 0.16 m, so the length of the lee slope was increased. The results of the sieve analysis of this experiment showed that the dune layers started to spread out from each other more than in the first experiment, but the sorting was still not clear despite the increased height of the dune.

The next step was to do another experiment but with a dune height of 0.30 m, almost double the second experiment, with a water depth of 0.32 m. The results of the sieve analysis for this experiment, shown in Figure 5, shows clear evidence of the vertical sorting within the deposited sand structure.

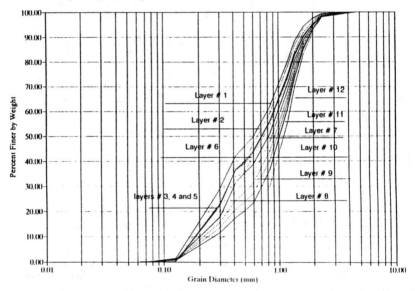

Figure 5 Size distribution curves for horizontal layers of delta type deposition in still water (delta height = 0.3 m)

The first six layers did not show a clear sorting process, maybe because the distance on the lee slope from the top of the dune to each layer of the first six layers was still insufficient to show the effect of the gravitational force on the sand grains, but as the lee slope distance was increased this gravitational force worked as in the lower five layers, which gave no doubt of this process.

From the above results it can be concluded that to get a sorting using the gravitational force only, the deposited sand height needs to be higher than that for running water.

5.2.3 Air experiments

For more evidence on the role of gravitational force in the sorting process, another set of experiments was conducted using the same sediment mixture. But the depositing, using the same technique as in the still water experiments of the sand, was in air. Two experiments were conducted. The deposited delta shape height of the first experiment was 0.16 m and the number of layers was 11 layers, each 14 mm thick. The sieve analysis results for each layer shown that all these layers fell on top of each other, indicating no evidence of vertical sorting in the dune structure. So the next step, as with the still water experiments, was to increase the dune height. The height of the deposited sand delta of the second experiment was 0.30 m, and the 11 layers were sampled (see Figure 6).

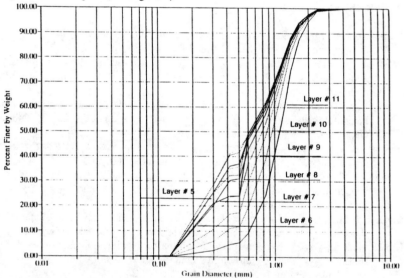

Figure 6 Size distribution curves for horizontal layers of delta type deposition in air (delta height = 0.3 m)

The top 6 layers, as in the third still water experiment, showed no vertical sorting of the sand. But as the lee slope length, which is a function in the height of the deposited sand, increased, the sorting process seemed to develop. From these two experiments with two different deposition environments, one can conclude that the hydrodynamic forces are important in the vertical sorting process within the dune structure.

6.2 CONCLUSIONS

The main conclusions drawn from the analysis of the results are as follows:
1- Extensive sampling of the bed showed that the sand grains of a sediment mixture tend to rearrange themsleves during the deposition of a dune in such way as to develop a vertical reduction in the grain diameter within the dune structure.
2- From the results of the still water and air experiments, it is clear that to develop vertical sorting within a deposited sand mixture by avalanche over the lee side needs the lee slope length to be much longer than that for the running water dune.
3- The role of hydrodynamic forces on the development of the vertical sorting process is more obvious than the gravitational force.

REFERENCES

Abdel-Motaleb, M., (1992). "Vertical sorting within dune structure", PhD. Dissertation, Colorado State University, Fort Collins, Co, U.S.A.
Brush, L.M., (1965a). "Experimental work on primary sedimentary structures", Society of Economic Paleontologists and Mineralogists Special Publication 12, p. 17-24, Tulsa, OK.
Brush, L.R., (1965b). "Sediment sorting in alluvial channels", Society of Economic Paleontologists and Mineralogists, Special Publication 12, p 25-33, Tulsa, OK.
Jopling, A.V., (1964). "Laboratory study of sorting processes related to flow separation", J. of Geophysical Research, vol. 69, N16, p 3403-3418.
Jopling, A.V., (1965). "Laboratory study of the distribution of grain sizes in cross-bedded deposits", Soc. Economic paleontologists and Mineralogists, Special Pub. 12, p 3-65, Tulsa, OK.
McKee, E.D., (1965). "Experiments on ripple lamination", Society of Economic Paleontologists and Mineralogists, Special Pub. 12, p 66-83, Tulsa, OK.
Ribberink, J.S., (1983) "Experiments with non-uniform sediment in case of bed-load transport", communications on hydraulics, Department of Civil Engineering, Delft University of Technology, Report no. 83-2.
Simons, D.B., and Richardson, E.V., (1966). "Resistance to flow in alluvial channels" USGS Professional Paper 422-J, 61 p.

ON MEASUREMENTS OF PARTICLE SPINNING MOTION

By Hong-Yuan Lee, [1] Member ASCE and In-Song Hsu [2]

ABSTRACT: Investigation of particle saltation motion is crucial to the development of the theory of bed load transport. Due to the combination effects of the bed roughness and the velocity gradient, particle spins during saltation process. This spinning motion will generate additional lift force and thus increase the saltation length and height. The increments can be as much as 12%. A high speed photographic technique is developed in this study to measure the saltation trajectories and corresponding velocities and spinning rates. Two kinds of materials, natural sand and plastic balls were used in the experiments. The longitudinal variations of the saltation velocities and the spinning rates were recorded. Statistical analyses were performed to investigate the probability density function of the spinning rates.

---

1. Professor, Dept. of Civil Engineering and Hydraulic Research Laboratory, National Taiwan University, Taipei, Taiwan, R.O.C.

2. Graduate student, Dept. of Civil Engineering and Hydraulic Research Laboratory, National Taiwan University, Taipei, Taiwan, R.O.C.

INTRODUCTION

Bed load transport can be classified into three different modes, namely rolling, sliding and saltation. About seventy-five percent of the bed load transport is in the mode of saltation motion. Hence, investigation of particle saltation motion is crucial to the development of the theory of bed load transport. Due to the combination effects of the bed roughness and velocity gradient, particle spins during saltation process. This spinning motion will generate additional lift force and thus increase the saltation length and height. The increments can be as much as 12% (Lee and Hsu 1994).

Due to experimental difficulties, the Magnus effects were usually assumed negelected in the previous saltation models. (Fernandez Luque and Beek 1976, Abbott and Francis 1977, Murphy 1985, Van Rijn 1984). Using Rubinow and Kellers' (1961) relation, Lee and Hsu (1994) successfully included the Magnus effect in their model. However the spin rate of the saltating particle is still left to be quantified. Observing the bright and dark spots in the photographs of the saltating paricles, Chepil (1945) roughly estimated that the spin rate of the saltating particle in the wind was between 200 to 1000 rev/sec. Using high speed camera, White and Schultz (1977) found the particle spin rate was between 100 to 300 rev/sec in wind sediment transport. Hui and Hu (1991) conducted series of flume tests and found the spin rate was about 40 rev/sec in water sediment transport. Due to lack of the experimental data, estimation of the additional lift force due to particle spinning is still very rough. More experimental works are needed in the future.

A high speed photographic technique is developed in this study to measure the saltation trajectories and corresponding velocities and spinning rates. The longitudinal variations of the saltation velocities and the spinning rates were recorded. Statistical analyses were performed to investigate the probability density function of the spinning rates.

EXPERIMENTAL SETUP

The experiments were conducted in a 12 m long, 0.3 m wide slope-adjustable recirculating flume. The experiments were conducted in a fully turbulent uniform flow. The corresponding water depth was 8.5 cm, channel bed slope was 0.004, the Reynolds number was 41,263, and the shear velocity was 0.046 m/sec. Plastic balls of size 4.15 mm were glued to the channel bed to increase the roughness of the bed surface. Two kinds of materials, natural sands with particle size equals 5.8 mm and specific gravity equals 2.64 and plastic balls with size equals 5.9 mm and specific gravity equals 2.8, were used in the experiments as the saltating particles. The particles were marked with crosses composed of blue and red colors, and the spinning rates were determined according to the rotation angles of the crosses.

A flash tube of maximum frequency 3,000 flash/sec was used as the light source and two 35 mm conventional cameras with 400 ASA color films were used to take pictures. The general configuration of the experimental setup is shown in Fig. 1. During the experimental process, the room was kept dark and the only light source is from the flash tube. The frequency of the flash tube was set at 50 Hz. The marked particles were released 4 m from the working section, and the saltating trajectories were recorded by the camera. The photographs were then projected to a big screen to measure the longitudinal variations of the saltation velocities and the corresponding spinning rates. A typical trajectory is shown in Fig. 2.

RESULTS

According to the experimental observations, the particle spinned with respect to an axis parallel to the channel bed most of the time, i.e., the spinning process was dominated by top-spin rather than screw-motion. The longitudinal variations of the saltation velocities, accelerations and the spinning rates are shown in Fig. 3. The horizontal component of the saltation velocity is slow at the rising limb and reaches a maximum value very quickly at the top of the trajectory and remains more or less constant from that moment on. The vertical component of the saltation velocity gains its maximum value from rebound of the channel bed and decreases along the trajectory. It reaches a zero value at the top of the trajectory and maintains a negative value at the falling limb. The horizontal component of the acceleration decreases along the trajectory. The vertical component of the acceleration reaches a minimum value at the top of the trajectory. This is because the relative velocity is small at the top of the trajectory and hence the corresponding lift force is minimum there. The spinning rate decreases along the trajectory. It is large at the rising limb, remains more or less constant at the central portion and slows down at the falling limb. It indicates that the Magnus effect is significant at the rising limb. Particle spinning is a combination effect of collision with the bed particles, velocity gradient and particle shape. Particles with irregular shape spin faster than the sphere particles. The average spinning rate for the natural sand and the plastic balls are 20 and 10 rev/sec respectively. The corresponding probability density functions are shown in Fig. 4. The distribution of the spinning rate of the sphere plastic balls is more uniform than that of the natural sand. This is due to complication of the flow pattern generated due to irregularity of the particle shape.

ACKNOWLEDGEMENTS

This study was supported by the National Science Council of Taiwan, the Republic of China. The authors would like to thank the staff of the Hydraulic Research Laboratory of National Taiwan University for support in conducting the experiments.

APPENDIX I-REFFERENCE

1. Abbott, J. E., and Francis, J. R. D., "Saltation and Suspension Trajectories of Solid Grains in a Water Stream," Proc. Royal Soc., London, England, Vol. 284, A 1321, 1977.

2. Chepil, W.S., "Dynamics of Wind Erosion I. Nature of Movement of Soil by Wind", Soil Science, 60,pp. 305-320,1945.

3. Fernandez Luque, R., and Beek, R. Van, "Erosion and Transport of Bed-load Sediment," Journal of Hydraulic Research, Vol. 14, No. 2, 1976.

4. Lee, H. Y., and Hsu, I. S., "Investigation of Saltating Particle Motions", Journal of Hydraulic Engineering, ASCE, 1994(in press).

5. Hui, Y., and Hu, E., "Saltation Characteristics of Particle Motions in Water", SHUILIXUEBAO, Vol. 12,1991, pp. 56-94 (in Chinese).

6. Murphy, P. J. and Hooshiar, Hamid, "Saltation in Water Dynamics." J. Hydr. Engr., ASCE ,Vol. 108,No. 11., 1982,p. 1251.

7. Rubinow, S. I., and Keller, J. B., "The Transverse Force on a Spinning Sphere Moving in a Viscous Fluid," Journal of Fluid Mechanics, Vol. 11, 1961,p. 454.

8. Van Rijn, L. C., "Sediment Transport, Part I: Bed Load Transport," J. Hydr. Engr., ASCE, Vol. 110, No. 10, 1984,pp. 1431-1456.

9. White, B. R., and Schultz, J. C., "Magnus Effect in Saltation," Journal of Fluid Mechanics, Vol. 81,1977,p. 507.

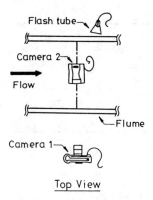

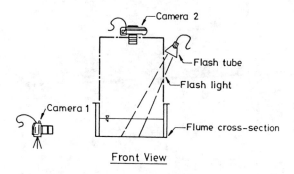

FIG.1 General Configuration of the Experimental Setup

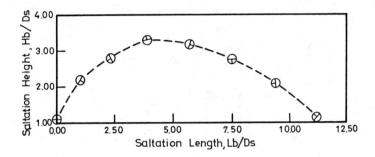

FIG.2 A Typical Salation Trajectory

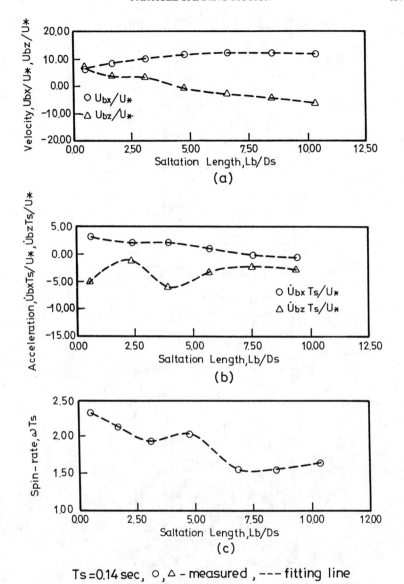

$T_s = 0.14$ sec, $\circ, \triangle$ - measured, --- fitting line

FIG.3 The Longitudinal Variations of the Saltation Velocities(a), Accelerations(b) and the Spinning rates(c)

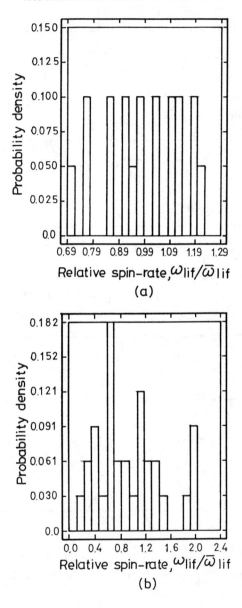

FIG.4 Probability Density Function of the Spinning Rates for (a) Plastic Ball and (b) Natural Sand

An Investigation of Turbulence
in Open Channel Flow
via Three-component Laser Doppler Anemometry

by
Mahalingam Balakrishnan[†] and Clinton L. Dancey[‡]

Abstract

A three-component laser Doppler anemometer is used to reexamine the distribution of single-point mean statistics in a nearly uniform fully-developed 2D turbulent open channel flow of water at a channel Re of 16100 and Fr = 0.17. Measurements of mean velocity, Reynolds stress, turbulent component intensities, 2D probability density functions and conditioned quadrant statistics are reported. Multi-element hot film measurements reported in the literature are used for limited comparisons. The viability of 3D LDA for nonintrusive, high spatial resolution measurements in open-channel flow is demonstrated.

Introduction

In recent years laser Doppler anemometry (LDA) has been increasingly applied to the experimental investigation of closed and open-channel water flows (Nezu and Rodi, 1986; Karlsson and Johansson, 1986; Wei and Willmarth, 1989; Schroder et.al., 1991). In contrast to pitot probes or hot-wire and film anemometry, LDA offers nonintrusive, multi-component, point-wise velocity measurements with potentially high spatial and temporal resolution. These attributes makes multi-component LDA particularly attractive for the investigation of turbulent water channel flow in general and channel flows with sediment and bedload transport in particular since the presence of modest amounts of suspended solids does not compromise the technique. Although high spatial resolution multi-component measurements are clearly desireable, to date, most LDA investigations in water channel flows have been limited to two-component velocity measurements, or three-component investigations requiring two sets of independently obtained measurements via a 2D-LDA system.

In the present paper a three-component LDA system (3D-LDA) is applied to the experimental investigation of open-channel water flow over a smooth surface. This technique permits the instantaneous measurement of the three velocity components and the determination of the complete Reynolds stress tensor, as well as higher order single point statistics. In the present paper measurements via 3D-LDA are compared to a limited set of measurements

---
[†] Graduate Research Assistant, Department of Mechanical Engineering
[‡] Assistant Professor, Department of Mechanical Engineering, Virginia Polytechnic Institute, Blacksburg, Va. 24061

obtained by Nezu (1977) using multi-element hot-film anemometry in a similar flow. In addition to reporting the usual single-point first and second order turbulence statistics, quadrant analysis of the 3D-LDA measurements are reported. This paper represents the first installment in a program to investigate open-channel flow with bedload transport.

The paper is organized as follows: first, the application of LDA to water channel flow is briefly reviewed and the adaptation of an existing 3D-LDA to a water flume is discussed. The experimental facilities and equipment are identified and the open-channel uniform flow operating conditions are specified. Following this the LDA results are presented and finally the paper concludes with technical items which must be addressed in the future and suggestions for system improvements.

Background

Despite the nonintrusive nature of laser anemometry and therefore its applicability to flows with sediment transport, most LDA studies of water channel flow have been limited to flows without sediment entrainment or suspended sediment transport. Because of space limitations only three of the more recent and significant contributions are discussed below.

Nezu and Rodi (1986) used a single transmission lens, two-channel LDA in forward scatter to measure the streamwise and normal components of the velocity in an open-channel flume over a range of Reynolds and Froude numbers. They demonstrated that the von Karman constant and the logarithmic constant could be treated as universal constants at least over the ranges studied. In addition, they presented turbulence intensity profiles and eddy viscosity distributions within the channel flow and, among other things, demonstrated that, when properly applied, LDA in water can yield accurate, high spatial resolution measurements throughout the wall layer in turbulent channel flow.

Karlsson and Johansson (1986) and Johansson and Karlsson (1988) extended Nezu and Rodi's 2D measurements to include the third spanwise velocity component in the near wall region in closed channel flow, again employing a 2D-LDA, in this case with backscatter collection, by acquiring two independent sets of measurements of the respective components and appropriate covariances. They reported measurements well within the inner wall layer as well as in the inertial subrange at two Reynolds numbers. Distributions of turbulence intensities and higher order single-point statistics including third order turbulent transport covariances were presented. Karlsson and Johansson (1986) parenthetically remark difficulty with simultaneous three-component measurements.

Wei and Willmarth (1989 and 1991) also report measurements of the streamwise and normal components of velocity in a narrow closed water channel using a 2D-LDA. However, Wei and Willmarth demonstrate the viability of using an off-axis 2D-LDA resembling the typical off-axis three-component LDA geometries. They report both high temporal and spatial resolution measurements and demonstrate through these measurements that the distributions of second-order turbulence quantities, including the rms turbulence levels and Reynolds stress, do not scale on inner viscous variables but are in fact Reynolds number dependent. In contrast to the previous 2D-LDA two-component investigations, a three-component LDA is described below which permits relatively high spatial resolution measurements within the wall layer. First and second-order turbulence statistics, as well as quadrant statistics, and multi-component probability density functions, pdfs, are reported, in this case, at a single channel Reynolds number.

Experimental Facilities

3D-LDA measurements were obtained near the smooth bottom wall of a tilting laboratory water flume. The flume is 20.5 m long, 0.6 m wide, and 0.3 m deep with a useful length of 15 m. In the present case, measurements were confined to the last 2 meters of the useful length. Constant head tanks are located at the inlet and exit of the channel and connected by two separately controlled centrifugal pumps. Flow straighteners are employed at the flume entrance and an adjustable gate located at the exit of the flume is available for flow depth adjustment. In the present case a fully-developed, uniform, turbulent open-channel flow with depth, H, of 10 cm. was maintained throughout the tests which gave a flow aspect ratio (depth/width) of 0.164. The area average flow velocity, V_m, was approximately 0.18 m/s throughout the tests, yielding a channel Reynolds number, $Re = V_m H/\nu$, of approximately 16100 and a Froude number, $Fr = \sqrt{V_m^2/gH}$, of 0.17. The friction velocity, u_*, was calculated to be 0.0096 m/s and the viscous length scale, ν/u_*, was 113 μm.

The coordinate system is shown in figures 1 and 2. x is the streamwise downstream coordinate, y is normal to the bottom smooth wall, and z is the spanwise coordinate. The corresponding velocity components are designated, u, v, and w, respectively. The origin of the coordinate system is located on the surface of the smooth bottom wall approximately 3.75 cm from the flume centerline, toward the LDA system. LDA optical access is available through the transparent side walls of the flume (see figures 1 and 2).

Instrumentation

A six beam, non-orthogonal, color-separated, fringe-mode, off-axis backscatter, three-component LDA system with counter-based signal processing was employed for the velocity measurements. This particular system uses three independent optical channels to measure three non-orthogonal components of the velocity. Two of these components are approximately coplanar with a coupling angle, the angle between the measured components, of approximately 30

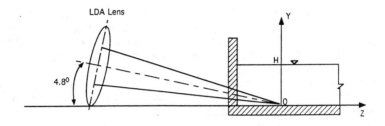

Figure 1. Flume Cross section and LDA Orientation

degrees. The third component is approximately orthogonal to the other two. The velocity components in the orthogonal coordinate system of the flume were computed during post-processing of the data. To improve optical access and facilitate near-wall measurements near the center-line of the flume the LDA system was tilted 4.8 degrees from the horizontal as illustrated in figure 1.

Accounting for 3.75X beam expansion and off-axis collection with 480 mm focal length lenses, the measurement volume is estimated to be approximately 87 µm in diameter (less than one viscous length) by approximately 318 µm (approximately 3 viscous lengths) in length.

A traverse system was employed to systematically position the LDA measurement volume within the flume in a prescibed sequence of measurement locations and all data acquisition and traverse movement was computer controlled. For the present experiment measurement volume was traversed normal to the bottom wall without any significant spanwise or streamwise movement. Silicon carbide was used as the LDA seed material.

3D-LDA in Water:

When optically probing from air into water adverse laser beam refraction effects are likely for off-axis 2D or 3D LDA with large coupling angles. In the present case it was necessary to reduce or eliminate these effects in order to maintain measurement volume coincidence of the three optical channels. (For 2D (Nezu and Rodi, 1986) or 5-beam 3D-LDA (Karlsson and Johansson, 1986)systems where beam transmission occurs through a single lens of relatively large f/# such effects are usually not serious.)

In the present case, the adverse effects of beam refraction were reduced by constructing an inexpensive, water-filled, acrylic vessel rigidly attached to the LDA optical table and coupled to the flume sidewall with a flexible rubber "bellows," (Lindsey and Owsenek, 1993). A schematic is shown in figure 2 for illustrative purposes. A similar approach was taken independently by Wei and Willmarth (1989). With this design, the beam transmission for each of the two separate optical axes is nearly normal to the acrylic vessel windows. This arrangement is similar to that of typical single-lens 2D LDA systems applied to water channel investigations. (Without this vessel, the transmission beams intersect the flume wall at approximately 15° off of the normal and the adverse beam refraction effects are intolerable.) The flexible rubber bellows permits traversing the LDA measurement volume in the three coordinate directions within the flume without a loss of measurement volume coincidence. The vessel is open to the atmosphere so that the water level within it can adjust as the LDA is traversed. This simple design permits good quality signals with adequate data rate, independent of LDA movement and measurement volume location within the flume.

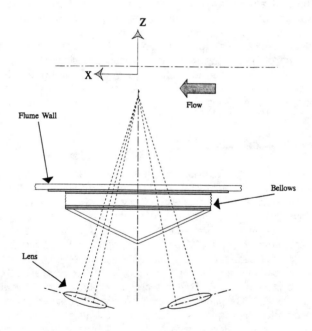

Figure 2. Plan view of Flume and LDA

Measurements:

Data Acquisition and Analysis:

The instantaneous measurement of the three non-orthogonal velocity components will be termed a triplet. In the present case, 2048 triplets were acquired at each measurement location above the bottom wall. Counter processors were used to determine the separate channel particle transit times for 8 fringes with 5/8 comparison at the 1% tolerance level. The separate channel velocities were computed during post-processing as were the orthogonal components and all statistical quantities. Bragg cell frequency shifting was employed on all channels to reduce angular bias effects and to shift the signal frequencies into more favorable filter bandwidths in order to increase the signal to noise ratio. Trial measurements were used to identify an acceptable channel coincidence time window of 1000 µs, to assure measurements on the same seed particles. This window size permitted good quality measurements at a low but tolerable data rate. At the full free-stream mean velocity, this window corresponds to approximately two particle transit times.

The data rate was low enough that on average the time between measurements was greater than twice the turbulence integral time scale, for all measurement locations, so that the individual triplets may be treated as statistically independent samples. This low data rate does however preclude time-resolved measurements of the velocity history and introduces biasing errors in the estimate mean quantities. In order to reduce this velocity biasing effect full three-component inverse velocity weighting was employed for all calculations of estimates of mean statistics, including the quadrant and pdf results (Madsen, 1994).

Measurement Uncertainty:

There are many potential sources of measurement uncertainty in LDA; among others one can identify: counter clock resolution/synchronization uncertainty, digital truncation error, particle lag, velocity gradient error, angular bias error, velocity bias error, fringe spacing uncertainty, statistical sampling error, and uncertainty in the parameters used to define the LDA system geometry (such as the angular orientation of the LDA beam geometry to the flume coordinate system and the angles between the several LDA optical channels). In the present case, it can be shown that the first five sources listed above are negligibly small compared to the last three. Systematic or fixed errors are dominated by the uncertainty in the LDA fringe spacings on the separate channels and the uncertainty in the angles which characterize the LDA geometry. Uncertainties due to random sources, in the present case, are predominantly associated with statistical sampling error.

A propagation of error analysis on the data yielded an estimated relative uncertainty in the mean streamwise velocity of ± 1.2%, in the streamwise velocity fluctuation variance of ± 4.8%, in the normal variance of ± 7.5%, the spanwise variance of ± 80%, and in the covariance, $<uv>$, of ± 10%. The relatively large estimated uncertainty in the spanwise direction reveals the weakness of non-orthogonal 3D-LDA. The "on-axis" or spanwise component, in this case, is very sensitive to errors in the measured non-orthogonal velocity components when the coupling angle of the system is small (Dancey, 1990). It is noted, however, that these estimates are conservative and that the scatter in the reported measurements are typically much less than these estimates suggest, particularly in statistics containing the spanwise velocity component. (It is also noted that in practice the relative uncertainty in the different measured and computed quantities varies with measurement location. Only representative estimates are presented here.)

Results:

Figures 3-10 present the distributions of selected single-point turbulence statistics computed from the complete data set. Although velocity bias correction has been applied to the plotted results, no attempt was made to remove "suspect" data from any of the measured samples, ie. no attempt was made to remove data "outliers".

The distributions of the nondimensional Reynolds and mean shear stress with y/H are shown in figure 3. Angle brackets, < >, are used throughout to denote average quantities and primes, ( )', are used for root mean square quantities.

For 2D uniform, fully-developed turbulent channel flow the Reynolds and mean shear stress are related by:

$$\frac{\nu \frac{\partial <U>}{\partial y} - <uv>}{u_*^2} = 1 - \frac{y}{H} \quad (1)$$

where u_* is defined by

$$u_* = \sqrt{\frac{<\tau_w>}{\rho}} \quad (2)$$

and $<\tau_w>$ is the average wall shear stress. ν is the kinematic viscosity and ρ is density. $<uv>$ represents the kinematic Reynolds stress in the x,y plane. The two nondimensional stress terms on the left-hand-side of equation 1, are computed from the measurements and shown separately in the figure. For comparison, the total theoretical stress, given by the right-hand-side is also shown in the figure. The data shown in figure 3 together with equation 1 were used to determine the best estimate of u_* (0.00960 ± 0.00008) by a least squares regression of the experimental stress data about the theoretical linear total stress profile. Although there is scatter in the measured Reynolds stress, due to the limited sample size, in general the data fit the theoretical linear profile well.

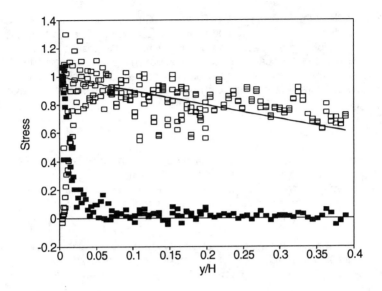

Figure 3. Stress Distribution, □ Reynolds Stress, ■ Viscous Stress

u. established from the stress profile is used in figure 4 to determine the streamwise mean velocity distribution in nondimensional wall coordinants, i.e.

$$u^+ = \frac{<U>}{u_*} \qquad (3)$$

and

$$y^+ = \frac{yu_*}{\nu} \qquad (4)$$

For $y^+ \geq 50$ the data are well predicted by the logarithmic distribution

$$u^+ = A + B \ln y^+ \qquad (5)$$

A least squares regression of the data about this line yielded A = 5.083 ± 0.240 and B = 2.532 ± 0.048, which are within 4 % of the "universal" values determined by Nezu and Rodi (1986) via 2D-LDA. It is parenthetically remarked that the measurements very close to the wall satisfy the linear law.

$$u^+ = y^+ \quad (0 < y^+ < 5) \qquad (6)$$

AN INVESTIGATION OF TURBULENCE

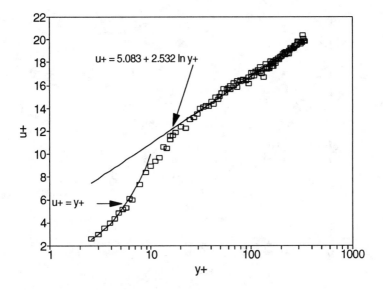

Figure 4. Streamwise Mean Velocity Distribution

Figures 5, 6, and 7 present the nondimensional turbulent intensity component distributions. In these figures the respective rms levels are nondimensionalized by u_*. Data from Nezu (1977) at two Reynolds numbers (10900 and 28900 based upon V_m and H) obtained with hot-film anemometry in an open-channel flow are also shown in the figures for comparison.

Figure 5 for the streamwise intensity reveals extraordinarily good agreement between the 3D-LDA measurements and Nezu's lower Re results, and fair agreement with the Re = 28900 results. Although Nezu's maximum measured intensity near the wall is lower, the two distributions obtained by these widely different techniques are nearly coincident. The measured maximum value, u'/u_* ≈ 2.9 in the present case, which occurs at y^+ ≈ 15 is, however, consistent with the level reported by Alfredsson and Johansson (1989). Furthermore the nondimensional RMS wall shear stress fluctuation intensity, $\tau'_x / <\tau_w>$, can be determined from figure 5, due to the linearity of the velocity profile in the viscous sublayer. A value of 0.40 is obtained which falls within the range 0.36-0.40 reported by Alfredsson and Johansson and is only 10% larger than the level reported by Moin and Spalart (1989) using a direct numerical simulation of channel flow at Re = 3300.

Nezu's measurements at the higher Re which are only available at larger y^+ do not compare quite as well. The 3D-LDA results at Re = 16100 show that

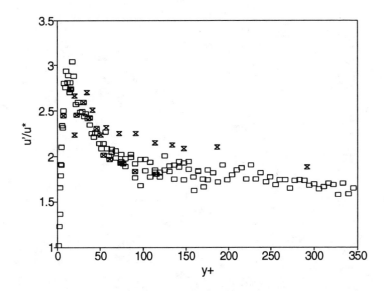

Figure 5. u'/u_* distribution: ▫ Data, ▨ Nezu (R=10900), ✖ Nezu (R=28900)

the streamwise intensity decreases more quickly with distance from the wall than reported by Nezu at Re = 28900. Nevertheless, and generally speaking, the measurements by these two different experimental techniques compare well.

The comparisons with Nezu shown in figures 6 and 7 for the v'and w' intensities are not as favorable. In the case of v', the peak intensity measured by 3D-LDA and by Nezu at Re = 28900 are comparable, and the trend with y^+ is similar; however, the Nezu measurements at Re = 10800 yield intensities nearly 20% lower than those reported here. Likewise, figure 7, shows that the spanwise intensity decreases relatively slowly from a maximum of 1.7 as y^+ increases. The Nezu results at Re = 28900 show more rapid decay (although the maximum intensity level is comparable), and the results at Re = 10800 show extremely rapid decay of the spanwise intensity compared to those measured here. At this writing the reasons for these anomalies are not clear. The observed behavior may be due in part to Reynolds number and probe spatial resolution effects. 3D-LDA measurements over a range of Re and Fr may be required to help resolve these apparent differences.

Although the spatial variations of mean turbulence statistics are necessary for the characterization of turbulent flow, and are the primary variables in modeling investigations, the individual instantaneous velocity contributions to the underlying

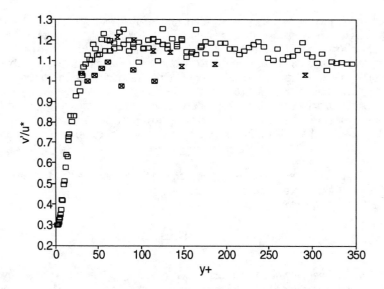

Figure 6. v'/u* distribution: See figure 5 for symbols.

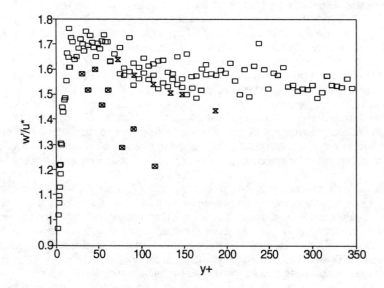

Figure 7. w'/u* distribution: See figure 5 for symbols.

distribution functions are becoming more relevant to the study of transient random events (such as sediment entrainment). Spatially resolved point-wise three-component measurements permit the determination of a wide range of detailed statistical measures of the velocity field, including multi-component joint probability density functions, pdfs, and other related functions, high order multi-component statistical correlations, conditioned statistics, and correlations of instantaneous velocity vector measurements with other auxiliary signals. Although such analyses are possible with the 3D measurements available in the present data set, since the flow is 2D in the mean velocities, the spanwise velocity component, w, is not as relevant as statistical information in the u,v plane.

In the present case, two distribution functions are of particular interest and figures 8 and 9 are representative of them. The joint pdf, $B(u/u',v/v')$, and the corresponding distribution of

$$\frac{uv}{<uv>} B \left[\frac{u}{u'}, \frac{v}{v'} \right] \qquad (7)$$

which is directly related to the production rate of turbulent kinetic energy were computed from the ensemble of measured velocity components. They are shown in figures 8 and 9 respectively at $y^+ \approx 15$ where the turbulent kinetic energy is at a maximum. Both figures are consistent with similar distributions reported by Lekakis (1988) using a triple hot-film probe in a turbulent pipe flow.

Figure 8 reveals the anticipated negative correlation associated with the Reynolds stress throughout the wall layer and the tilting of the pdf into quadrants 2 and 4 (u<0. v>0 and u>0, v<0 respectively) in the u,v plane. Figure 9 reveals the relatively large positive contributions to the turbulent production rate (and <uv> itself) associated with Reynolds stress events in the 2nd and 4th quadrants (ejection and sweeps). As Lekakis points out, the two prominant peaks in quadrants 2 and 4 imply that events which contribute most to <uv> and the production rate are not the most probable (u,v) events, but may be relatively infrequent. The contributions of particular events (which may be relatively infrequent) may be determined by the introduction of an indicator function.

The information carried in figures 8 and 9 (and similar figures throughout the wall layer) together with a user-defined indicator function make the estimation of relevant conditioned statistics possible. These distribution functions permit the determination of the Reynolds stress and turbulent production rate contributions associated with individual velocity events which satisfy the indicator function criteria. One such useful indicator function associated with the "quadrant" method selects those events from the ensemble of measurements which lie in the separate quadrants of the u,v plane and exceed a specified instantaneous stress level (octant maps of the u,v,w plane are possible with 3D-LDA). That is:

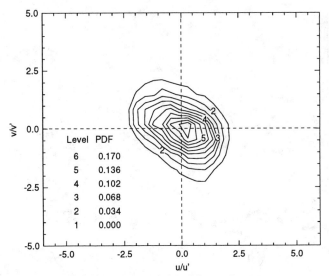

Figure 8. Probability Density, $B(u/u',v/v')$, at $y^+ = 15$

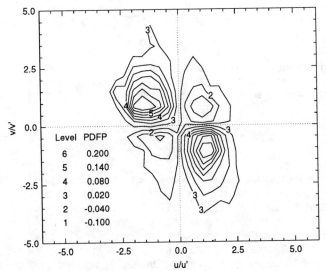

Figure 9. $(uv)/<uv>\ B(u/u',v/v')$ at $y^+ = 15$

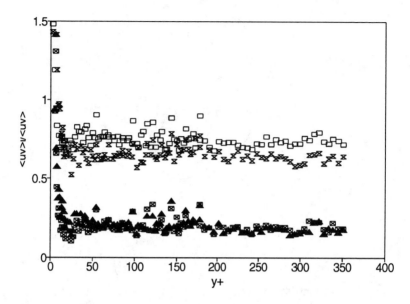

Figure 10. Quadrant Map: □ Q 2, X Q 4, ▲ -Q 1, ⊠ -Q 3

$$I_i(t_j, K) = \begin{bmatrix} 1 : \left|\dfrac{u_j v_j}{<uv>}\right| > K \; ; \; i = 1, 2, 3, 4 \\ 0 : \text{otherwise} \end{bmatrix} \quad (8)$$

where j indicates the particular measurement triplet, i identifies the specified quadrant, and K is the criterion level, ≥ 0. The conditioned (quadrant) Reynolds stresses, $<uv>_i$, where i = 1, 2, 3, or 4 for the respective quadrants are obtained from

$$<u,v>_i = \dfrac{1}{N} \sum_{j=1}^{N} u_j v_j \, I_i(t_j, K) \quad (9)$$

where N is the number of measurements in the ensemble. Figure 10 is the result of such quadrant analysis with K = 0 (and with velocity bias correction) for 5 <= y^+ <= 350. Nezu (1977) and Lekakis (1989) report similar measurements in the near wall region via hot-film anemometry. The relative contributions from the respective quadrants shown in figure 10 are consistent with the results of Nezu and Lekakis. Quadrants 2 and 4 make large contributions (on the order of 60-80%) to the Reynolds stress throughout the wall layer, while quadrants 1 and 3 are relatively weak contributors except within the sublayer.

Such conditioned statistical analysis together with joint pdf construction in strongly three dimensional flows will aid in the identification of isolated or rare intense stress component episodes. It is expected that three dimensional velocity event correlation with other auxiliary signals, such as high speed video images, may permit the isolation of those stress episodes which contribute most to wall transport processes (such as sediment entrainment events). The previous representative illustrations demonstrate the applicability of 3D-LDA to such demanding research investigations.

Summary

A three-component laser Doppler anemometer was applied to the measurement of the velocity field near the center-line of a rectangular cross section smooth walled open-channel flume at a Reynolds number of 16100. The measurements reported in this paper include the streamwise mean velocity, the three components of the turbulence fluctuation intensities, the dominant Reynolds stress in the (u,v) plane, and joint probability density functions within the wall region, $5 \leq y^+ \leq 350$. The results of quadrant analysis in the (u,v) plane are also reported throughout this region. It is demonstrated that the mean velocity and the streamwise and normal components of the turbulence intensities compare well to other measurements obtained with a multi-element hot-film anemometer, in a similar Reynolds number range. The mean velocities in the normal and spanwise directions were found to be zero, demonstrating two-dimensionality of the flow and the correlation coefficients between fluctuating velocities in the spanwise plane were also found to be zero, as expected. The spanwise component of the turbulence intensity did not compare as favorably, however. The measured magnitude and qualitative behavior of the spanwise intensity with distance from the wall is not consistent with the limited number of other reported measurements. The results obtained here indicate a relative maximum intensity of approximately $w'/u_* = 1.7$ with a slow decrease (toward v') with distance from the smooth wall. These results in the spanwise component may be due, at least in part, to the accumulation of systematic or fixed uncertainties associated with the measurement of the LDA fringe spacings and geometrical parameters, since the spanwise component is very sensitive to errors in the measured components for non-orthogonal 3D-LDA. The accuracy of the spanwise measurement could be improved with a carefully designed calibration experiment to reduce this fixed error contribution. Nevertheless, the joint pdf's and quadrant results compare very closely to other reported measurements in the wall region of turbulent channel and pipe flows.

Enhanced performance of the LDA system is possible through improvements in the optical system and signal processors (and the introduction of calibration experiments to reduce fixed errors). The signal quality and data rate may be improved in the present case through the introduction of simple spherical optical elements in the acrylic water-filled box which couples the LDA system to the flume side wall. This acrylic box was introduced in the present case to reduce adverse beam refraction effects which limit spatial resolution and measurement volume movement. This could be improved further by the introduction of simple thin spherical glass shells to replace the plane windows in this optical box, in the manner of Wei and Willmarth (1989). Such elements will permit better laser beam focussing and scattered light collection, as well as improved spatial resolution of the measurement volume. Data rate increases, which would permit higher temporal resolution are possible with this modification, and could be increased further through the use of frequency domain based signal processors, which now have adequate dynamic range and data rate capability.

Acknowledgements

Research supported by the U.S. Geological Survey (USGS), Department of the Interior, under USGS award number. The views and conclusions contained in this document are those of the authors and should not be interpreted as necessarily representing the official policies, either expressed or implied, of the U.S. Government. The authors would also like to acknowledge C. Lindsey, B. Owsenek, and C.F. Madsen for their contributions to this work.

References

Alfredsson, P.H. and A.V. Johansson, "Turbulence Experiments-Instrumentation and Processing of Data," Advances in Turbulence 2, Fernholz and Fiedler, eds., Spring-Verlag, Berlin, 1989, pp. 230-243.

Dancey, C.L. "Measurement of Second Order Turbulence Statistics in an Axial-Flow Compressor via 3-Component LDA," AIAA Paper 90-2017, 1990.

Johansson, T.G. and R.I. Karlsson, "The Energy Budget in the Near-Wall Region of a Turbulent Boundary Layer," 4th International Symposium on Applications of Laser Anemometry to Fluid Mechanics, Lisbon, Portugal, 1988.

Karlsson, R.I. and T.G. Johansson, "LDV Measurements of Higher Order Moments of Velocity Fluctuations in a Turbulent Boundary Layer," 3rd International Symposium on Applications of Laser Anemometry to Fluid Mechanics, Lisbon, Portugal, 1986.

Lekakis, I.C., "Coherent Structures in Fully Developed Turbulent Pipe Flow," PhD Dissertation, Department of Mechanical Engineering, University of Illinois, 1988.

Lindsey, C.A. and B. Owsenek, "The Design of a Three Component LDA System for Application to an Open Channel Flow," presented at the 1993 ASME Fluids Engineering Conference, Washington, D.C. June, 1993. FED-Vol. 153 ASME, 1993, pp. 11-14.

Madsen, Carl F., "An Investigation of Velocity Bias with a Three-component LDA in Open Channel Flow." Master's of Science Thesis, Department of Mechanical Engineering, Virginia Polytechnic Institute and State University, Blacksburg, Va., 1994.

Moin, P. and P.R. Spalart, "Contributions of Numerical Simulation Data Bases to the Physics, Modeling, and Measurement of Turbulence," Advances in Turbulence, George and Arndt, eds., Spring-Verlag, Berlin, 1989, pp. 11-20.

Nezu, I., "Turbulent Structure in Open-Channel Flows," Doctoral Dissertation, Department of Civil Engineering, Kyoto University, Kyoto, Japan, 1977.

Nezu, I. and W. Rodi, "Open-Channel Flow Measurements with a Laser Doppler Anemometer," Journal of Hydraulic Engineering, Vol. 112, No. 5, May, 1986, pp. 335-355.

Schroder, M., Stein, C.J. and G. Rouve, "Application of the 3D-LDV Technique on Physical Model of Meandering Channel With Vegetated Flood Plain," 4th International Conference on Laser Anemometry, Advances and Applications, Cleveland, Ohio, Vol. 2, 1991, pp. 511-520.

Wei, T. and W.W. Willmarth, "Reynolds-number effects on the structure of a turbulent channel flow," Journal of Fluid Mechanics, Vol. 204, 1989, pp.57-95.

Wei, T. and W.W. Willmarth, "Examination of v-velocity fluctuations in a turbulent channel flow in the context of sediment transport", Journal of Fluid Mechanics, Vol. 223, 1991, pp.241-252.

Hot-Film Response in Three-Dimensional Highly Turbulent Flows

P. Prinos[1]

Abstract

A methodology is described for accurate turbulence measurements with hot film or hot wires even at turbulent intensities as high as 100%. The assumptions involved and the approximations introduced are clearly stated. Corrections to mean velocities due to high turbulence levels are shown to increase with increasing turbulence intensity.

Introduction

It has been well established (Jorgensen, 1971, Perry, 1982, Tavoularis, 1983) that the output voltage E, of a cylindrical hot-film or hot-wire with large length-to-diameter ratio, operating at a constant temperature T_w, in a flow stream with velocity Q and temperature T, is given by the modified King's law

$$\frac{E^2}{T_w - T} = A + BU_e^n \tag{1}$$

where A, B and n are calibration constants depending on wire and fluid material properties and on wire dimensions, and U_e is the effective cooling velocity. If the flow direction is perpendicular to the wire axis, $U_e = Q$; otherwise, U_e represents the hypothetical flow velocity perpendicular to the wire that would give the same voltage output as the actual flow velocity at the particular orientation. While the validity of equation (1) has been verified for commonly encountered laboratory flows, the relationship between Q and U_e is still under scrutiny. Q can be decomposed into a normal component, Q_n, a tangential component, Q_t, and a binormal component, Q_b.

---
[1] Asst. Professor, Hydraulics Lab., Dept. of Civil Eng., Univ. of Thessaloniki, Thessaloniki, Greece

It is clear that Q_n and Q_b should have a major contribution to the cooling of the wire, while the contribution of Q_t should be of secondary importance. A commonly used expression is:

$$U_e^2 = Q_n^2 + k^2 Q_t^2 + h^2 Q_b^2 \qquad (2)$$

where k^2 and h^2 are calibration constants to account for the hot-film length-to-diameter ratio, the mounting type, the size and spacing of prongs and the shape and proximity of the probe body. For infinite sensors $k^2 = 0$, $h^2 = 1$, while for a range of commercial sensors $k^2 = 0.01$ to 0.20 and $h^2 = 1.0$ to 1.2 (Samet and Einav, 1985, Mobarak et al, 1986). It has been shown (Samet and Einav, 1985) that these coefficients also depend on the Reynolds number of the flow past the sensor. In terms of the angles ψ and β, the relationship between U_e and Q can be written as:

$$|U_e/Q| = (\cos^2\psi + k^2 \sin^2\psi \cos^2\beta + h^2 \sin^2\psi \sin^2\beta)^{1/2} \qquad (3)$$

Figure 1 illustrates the general orientation of a single, cylindrical, sensor, mounted perpendicular to the probe axis with respect to an arbitrary velocity vector Q. The angles ψ and β are termed as the yaw and roll angles respectively. Figure 2 shows the calibration of a single hot wire probe (TSI, model 1260 miniature straight probe) which indicates the validity of equation (1) with excellent accuracy. Typical yaw and roll sensitivities of the same probe are shown in figure 3. These results are used to evaluate the coefficients k and h of equation (2). The optimal values $k^2 = 0.03$ and $h^2 = 1.11$ resulted in the minimum errors for three roll angles and two velocities. The error $(Q_m - Q)/Q$, where Q_m is the measured value using equation (3), is illustrated in figure 4. It can be seen that the errors are generally smaller than 1% for $-30 < \psi < 30$ and exceed 2.5% only for $\beta = 45$ and $60 < \psi$.

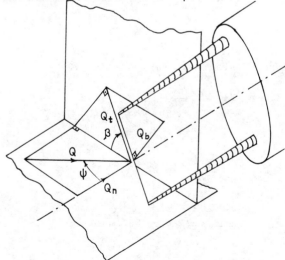

Figure 1: Velocity components and the yaw and roll angles for a single probe

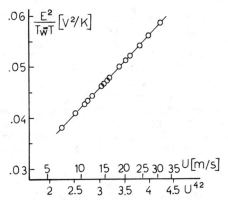

Figure 2: Calibration of a single wire probe

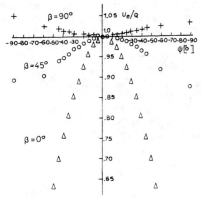

Figure 3: Calibration of single wire probe to yaw and roll

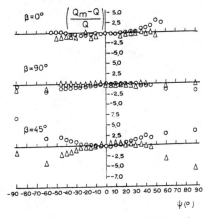

Figure 4: Percentage error in mean velocity using equ. (2)

In a stationary turbulent flow, the instantaneous velocities U_e, Q_n, Q_t, Q_b can be decomposed respectively into mean $\bar{U}_e$, $\bar{Q}_n$, $\bar{Q}_t$, $\bar{Q}_b$ and fluctuating components u_e, q_n, q_t, q_b, where, by definition, the quantities denoted by lower case letters have zero averages. Substituting these expressions into equation (2) and expanding the squares, we have

$$\bar{U}_e^2 + 2\bar{U}_e u_e + u_e^2 = \bar{Q}_n^2 + 2\bar{Q}_n q_n + q_n^2 + k^2(\bar{Q}_t^2 + 2\bar{Q}_t q_t + q_t^2) + h^2(\bar{Q}_b^2 + 2\bar{Q}_b q_b + q_b^2) \quad (4)$$

which, upon averaging, is reduced to

$$\bar{U}_e^2 + \overline{u_e^2} = \bar{Q}_n^2 + k^2 \bar{Q}_t^2 + h^2 \bar{Q}_b^2 + \overline{q_n^2} + k^2 \overline{q_t^2} + h^2 \overline{q_b^2} \quad (5)$$

For the general mean velocity orientation and/or in the presence of turbulence fluctuations, it is impossible to resolve the actual from the effective cooling velocity, except by performing measurements at different orientations of the hot film or with multiple sensors. Such procedures usually assume low turbulence levels and breakdown for turbulence intensities above about 20%. Some response equations have been developed in the past for high turbulence levels. In order to simplify the equations, the following novel procedure has been developed for highly turbulent flows with turbulent intensities higher than 20%. The assumptions involved and the approximations introduced are clearly stated and hence the conditions of applying such a procedure are easily identified.

Methodology

We assume that the probe is inserted in a stationary flow such that its axis remains fixed, while it is permitted to roll about this axis. If we decompose the velocity vector Q into an axial component U and a transverse component V, the component W normal to both U and V is, by definition, equal to zero. The components in the probe-oriented and wire-oriented systems are related as

$$Q_n = U, \quad Q_t = V\cos\beta, \quad Q_b = V\sin\beta$$

It is obvious from Equation (3) that, for fixed Q and ψ, U_e is a function of β. For simplicity, and, since for the particular sensors used it was found that $k^2 = 0.03 << h^2 = 1.11$, it is possible to neglect the second term in the right hand side of Equation (3), except for rare situations where $\psi = \pi/2$ and $\beta = 0$.
Then, it is obvious that the minimum and maximum values of U_e occur when $\beta = 0$ and $\pi/2$ respectively. Hence

$$U_{emin} \approx U, \quad U_{emax} \approx U^2 + h^2 V^2 \quad (6)$$

In turbulent flows the equations for the mean velocity are complicated by the presence of fluctuation statistics. However, if the turbulence does not deviate significantly from isotropy ($\overline{u^2} = \overline{v^2} = \overline{w^2}$) or, at least from axisymmetry ($\overline{v^2} = \overline{w^2}$), it seems reasonable to assume that U_{emin} and U_{emax} will be reached at two perpendicular roll orientations such that $Q_b = 0$ and $Q_b = V$ respectively. Since the contribution of tangential velocities can be neglected ($k^2 \ll h^2$), it is possible to derive the following approximate relationships

$$\bar{U} \approx (\bar{U}_{emin}^2 + \overline{u_{emin}^2} - \bar{u}^2 - h^2 \overline{w^2})^{1/2} \tag{7}$$

and

$$\bar{V} \approx \frac{1}{h}(\bar{U}_{emax}^2 - \bar{U}^2 + \overline{u_{emax}^2} - \bar{u}^2 - h^2 \overline{v^2})^{1/2} \tag{8}$$

Equations (7) and (8) would provide the mean velocity vector direction and magnitude from two positions of the hot wire, provided that turbulence parameters are negligible, known or possible to estimate.

Equations for turbulent fructuations can be derived by subtracting appropriate mean equations from the corresponding instantaneous equations. For example, subtracting Equation (5) from Equation (4), it is possible to derive the relationship

$$2\bar{U}_e u_e + \bar{u}_e^2 - u_e^2 = 2\bar{Q}_n q_n + q_n^2 - \bar{q}_n^2 + k^2(2\bar{Q}_t q_t + q_t^2 - \bar{q}_t^2) + h^2(2\bar{Q}_b q_b + q_b^2 - \bar{q}_b^2) \tag{9}$$

which shows that the fluctuating anemometer signal contains contributions from all three velocity fructuation components. Equation (9) can be simplified by neglecting the tangential term, as before. Further simplification can be obtained if equation (9) is written for the roll angle that provides U_{emin}. Then $Q_b = W = 0$ and $Q_n = U$, so that

$$2\bar{U}_{emin} u_{emin} + u_{emin}^2 - \bar{u}_{emin}^2 \approx 2\bar{U}u + u^2 - \bar{u}^2 + h^2(w^2 - \bar{w}^2) \tag{10}$$

Squaring both sides of Equation (10), averaging, omitting the vanishing terms and grouping the remaining ones, we have

$$4\bar{U}_{emin}^2 \overline{u_{emin}^2} + 4\bar{U}_{emin} \overline{u_{emin}^3} + \overline{u_{emin}^4} - (\overline{u_{emin}^2})^2 \approx 4\bar{U}^2 \overline{u^2} + 4\bar{U}(\overline{u^3} + h^2 \overline{uw^2})$$

$$+ \overline{u^4} - (\overline{u^2})^2 + h^4(\overline{w^4} - (\overline{w^2})^2) + 2h^2(\overline{u^2 w^2} - \overline{u^2}\,\overline{w^2}) \tag{11}$$

All statistical properties of u_e, $\bar{U}_e$ are measurable. The mean velocity $\bar{U}$ can be eliminated in terms of turbulent statistics using Equation (7). Then, Equation (11) becomes an almost exact relationship among turbulent moments of u and w. In order to solve it, for example for the variance $\overline{u^2} = u'^2$ (primes indicate r.m.s. values), a turbulence model must be employed.

It is reminded that, for small turbulence intensities ($|u|, |w| \ll \bar{U}$), $u = u_{emin}$, in which case the solution is readily available. Thus, it seems plausible to assume that the dimensionless statistical properties of u are the same as those of u_{emin}, when such properties appear in correction terms. This assumption is applied to the dimensionless third and fourth mometns, known as skewness and flatness factors respectively,

$$\frac{\overline{u}^3}{u^3} \approx \frac{\overline{u}_{emin}^3}{u_{emin}^3} \equiv S \qquad (12)$$

$$\frac{\overline{u}^4}{u^4} \approx \frac{\overline{u}_{emin}^4}{u_{emin}^4} \equiv f \qquad (13)$$

For Gaussian random processes $f=3$, $S=0$, but measurements show some deviations of u_e from Gaussianity. In the absence of any prior information about w, we shall also assume that

$$\frac{\overline{w}^4}{w^4} \approx f \qquad (14)$$

The joint statistics of the turbulent fluctuations u and w depend on the flow type. In isotropic turbulence, w is uncorrelated with u, while in sheared flows the correlation coefficient

$$\rho \equiv \frac{\overline{uw}}{u'w'} \qquad (15)$$

may take values between -1 and +1 and is typically of the order ±0.4. Introducing a plausible model, we assume that w and u are jointly quasi-Gaussian as far as fourth moments are concerned, i.e.

$$\frac{\overline{u^2 w^2}}{u'^2 w'^2} \approx \frac{1}{3}(1+2\rho^2)f \qquad (16)$$

We further employ an assumption for the third moment, i.e.

$$\frac{\overline{uw^2}}{u'w'^2} \approx |\rho|^{3/2} S \qquad (17)$$

Finally, we introduce as parameters the measured turbulent intensity

$$t \equiv \frac{u'_{emin}}{\overline{U}_{emin}} \qquad (18)$$

and the ratio of r.m.s. turbulent fluctuations

$$a \equiv \frac{hw'}{u'} \qquad (19)$$

The hot film binormal sensitivity coefficient h(h = 1.05 for the hot film used) has been included in the definition for convenience. Using all assumptions and simplifications, Equation (11) can be written in terms of the quantity.

$$x \equiv \frac{u'}{U_{emin}} \quad (20)$$

as

$$t^2 + St^3 + \frac{1}{4}(f-1)t^4 = [1+t^2-(1+a^2)x^2]x^2$$

$$+[1+t^2-(1+a^2)x^2]^{1/2}(1+a^2|\rho|^{3/2})Sx^3$$

$$+\{\frac{1}{4}(f-1)(1+a^2)+[\frac{f}{2}(1+2\rho^2)-1]\frac{a^2}{2}\}x^4 \quad (21)$$

Equation (21) can be solved numerically for x in terms of the measurable quantities t, S and f for commonly encountered ranges of the parameters a and ρ. Then the mean velocity U can be computed from equation (7) as

$$\bar{U} = \bar{U}_{emin}[1+t^2-(1+a^2)x^2]^{1/2} \quad (22)$$

Once U has been obtained, the transverse mean velocity componet V can be estimated using equation (8). An additional assumption must be made for the ratio hv'/u'; in the absence of other information, one may assume that it is equal to a, as would be the case in axisymmetric turbulence, where v' = w'.

General Remarks

Figure 5 presents the ratio u'/u'_{emin}, obtained by solving Equation (21) using the Newton-Raphson technique. The figure includes the case of Gaussian, isotropic turbulence (f=3, S=0, a=1, ρ=0) as well as few cases with values of f and S typical of turbulent flows and a and ρ spanning the most likely ranges to be encountered in highly turbulent flows. The results are presented for effective cooling velocity intensities between 0 and 0.85, which covers flows with low and high turbulence intensities. This figure illustrates the following interesting points:
a) The r.m.s. axial velocity u' can be higher or lower than the measured u'_{emin}, depending on the combination of values of S, f, a, ρ.
b) In all cases, the difference between u' and u_{emin} is relatively small, rarely exceeding ±10%. Considering that other systematic and random errors might exceed such levels, it appears that corrections to u'_{emin} are not essential and can be omitted.
c) It has been illustrated that hot films or hot wires can give reasonably accurate turbulence measurements, even at turbulent intensities as high as 100%.

The ratio U/U_{emin}, based on equation (22) and the solution of equation (21), is shown in Fig. 6. The following observations can be made:
a) U is always smaller than U_{emin}. Their difference increases monotonically with increasing turbulence intensity.
b) Corrections to U_{emin} are generally less than 10% for measured turbulence intensities $t < 40\%$, but may increase up to 30-40% at $t = 60\%$ and may exceed 60% at $t = 80\%$.
c) The errors in U are higher in Gaussian isotropic turbulence than in non-isotropic shear flow, where usually $u' > w'$ and $\rho = 0$.
d) The above procedure presents self consistent corrections for hot-wire measurements of mean velocity and flows with measured turbulence intensity up to 80%.

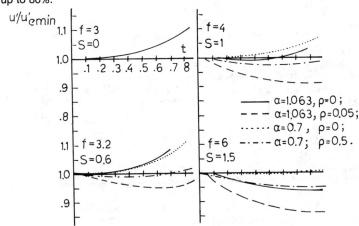

Figure 5: Ratio of rms axial and effective cooling velocities for various parameters

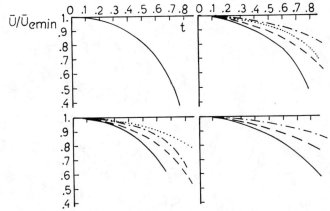

Figure 6: Ratio of mean axial and effective cooling velocities for various parameters

Conclusions

A methodology has been developed for accurate measurements of mean and turbulence characteristics with hot films or hot wires in turbulent flows with intensities as high as 100%. The method is based on measurements with a single probe at two positions where the roll angle is 0 and $\pi/2$ respectively. Turbulence intensities are calculated based on the assumptions of equal flatness and skewness coefficients at all poritions, as well as on some approximations for the third and fourth moments. Corrected estimates of mean velocities due to high turbulence levels are obtained which are shown to be up to 30-40% for 60% turbulence intensity and may exceed 60% for turbulence intensities as high as 80%.

References

Acrivlellis, M. (1977), Hot-wire measurements in flows of low and high turbulence intensity, DISA Inf., no. 22, pp. 15-20.

Acrivlellis, M. (1978), An improved method for determining the flow field of multidimensional flows of any turbulence intensity, DISA Inf., no 23, pp. 11-16.

Jaroch, M. (1985), Development and testing of pulsed-wire probes for measuring fluctuating quantities in highly turbulent flows, Experiments in fluids, 3, pp. 315-322.

Jorgensen, F. (1971), Directional sensitivity of wire and fiber-film probes, DISA Inf., no 11, pp 31-37.

Mobarak, A., Sedrak, M. F and El Telbany M.M.M (1980), On the direction sensitivity of hot-wire probes, Dantec Inf., no 2, pp 7-9.

Perry A.E. (1982), Hot-Wire Anemometry, Clarendon Press, Oxford.

Samet, M. and Einav, S. (1985) Directional sensitivity of unplated normal-wire probes, Rev. Sci. Instrum., 56(12), pp. 2299-2305

Rodi, W. (1975), A new method of analysing hot-wire signals in highly turbulent flow and its evaluation in a round jet, DISA Inf., no. 17, pp. 9-18.

Tavoularis, S. (1983), Simple corrections for the temperature sensitivites of hot-wires, Rev. Sci. Instrum., vol. 54, pp. 741-743.

EXPERIMENTAL STUDY ON TURBULENT STRUCTURES IN UNSTEADY OPEN-CHANNEL FLOWS

Iehisa NEZU[1], M. ASCE, Akihiro KADOTA[1] and Hiroji NAKAGAWA[1]

ABSTRACT
 Turbulence measurements in *unsteady open-channel flows* over smooth beds were conducted accurately by the simultaneous use of two-component LDA system and water-wave gauges. In the inner region near the bed, the normalized turbulent structures were almost similar to those in steady and uniform open-channel flows. In contrast, in the outer region near the free surface, especially in depth-varying zone that could first be measured with LDA, the effects of low frequency component like fluctuations of free surface were observed in the distributions of turbulence intensities. As the result, the turbulence intensities did not indicate the universal distribution.

INTRODUCTION
 In steady open-channel flows, anisotropic and inhomogeneous turbulence is observed due to the existence of free surface. Therefore, open-channel flows have their own peculiar distribution of turbulence intensities which are different from pipe flow with no free surface. Especially in the critical open-channel flow in which the Froude number Fr becomes about unity, the turbulence intensities do not show universal distributions. This may be caused by Bradshaw's (1967) hypothesis that the *inactive motion* composed of low frequency component such as the depth or pressure fluctuations becomes dominant although the Reynolds stress shows the same triangular distribution as pipe flow. In *unsteady open-channel flows*, the effect of *inactive motion* are expected to be more dominant for the distributions of turbulence intensities. A comprehensive review is available in the IAHR monograph by Nezu and Nakagawa (1993a).
 There have been a number of studies on flood flow hydraulics. Most of them, however, have used some assumptions of quasi-steady flow. That is to say, they were analyzed on the basis of a peak discharge of flood flow and ignored the *turbulence*. Taking these problems into account, Hayashi et al. (1988) have measured the turbulent structure by making use of hot-film anemometers and suggested that the turbulence become larger in the rising stage than in the falling one. Tu and Graf (1992) have studied the velocity distributions in unsteady open-channel flows over gravel beds by

---

[1] Dept. of Civil and Global Environment Eng., Kyoto University, Kyoto 606, Japan.

using propeller current meters and examined the unsteadiness effects on them. Nezu and Nakagawa (1993b) have first measured the turbulent structures in unsteady open-channel flows over smooth bed by making use of two-component laser Doppler anemometer (LDA) and suggested that the Coles' wake strength parameter Π which indicates the magnitude of deviation from log-law distribution increased in the rising stage, whereas it decreased in the falling stage. It was concluded that the unsteadiness effects on turbulent structures in open-channel flows might be more significant in the outer region near the free surface.

In this study, the more detailed turbulence measurements in the outer region, especially in depth-varying zone that can be first measured with LDA, were conducted accurately in unsteady open-channel flows. The unsteadiness effects on mean velocity, turbulence intensities and bed shear stress are discussed and compared with the theory.

THEORETICAL CONSIDERATION

The momentum equation for unsteady and two-dimensional open-channel flows can be described by

$$\frac{\partial U}{\partial t} + \frac{\partial U^2}{\partial x} + \frac{\partial UV}{\partial y} = -\frac{1}{\rho}\frac{\partial p}{\partial x} + \frac{1}{\rho}\frac{\partial \tau}{\partial y} \quad (1)$$

in which, U and V are the mean velocity components in the streamwise (x) and vertical (y) directions, respectively. p is the total pressure including the gravity and τ is the total shear stress. The total shear stress τ is defined as follows :

$$\frac{\tau}{\rho} \equiv \nu \frac{\partial U}{\partial y} - \overline{uv} \quad (2)$$

where, ν is the kinematic viscosity. u and v are turbulence components corresponding to U and V, respectively. In open-channel flow, the pressure gradient $\partial p/\partial x$ in Eq. (1) is given by the water surface slope S_s, as follows :

$$-\frac{1}{\rho}\frac{\partial p}{\partial x} = gS_s, \qquad S_s \equiv \sin\theta - \cos\theta\,\frac{\partial h}{\partial x} \quad (3)$$

where, h is the flow depth and $S_b \equiv \sin\theta$ is the bed slope. In strong unsteady open-channel flows, it can be considered that the first term in the left hand side of Eq. (1) is greater than the other inertial terms. Therefore, the bed shear stress τ_b is obtained by integrating Eq. (1) from the bed $y=0$ to the water surface $y=h$, as follows :

$$\frac{\tau_b}{\rho} = gS_s h - \int_0^h \frac{\partial U}{\partial t}dy \cong gS_s R - \frac{1}{B}\frac{\partial Q}{\partial t} \quad (4)$$

in which, R is the hydraulic radius and is adopted for the side-wall effects. B is the channel width and Q is the water discharge.

The depth gradient $\partial h/\partial x$ can be approximated by $-1/c \cdot \partial h/\partial t$. Thus, the water surface slope S_s is given by $S_b - 1/c \cdot \partial h/\partial t$. c is the celerity of flood waves and can be expressed by $U_m + \sqrt{gh}$; U_m is the bulk mean velocity. Therefore, the first term in the right hand side of Eq. (4) becomes larger in the rising stage than the falling one, whereas the second term varies in the opposite side of first term.

EXPERIMENTAL EQUIPMENT AND HYDRAULIC CONDITIONS

The experiments were conducted in a 10m long, 40cm wide and 50cm deep tilting flume. In this water flume, the discharge $Q(t)$ can be automatically controlled by a personal computer in which the rotation speed of a water-pump motor involving a transistor inverter is controlled by the feedback from the signals of an electromagnetic flow-meter. If any hydrograph is input into the computer, the corresponding discharge can be reproduced accurately in this circulation system.

Table 1 Hydraulic Conditions.

Case	T_d (sec)	Q_b (l/s)	Q_p (l/s)	h_b (cm)	h_p (cm)	Re_b	Re_p	Fr_b	Fr_p	S_b
SC3T1	60	5.00	15.48	4.05	6.60	12748	39469	0.49	0.73	1/600
SC3T2	90	5.00	15.87	4.10	6.80	12748	40463	0.48	0.71	1/600
SC3T3	120	5.00	16.19	4.10	6.90	12748	41279	0.48	0.71	1/600
SD3T1	60	2.50	7.34	4.05	5.85	6374	18714	0.24	0.41	1/1000
SD3T2	90	2.50	7.62	4.00	6.00	6374	19428	0.25	0.41	1/1000
SD3T3	120	2.50	7.74	4.00	6.00	6374	19734	0.25	0.42	1/1000

Two components of instantaneous velocities ($\tilde{u}, \tilde{v}$) were measured with a three-beam LDA system operated in the forward-scattering differential mode (Dantec-made). The flow depth $h(t)$ was measured by water-wave gauges (Kenek-made). The LDA was set in 7m distance from the upstream so that the flow was fully developed. The water-wave gauge was set just at the downstream side of the LDA. In the present experiments, the instantaneous velocity in the depth-varying zone which corresponds to the region ($h_b<y<h_p$) was also measured in order to discuss the free-surface effects; h_b and h_p are the depth at base and peak flows, respectively. All of the output signals of LDA and water-wave gauge were recorded in a digital form with sampling frequency f=200Hz and sampling time 250-370 sec, depending on the hydraulic condition. Statistical analyses were conducted then by a large digital computer in the Data Processing Center, Kyoto University.

As shown in Table 1, the duration time T_d from the base discharge to the peak one in flood was set to 60, 90 and 120 sec in three series. The base discharges Q_b were set to 2.5 and 5.0 (l/s) in two series. Therefore, total number of experimental cases was six in the combination of T_d and Q_b. In each case, the peak discharge Q_p was chosen as Q_p/Q_b=3. The depth at peak flow h_p satisfied the criterion of the aspect ratio $B/h_p>5$ that the velocity-dip phenomena caused by the secondary currents do not occur in the center line of channel, as pointed out by Nezu et al. (1993b).

UNSTEADINESS PARAMETER

A few kinds of unsteadiness parameters which indicate the effects of unsteadiness on the hydrodynamic and turbulent structures have ever been proposed. For example, Hayashi (1951) considered that the ratio of depth-varying acceleration to gravity described the characteristics of unsteady flow, and he introduced the following parameter σ in one-dimensional (1-D) momentum equation.

$$\sigma = \sqrt{-\ddot{h}_p / (gS_b)} \tag{5}$$

where, $\ddot{h}_p$ is defined as $\partial^2 h_p / \partial t^2$. However, it is difficult to evaluate the $\ddot{h}_p$ at any sections because the whole profile of flood wave can not be measured simultaneously. Takahashi (1969) also introduced the following parameter λ by abbreviating the longitudinal and vertical accelerations.

$$\lambda \equiv V_s / (cS_b) \tag{6}$$

where, $V_s = (h_p-h_b)/T_d$. The parameter λ implies the ratio of the rising speed of water surface to the vertical component of celerity of flood waves. The unsteadiness parameters σ and λ contain the bed slope S_b. This means that if the bed slope S_b changes slightly, turbulent structures in unsteady open-channel flows with almost the same hydraulic conditions change significantly. Due to these shortcomings, σ and λ are not suitable for discussing the effects of unsteadiness on turbulent structures in unsteady open-channel flows. On the other hand, Tu and Graf (1992) applied the Clauser's pressure-gradient parameter β to unsteady open-channel flow over rough bed and proposed the following unsteadiness parameter β.

$$\beta \equiv K\left(-\frac{h}{U_m^2}\frac{dU_m}{dt}\right) \quad (7)$$

in which, K depends on both the relative roughness and the mean velocity distribution. This parameter β, however, has an assumption that the mean velocity profiles in the whole depth may be expressed in the form of power law and it was defined at arbitrary time of flood period.

As mentioned previously, there are some difficulties in order to apply the existing unsteadiness parameters to turbulent structures in unsteady open-channel flows. In the present study, taking the importance of pressure gradient $-dp/dx$ into account, the same unsteadiness parameter α as used by Nezu and Nakagawa (1993b) is also adopted in the following.

By using Eq. (3) and assuming the hydrostatic pressure distribution, the pressure gradient $-dp/dx$ leads to:

$$\frac{1}{\rho g}\frac{dp}{dx} = -S_b + \frac{1}{U_c}\frac{\partial h}{\partial t}\left(1 - \cos\theta\frac{U}{c}\right) \quad (8)$$

where, U_c is the convection velocity of turbulent eddies; it is roughly approximated as $(U_b+U_p)/2$. U_b and U_p are the bulk mean velocities at the base and peak time, respectively. The second term of right hand side in Eq. (8) shows the unsteadiness. Because the celerity of flood wave $c > U$; e.g., Kleitz-Seddon's law, the unsteadiness parameter α is defined, as follows:

$$\alpha \equiv \frac{1}{U_c}\frac{\partial h}{\partial t} \approx \frac{1}{U_c}\frac{h_p - h_b}{T_d} = \frac{V_s}{U_c} \quad (9)$$

The unsteadiness parameter α implies the ratio of the rising speed of water surface V_s to the convection velocity U_c.

DEFINITION OF MEAN VELOCITY COMPONENT

The most difficult and important aspect to investigate turbulent structures in unsteady flow is how to determine the mean velocity component $U(t)$ from the instantaneous velocity $\tilde{u}(t) \equiv U(t) + u(t)$.

In the investigations on oscillatory pipe flow and unsteady boundary layers, three kinds of methods to determine the mean velocity are often used. One is (a) the ensemble-average method, the other is (b) the moving time-average method and the other is (c) the Fourier component method. From the theoretical point of view, the ensemble method (a) is the most suitable. However, in order to apply the method (a) to unsteady flows, the same phase of flows must be accurately used as a trigger of sampling. Furthermore, it is difficult to apply the method (a) to the unsteady open-channel flows because the measurement of sufficient number of waves for ensemble averaging takes much time and because the measurements of depth fluctuations as the trigger by using water-wave gauge is not so accurate. On the contrary, in the moving time-average method (b), there are shortcomings that the mean velocities are affected by the hydraulic conditions and the limits in time series of mean velocities can not be determined. Nezu and Nakagawa (1991) have examined the applicability of these three methods to unsteady open-channel flows and suggested that the Fourier component method (c) was the best among them. In the method (c), the mean velocity component is defined in the following way.

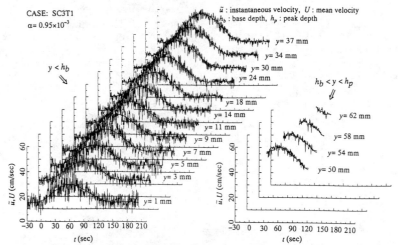

Fig. 1 Time Series of Instantaneous and Mean Velocity.

The instantaneous velocity $\tilde{u}_i$ ($i=1, 2, \cdots, n$) is transformed into a frequency domain by using a discrete Fourier transform (DFT). Only the frequency components less than $(m-1)/2$ are used as mean velocity component, as follows:

$$U_i = \frac{1}{2} a_0 + \sum_{k=1}^{(m-1)/2} \left(a_k \cos \frac{2\pi k i}{n} + b_k \sin \frac{2\pi k i}{n} \right) \quad (10)$$

$$a_k = \frac{1}{n} \sum_{i=1}^{n} \tilde{u}_i \cos \frac{2\pi k i}{n} \quad (k=0, 1, 2, \cdots, m-1/2) \quad (11a)$$

$$b_k = \frac{1}{n} \sum_{i=1}^{n} \tilde{u}_i \sin \frac{2\pi k i}{n} \quad (k=0, 1, 2, \cdots, m-1/2) \quad (11b)$$

The cutoff frequency of Fourier components for mean velocity was reasonably chosen so as to be much smaller than the burst frequency of turbulence. Therefore, in this study, the number of Fourier components m was adopted as 7, as used by Nezu and Nakagawa (1991).

Fig. 1 shows an example of time series of instantaneous velocity $\tilde{u}(t)$ and mean velocity $U(t)$ which was calculated by the Fourier component method; the solid curve in Fig. 1 indicates $U(t)$. The right side of this figure corresponds to the depth-varying zone, i.e., the region of $h_b < y < h_p$. The Fourier component method is applicable not only for the wall region near the bed, but also for the depth-varying zone near the free surface.

RESULTS AND DISCUSSION
Mean Velocity Profiles

The mean velocity profiles that are normalized by the inner variables, i.e., the friction velocity U_* and the kinematic viscosity ν, are shown in Fig. 2. The left side of this figure corresponds to the rising stage ($0<T<1$) and the right side is the falling one ($1<T<2$). T is the time normalized by the duration time T_d. The mean velocity profile at each non-dimensional time T is shifted by 10 units. In the inner region except for buffer layer, i.e., in the region of $y^+ \equiv U_* y/\nu > 30$ and $y/h < 0.2$, the mean velocities show the linear distribution for all non-dimensional time T. This verifies that the mean velocities in the inner region obey the log-law, irrespective of unsteadiness. Therefore, the friction velocity U_* can be evaluated reasonably from the log-law because the Karman constant κ is universal to be equal to 0.41, irrespective of pipes, open channels and boundary layers (see, Nezu and Nakagawa 1993a).

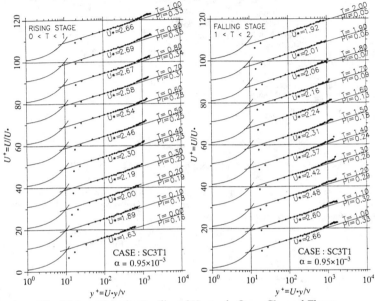

Fig. 2 Mean Velocity Profiles of Unsteady Open-Channel Flows.

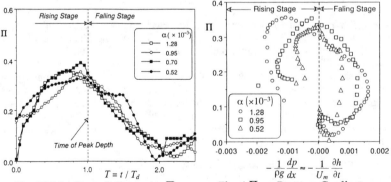

Fig. 3 Wake Strength Parameter Π.   Fig. 4 Π vs. Pressure Gradient.

On the other hand, in the outer region near the free surface, i.e., $y/h > 0.2$, the deviation from the log-law which is named 'wake' can not be neglected any longer around the time of peak depth ($T=1$) although it is small deviation in comparison with boundary layers. Nezu and Rodi (1986) have suggested that the Coles' wake function $w(\xi)$ used in the boundary layers could also be applied in open-channel flows, as follows:

$$\frac{U}{U_*} = \frac{1}{\kappa} \ln\left(\frac{U_* y}{\nu}\right) + A_s + \frac{\Pi}{\kappa} w(\xi), \quad w(\xi) = 2 \sin^2\left(\frac{\pi}{2}\xi\right) \quad (12a,b)$$

where, Π is the wake strength parameter which indicates the magnitude of deviation from the log-law and it is a function of Reynolds number Re in uniform open-channel flows. A_s is the integration constant and $\xi = y/h$ is the normalized vertical coordinate. The variations of Π against the non-dimensional time T are shown in Fig. 3. The Π-value attains a maximum before the time of peak depth ($T=1$). The Π-value increases slowly in the rising stage, whereas it decreases rapidly in the falling stage. These features become more significant as the unsteadiness is larger, i.e., the value of unsteadiness parameter α is larger. It should be noted that as the Reynolds number Re in the steady flow at $T=0$ is larger, the Π-value increases, as pointed out by Nezu and Rodi (1986). However, the Π-value in the flood period ($0<T<2$) has nothing to do with the Reynolds number Re, but it seems to be a function of pressure gradient as pointed out in the boundary-layer experiments. Fig. 4 shows the variations of Π-value against the pressure gradient. Of particularly significance is the clockwise loop of Π vs. T. The area of loop is larger as the effect of unsteadiness becomes stronger. In the flow with adverse pressure-gradient ($-dp/dx<0$), the wake strength parameter Π is larger than favorable pressure-gradient ($-dp/dx>0$).

Turbulence Characteristics

Figs. 5 and 6 show the distributions of turbulence intensity u', v' and Reynolds stress $-\overline{uv}$ normalized by the friction velocity U_*, respectively. The broken curves in Fig. 5 are the following semi-theoretical one (see, Nezu and Nakagawa 1993a).

$$u'/U_* = 2.30 \exp(-\xi), \quad v'/U_* = 1.27 \exp(-\xi) \quad (13a,b)$$

These curves are universal, independent of Reynolds number Re in uniform open-channel flows. The straight line in Fig. 6 is the triangular distribution except for viscous effects very near the bed. In Fig. 5, a little deviation from Eq. (13a) can be observed in the falling stage whereas the turbulence intensities in the base flow ($T=0$) and rising stage coincide well with Eq. (13). In the outer region very near the free surface, the streamwise turbulence intensity u' draws close to the vertical one v' in the falling stage and thus the isotropic tendency can be observed. In contrast, the Reynolds stress indicates almost linear distributions although the data are somewhat scattered.

By using Eq. (13) and the triangular distribution of $-\overline{uv}$, the correlation coefficient of Reynolds stress in uniform flow is given by

$$R \equiv \frac{-\overline{uv}}{u'v'} = \frac{0.342(1-\xi)}{\exp(-2\xi)} \quad (14)$$

The distributions of correlation coefficient R in unsteady open-channel flows are shown in Fig. 7. The calculated curve of Eq. (14) is also included in this figure. The experimental value of R deviates from Eq. (14) near the free surface in the falling stage. These important features suggest strongly that the aforementioned *inactive motion* causes such distribution of turbulence intensities because the effect of depth fluctuations may be more significant than steady flows.

The ratio of average turbulence intensity in the rising stage to that in the falling stage u'_r/u'_f and v'_r/v'_f are shown in Fig 8. In the same manner as Fig. 8, the ratio of Reynolds stress $-\overline{uv}_r/-\overline{uv}_f$ is also shown in Fig. 9. In these figures, the vertical coordinate y is normalized by the base depth h_b. Therefore, the region $y/h_b>1$ corresponds to the depth-varying zone near the free surface. It can be seen from the figures that the ratio u'_r/u'_f and v'_r/v'_f in the region of $y/h_b>1.2$ become smaller than unity whereas they become larger near the bed. It can be indicated that the turbulence becomes larger in the falling stage near the free surface.

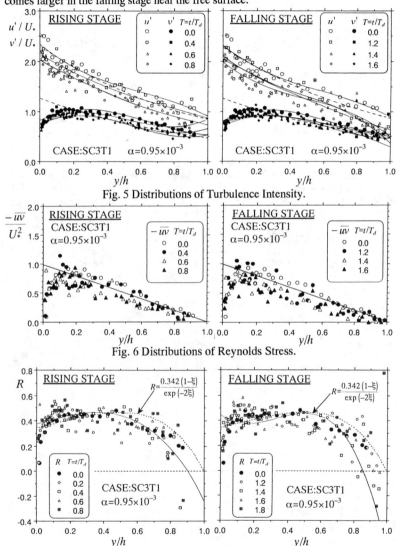

Fig. 5 Distributions of Turbulence Intensity.

Fig. 6 Distributions of Reynolds Stress.

Fig. 7 Correlation Coefficient R of Reynolds Stress.
(The solid curves are experimental ones)

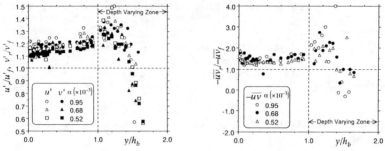

Fig. 8 Ratio of Average Turbulence Intensity in Rising Stage to Falling Stage.

Fig. 9 Ratio of Average Reynolds Stress in Rising Stage to Falling Stage.

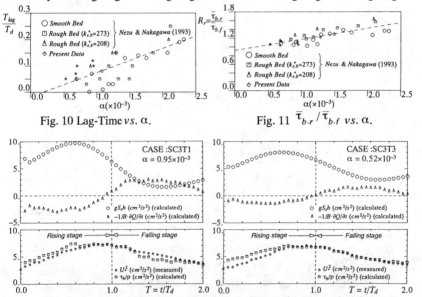

Fig. 10 Lag-Time vs. α.

Fig. 11 $\overline{\tau}_{b.r} / \overline{\tau}_{b.f}$ vs. α.

Fig. 12 Comparison of Measured Bed Shear Stress with Theoretical One.

Unsteadiness Effect on Bed Shear Stress

The friction velocity U_* and bed shear stress τ_b are the most important values in the turbulence theory and sediment hydraulics because the turbulence characteristics are normalized by them in the form of universal function as well as the scaling laws of coherent vortices. The lag-time T_{lag} between the maximum values of τ_b and h against the unsteadiness parameter α is shown in Fig. 10. In this figure, the experimental data over smooth and rough bed which were obtained by Nezu et al. (1993b) are also included. As seen in Fig. 10, the lag-time T_{lag} normalized by the duration time T_d have the linear relation for the unsteadiness parameter α. The T_{lag} become larger as the unsteadiness parameter α increases, irrespective of bed roughness. These findings implies that the α is more suitable parameter of unsteadiness than σ and λ in Eqs. (5) and (6). Fig. 11 shows the ratio of average bed shear stress in the rising stage to the falling one $R_r = \bar{\tau}_{b,r} / \bar{\tau}_{b,f}$ as a function of α. The linear relation between R_r and α can be observed. This indicates that the τ_b for the rising stage becomes much larger than the falling stage as the α becomes larger. The same features as Figs. 10 and 11 have been observed in the actual rivers. Fig. 12 shows the time variation of τ_b evaluated by Eq. (4). The variation of each terms in Eq. (4) are also included in this figure. The bed shear stress evaluated from log-law coincide well with the theory. Furthermore, The variation of first term in the right hand side of Eq. (4) shows the other side of second term. This explains that the τ_b attains maximum before the time of peak depth.

CONCLUSION

In this study, the turbulent structures in unsteady open-channel flows over smooth bed were measured accurately by making use of two-component LDA. Of particular significance was that the unsteadiness effect on turbulent structure in open-channel flows is relatively large in the outer region near the free surface. The *active-inactive* hypothesis proposed by Bradshaw (1967) may be applicable to explain the turbulent structures very near the free surface in unsteady open-channel flows.

ACKNOWLEDGMENT

The authors wish to thank Mr. T. Toda, graduate student of their Department, for the present experiments and thank for the financial support from the Asahi Glass Foundation.

REFERENCES

1) Bradshaw, P. (1967) : 'Inactive' Motion and Pressure Fluctuations in Turbulent Boundary Layers, J. Fluid Mech., vol.30, pp.241-258.
2) Hayashi, T. (1951) : Mathematical Theory and Experimental Study of Flood Waves, Trans. of JSCE, No.18, pp.13-26.
3) Hayashi, T., Ohashi, M. and Oshima, M. (1988) : Unsteadiness and Turbulence Structure of a Flood Wave, 20th Symp. on Turbulence, pp.154-159, (in Japanese).
4) Nezu, I. and Nakagawa, H. (1991) : Turbulent Structures over Dunes and its Role on Suspended Sediments in Steady and Unsteady Open-Channel Flows, Proc. of Int. Symp. on Transport of Suspended Sediments and Its Mathematical Modelling, IAHR, Firenze, pp.165-189.
5) Nezu, I., Nakagawa, H., Ishida, Y. and Kadota, A. (1993a) : Bed Shear Stress in Unsteady Open-Channel Flows., Proc. of 93 ASCE Hydraulic Conference, San Francisco.
6) Nezu, I., Tominaga, A. and Nakagawa, H. (1993b) : Field Measurements of Secondary Currents in Straight Rivers, J. Hydraulic Eng., ASCE, vol.119, No.5., pp.596-614.
7) Nezu, I. and Nakagawa, H. (1993a) : Turbulence in Open-Channel Flows, IAHR-Monograph, Balkema Publishers, Rotterdam.
8) Nezu, I. and Nakagawa, H. (1993b) : Basic Structure of Turbulence in Unsteady Open-Channel Flows, Proc. of 9th Symp. on Turbulent Shear Flows, Session 7.
9) Takahashi, T. (1969) : Theory of One-Dimensional Unsteady Flows in an Prismatic Open-Channel, Annual of Disaster Prevention Research Institute, Kyoto University, (in Japanese).
10) Tu, H. and Graf, W. H. (1992) : Velocity Distribution in Unsteady Open-Channel Flow over Gravel Beds, J. Hydrosc. and Hydr. Eng., JSCE, vol. 10/1, pp.11-25.

RESPONSE OF VELOCITY AND TURBULENCE TO ABRUPT CHANGES FROM SMOOTH TO ROUGH BEDS IN OPEN-CHANNEL FLOWS

Iehisa NEZU[1]; Member, ASCE, and Akihiro TOMINAGA[2]

ABSTRACT

Response of velocity distribution and turbulence characteristics to abrupt changes of bed roughness have been investigated experimentally with two-component laser Doppler anemometer (LDA). Of particular significance is the overshooting property of bed shear stress immediately downstream of roughness discontinuity in open-channel flows. These variations of open-channel turbulence due to the abrupt changes of bed roughness were well predicted numerically by making use of the k-ϵ model.

INTRODUCTION and REVIEW

Bed and side-wall protection works are often constructed to protect river bed and banks from local scours due to high shear stress in rivers. In many cases, river bed consists of smooth protected bed and rough non-protected sand bed. Therefore, it is one of the most important topics in hydraulic engineering to investigate turbulent structure and its response to abrupt changes of bed roughness in open-channel flows. This is true of bank walls of rivers. These experimental data are also necessary to develop a refined turbulence modelling in order to predict such a complicated response of turbulence over roughness discontinuity in open-channel flows.
When the bed roughness changes abruptly from smooth to rough beds and reversely from rough to smooth beds, a sub-boundary layer develops from the transition point between the smooth and rough beds, as shown in Fig. 1. The velocity distributions and turbulence characteristics vary in the streamwise direction due to the roughness discontinuity, as well as in the vertical direction due to the vertical shear stress. As far as the authors know, Plate (1971) have first reviewed comprehensively the response of velocity to abrupt changes of roughness in *air boundary layers* including model forest canopy. Of particular significance is the finding of overshooting property of wall shear stress. Fig. 2 shows the variation of wall shear stress downwind

---

1. Dept. of Civil and Global Environment Eng., Kyoto University, Kyoto 606, Japan.
2. Dept. of Civil Eng., Nagoya Institute of Tech., Nagoya 466, Japan.

of smooth to rough transition, which is cited from Plate (1971). The ratio of wall shear stress τ_2/τ_1 increases abruptly just after the roughness changes, and decreases then monotonously in the streamwise direction. This property is called here the *"overshooting"*. Yeh and Nickerson (1970) have measured velocity distributions in *air boundary-layer* roughness discontinuity like Plate (1971), by making use of 3.2mm diameter Pitot tube and single hot wire. Fig. 3 shows the typical results of friction velocity $U_* = (\tau_w/\rho)^{1/2}$ in the case of smooth to rough changes in air tunnel. The overshooting property is seen clearly.

Antonia and Luxton (1971,72) have conducted velocity measurements over roughness discontinuity in *air* boundary layers by using hot-wire anemometers. Their roughness elements were not *Nikuradse-type* sand roughness, but two-dimensional stripe ones. Therefore, these data may not be so applicable to river engineering.

In the end of the 80's, turbulence measurements over roughness discontinuity in *water* channel flow have been conducted using hot-film anemometers, e.g. Kanda and Muramoto (1989). Nezu and Nakagawa (1991) have first conducted accurate velocity measurements over roughness discontinuity in open-channel flows (Fig. 1) with laser Doppler anemometer (LDA). They found the overshooting property of turbulence as well as of wall shear stress in open-

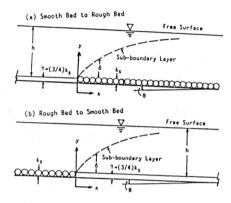

Fig. 1. Configuration of Roughness Changes

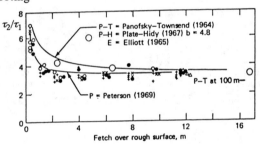

Fig. 2. Overshooting Property of Wall Shear Stress reviewed by Plate(1971).

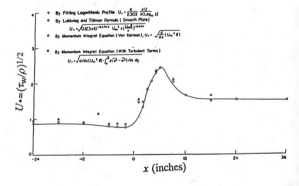

Fig. 3. Overshooting Property of Friction Velocity obtained by Yeh and Nickerson (1970).

channel flows, in the same way as in *air* boundary layers like in Figs. 2 and 3. Further, Nezu et al. (1993) tried to reveal turbulence structure and the associated bursting phenomena over abrupt changes of wall roughness in open-channel flows.

In the present study, accurate turbulence measurements over the same roughness discontinuity as used in open-channel flows of Nezu et al.(1993) have been intensively carried out with a two-component LDA system in order to make clear these important overshooting properties of bed shear stress and turbulence. In order to reveal an evolution mechanism of vortex generated upstream of the roughness discontinuity, space-time correlations of velocity components were carried out by simultaneous measurements using both LDA (as a *detection probe*) and hot-film anemometers (as a *sampling probe*). Both the conventional and conditional-sampling techniques were used to detect bursting vortex near the bed upstream of the roughness discontinuity.

Lastly, numerical calculations were conducted to predict such complicated variations of velocity and turbulence profiles by making use of the k-ε turbulence model. This calculation model could explain the overshooting property of bed shear stress in open-channel flows.

EXPERIMENTAL SET-UP AND PROCEDURES

Experimental Flume

The experiments were conducted in a 10m long, 40cm wide and 50cm deep tilting flume. In this water flume, the discharge can be automatically controlled by a personal computer in which the rotation speed of a water-pump motor involving a transistor inverter is controlled by the feedback from the signals of an electromagnetic flowmeter. This flume system is the same as used by Nezu's group (1991, 93b, 94).

LDA System

Two components of instantaneous velocity at arbitrary point in the central line of channel were measured with a three-beam laser Doppler anemometer (LDA) system operated in the forward-scattering differential mode (Dantec-made). The 10mW He-Ne was used as a laser light, together with the Bragg cell module and optic-axis rotation module. The three-dimensional (3-D) accurate traversing mechanism for LDA system was set 6m downstream of the channel entrance so that the inflow was fully developed. It was possible to traverse the LDA in the 2.5m distances of streamwise direction.

Simultaneous Measurements with LDA and CTA Combination

In order to reveal an evolution mechanism of vortex generated upstream of the roughness discontinuity, it is very useful to investigate space-time correlations from simultaneous measurements with both the laser Doppler anemometer (LDA) and hot-film anemometer (CTA). The same two-component LDA was used as a detection probe of coherent and bursting vortex; this LDA was fixed near the bed upstream of the roughness discontinuity. On the other hand, X-type hot-film CTA was used as a sampling probe of the corresponding evolving vortex. The LDA-CTA combination is very useful to obtain the space-time correlation of velocity, because the laser light (*detection signals*) set upstream ***never*** disturbs the downstream flow (*sampling signals*) of the hot-film probe.

These LDA-CTA combination techniques have been successfully developed by Nezu and Nakagawa (1989, 93b) in order to obtain quantitative information about the generation and interaction mechanisms between the separated vortex and kolk-boil vortex behind dune in corresponding open-channel flows. The hot-film probe in water was traversed automatically in the x-y plane by stepping motors controlled by a computer. All of the output signals of LDA and CTA were recorded in a digital form with sampling frequency f=200Hz and sampling time 150-200s. Then, statistical analyses and conditional sampling procedures were conducted with a large digital computer in the Data Processing Center, Kyoto University.

Hydraulic Conditions

Three kinds of spherical glass beads of k_S=2, 4 and 12mm in diameter were used for bed roughnesses. The Froude number Fr was varied from 0.2 to 0.8. The variations of water depth, h, in the streamwise direction, x, became significant over the abrupt changes of bed roughness at higher Froude numbers of 0.6 and 0.8, as shown in Fig. 4. Consequently, simultaneous measurements with LDA and CTA combination techniques were conducted at Fr=0.2 and Reynolds number Re=9.2x10^3, to avoid complexity of flows. The smooth (k_S=0) to rough (k_S=4 and 12mm) beds, its inversely rough to smooth beds, and the rough (k_S=4mm) to rough (k_S=12mm) beds and its inverse bed configurations were set up, as shown in Figs. 1 and 5.

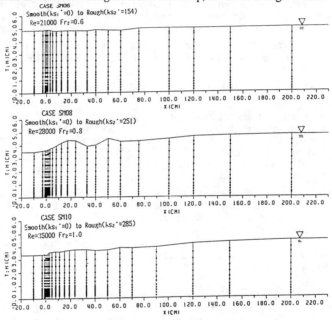

Fig. 4. Free-surface Profiles and Measuring Points over Roughness Discontinuity.

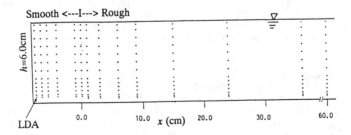

Fig. 5 Free-surface Profiles and Measuring Points of LDA-CTA Combination.

RESULTS and DISCUSSION

Variations of Mean Velocity Distribution

Fig. 6 shows an example of streamwise variations of the semi-log plot of mean velocity $U(x, y)$ normalized by the bulk mean velocity U_m as a function of the streamwise coordinate x/h. The hydraulic condition is Fr=0.4 and Re=1.8x10^4. The bed roughness changes abruptly from smooth (k_s^+=0) to completely rough (k_s^+=385) beds at x=0. The solid curves shown in this figure are the results calculated from the k-ε model by Nezu et al. (1993). The broken curve indicates the velocity distribution of uniform flow before the roughness discontinuity, i.e., $x<0$ (*smooth bed*). The velocity near the bed responds immediately to roughness changes. Consequently, a sub-boundary layer develops from x=0, in the same mechanism as boundary layers of air tunnel, e.g., *see* Plate (1971). The decelerated velocity in the sub-boundary layer coincides well with the log-law distribution, which is described by

$$U/U_* = (1/\kappa) \ln(y/k_s) + A_r \qquad (1).$$

It should be noted that the velocity in the outer region outside of the sub-boundary layer is accelerated due to the equation of continuity. This situation of open-channel flow is quite different from that of boundary-layer flows. The calculated values are in a good agreement with the observed data measured with the present LDA system. The friction velocity U_* and bed shear stress $\tau = \rho U_*^2$ can be evaluated from (1) with the Karman constant κ=0.41, as pointed out by Nezu and Nakagawa (1993a).

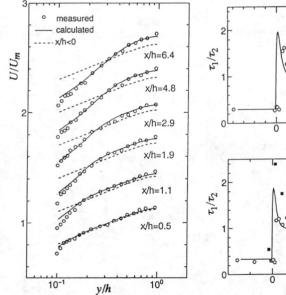

Fig. 6 Variations of Mean Velocity. (*Smooth to Rough Beds*)

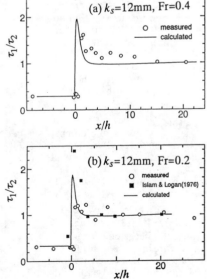

Fig. 7 Overshooting Property of Wall Shear Stress in Open Channel Flows.

Overshooting of Bed Shear Stress

Figs. 7 (a) and (b) show some examples of streamwise variation of bed shear stress $\tau(x)/\tau_2$, in which τ_2 is the value of bed shear stress sufficiently downstream of the roughness discontinuity. The solid line indicates the results calculated from the k-ε model. Of particular significance is the evidence of the overshooting property in open-channel flows in the same manner as air boundary layers shown in Figs. 2 and 3. The bed shear stress $\tau(x)$ overshoots immediately after the roughness discontinuity. The maximum overshooting value τ_{max} tends to increase with an increase of the Froude number Fr and the bed roughness k_s^+, as shown in Fig. 8.

The similar overshooting property have been observed in air boundary layers, e.g., Yeh and Nickerson (1970), Plate (1971), Antonia and Luxton (1971) and Smits and Wood (1985) and others, and also in air closed channel flows by Islam and Logan (1976). For example, the data of Islam and Logan are replotted in Fig. 7 and agree fairly well with the present open-channel data.

Turbulence Intensity

Fig. 9 shows some comparisons between the observed and calculated variations of turbulence intensity u' in the case of smooth to rough changes. Since only the turbulent kinetic energy k is calculated from the k-ε model, u' was evaluated approximately so that $u' = 0.744(2k)^{1/2}$ in the equilibrium region according to Nezu (1977). The value of u'/U_{max} increases near the bed immediately after the roughness discontinuity. This increase develops toward the free surface, which corresponds to the development of sub-boundary

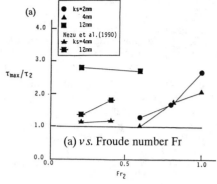

(a) $vs.$ Froude number Fr

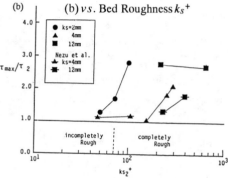

(b) $vs.$ Bed Roughness k_s^+

Fig. 8 Maximum Overshooting Value τ_{max}.

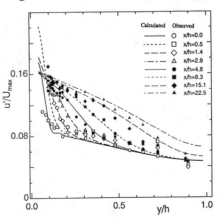

Fig. 9 Streamwise Variations of Turbulence Intensity u'/U_{max}.

layer, Nezu and Nakagawa (1991). The calculated values predict well the development process of the enhanced turbulence intensity. Of particular significance is the finding of overshooting property of u'/U_{max} in the same manner as bed and wall shear stresses. The similar results of v' to u' were also obtained, Nezu et al.(1991).

Reynolds Shear Stress
Fig. 10 shows an example of Reynolds shear stress $-\overline{uv}$ normalized by the maximum main velocity U_{max}. Fig. 10 (a) shows the data measured with the LDA, whereas (b) the data calculated from the k-ε model. The agreement between them is fairly good. The overshooting property is recognized clearly near the bed in both the observed and calculated values. The Reynolds stress is a *key* property to reveal turbulent structures and the associated sediment transport in shear flows. The value of $-\overline{uv}$ is reduced to bed shear stress τ/ρ near the bed except for the viscous term. The properties of Fig.10 coincide well with bed shear stress of Fig. 7.

Property of Bursting Events
Fig. 11 shows an example of the fractional contribution of each bursting event to the Reynolds stress, which were measured with the fixed LDA before the roughness changes, i.e., over *smooth* bed, Fig.5. These were analyzed using the conditional quadrant theory. More detailed information is available in the IAHR-monograph of Nezu and Nakagawa (1993a).

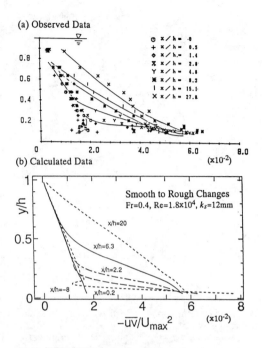

Fig. 10 Streamwise Variations of Reynolds Stress.

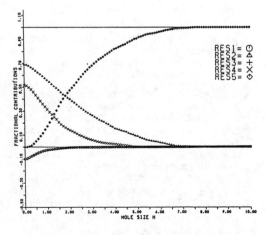

Fig. 11 Contributions of Bursting Events to Reynolds Stress.

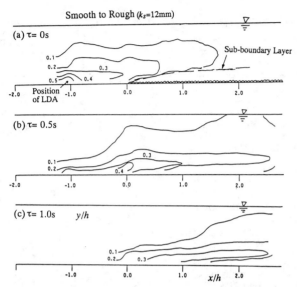

Fig. 12 **Conventional** Space-Time Correlation Contours $C_{uu}(x, y; \tau)$.

The ejection motion (RS2) is the strongest, and the sweep (RS4) is the second one, irrespective of the hole size H of conditional quadrant theory. The negative contributions from the outward (RS1) and inward (RS3) interactions are much smaller than those from the ejection and sweep motions. Therefore, the bursting period T_B can be reasonably evaluated from the half-value-threshold method of H. As the results, $T_B U_{max}/h$ was nearly constant of 1.5-3.0, irrespective of the bed roughness changes. In contrast, the inner-variable description, $T_B U_*^2/\nu$, changed from 100 to 500, depending on the bed roughness, i.e., friction velocity U_*. These findings suggest strongly that the bursting period should be scaled with the outer variables rather than the inner variables, even in the case of abrupt changes of roughness. These results coincide well with previous data that are in detail reviewed in the IAHR-monograph of Nezu and Nakagawa (1993a).

Conventional Space-Time Correlations and Evolution of Mean-Scale Eddies

Fig. 12 shows an example of conventional space-time correlation contours in the case of smooth to rough (k_S=12mm) changes at x=0. The contours C_{uu} of the u-u combination imply an evolution of mean-scale eddies. The mean-scale eddy appearing upstream of the roughness discontinuity is convected downstream by its mean velocity, but it seems that the eddy may not penetrate into the sub-boundary layer generated due to the abrupt changes of roughness.

Fig. 13 shows an example of conditional space-time correlation contours of $<u>_e$ which were analyzed using the conditional quadrant theory developed by Nezu and Nakagawa (1993). The value of $<u>_e$ implies the conditional-sampling ensemble-averaged value of u/u' from the sampling hot-film probe when the ejection motions are detected at τ=0 by the detecting device, i.e., LDA. Therefore, Fig. 13 indicates an evolution of ejection motions in regard to the u-component. Of course, ejection motions have the coherent structure of $u<0$ and $v>0$.

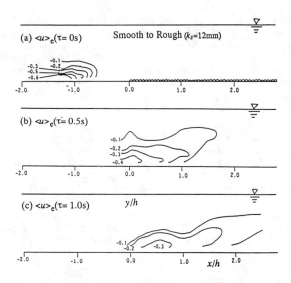

Fig. 13 **Conditional** Space-Time Correlation Contours $<u>(x, y; \tau)$ of u-component in Ejection Motions.

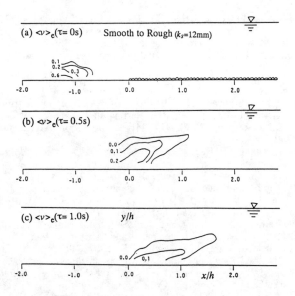

Fig. 14 **Conditional** Space-Time Correlation Contours $<v>(x, y; \tau)$ of v-component in Ejection Motions.

Fig. 14 shows the iso-lines of $<v>_e$ in regard to the corresponding v-component of ejection motion. Of particular significance is the inclination of ejection motions towards the bed. The same results were obtained in the sweep motions.

Although the conventional space-time correlation C_{uu} and the conditional-sampling one $<u>_e$ are quite different from each other in the analyzing techniques, they show similar evolution processes of coherent eddies. The ejection motions generated upstream of the roughness discontinuity may not penetrate into the sub-boundary layer developed at $x=0$. More detailed discussion will be given anywhere.

CONCLUSION

Response of velocity distribution and turbulence characteristics to abrupt changes of bed roughness have been investigated experimentally with the combination of two-component laser Doppler anemometer (LDA) and X-type hot-film anemometer (CTA). Of particular significance is the overshooting property of bed shear stress and turbulence characteristics immediately downstream of roughness discontinuity in open-channel flows. These variations of open-channel turbulence due to the abrupt changes of bed roughness were well predicted numerically by making use of the k-ε model. Finally, the conventional and conditional space-time correlation structures of bursting motions were revealed using the LDA-CTA combination techniques.

ACKNOWLEDGEMENTS

The present experiments were conducted by the help of the first author's graduate students, N. Kawashima and Y. Suzuki. The financial support was obtained from the Asahi Glass Foundation. These help and support are hereby acknowledged.

REFERENCES

1) Antonia, R.A. and Luxton, R.E. (1971);J. Fluid Mech., vol.48, pp.721-761.
2) Antonia, R.A. and Luxton, R.E. (1972);J. Fluid Mech., vol.53, pp.737-757.
3) Islam, O. and Logan, E. (1976); Trans. of ASME, No. 76/FE-4, pp.626-634.
4) Kanda, K. and Muramoto, Y. (1989); 33rd Japanese Conference on Hydraulics, pp.499-504 (in Japanese).
5) Nezu, I. (1977): *Turbulent Structure in Open-Channel Flows*, Ph.D Dissertation, Kyoto University.
6) Nezu, I. and Nakagawa, H. (1989); Proc. of Workshop on Instrumentation for Hydraulic Laboratories, CCIW/IAHR, Burlington, Canada, pp.29-44.
7) Nezu,I. and Nakagawa, H. (1991): 24th Congress of IAHR, Madrid, vol.A, pp.235-242
8) Nezu, I. and Nakagawa, H. (1993a); *"Turbulence in Open Channel Flows"*, IAHR-Monograph, Balkema Publishers, Rotterdam.
9) Nezu, I. and Nakagawa, H. (1993b): Proc. of 5th Int. Symp. on Refined Flow Modelling and Turbulence Measurements, Paris, pp.603-612.
10) Nezu, I. and Nakagawa, H. (1994): an invited paper submitted to J. Flow Measurement and Instrumentation, Butterworth-Heinemann, UK.
11) Nezu, I., Kadota, A. and Nakagawa, H.(1994); Symposium on Fundamentals and Advancements in Hydraulic Measurements and Experimentation, ASCE, Buffalo.
12) Nezu, I., Tominaga, A. and Nakagawa, H. (1993); Proc. of 5th Int. Symp. on Refined Flow Modelling and Turbulence Measurements, Paris, pp.629-636.
13) Plate, E.J. (1971), *Aerodynamic Characteristics of Atmospheric Boundary Layers*, U.S. Atomic Energy Commission, Critical Review Series.
14) Smitz, A.J. and Wood, D.H.(1985): Ann. Rev. Fluid Mech., vol.17, pp.321-358.
15) Yeh, F.F. and Nickerson, E.C. (1970), *Air Flow over Roughness Discontinuity*, Technical Report No.8, Colorado State University, Fort Collins, Colorado.

DEVELOPMENT OF A VISUAL METHOD TO TRACK THE MOVEMENT OF HYDROGEN BUBBLES IN A LABORATORY FLUME.

Athanasios N Papanicolaou[1] and Panayiotis Diplas[2], Member, ASCE

Abstract

The current study proposes a visual method for tracking the hydrogen bubbles position with time in a 2-D space. The experiment is filmed from above with a video camera. The obtained movie is imported into a video recorder for its timing process and then is digitized using a SGI Indigo workstation to create a movie file. A video grabber is used to transfer fifteen frames per second to a Sun workstation for their image development. From this process the spatial coordinates of hydrogen bubbles can be obtained in sequential frames and plots of several bubbles' pathlines can be drawn.

---

[1] Graduate student, Department of Civil Engineering, Virginia Polytechnic Institute and State University, Blacksburg, VA 24061, USA.

[2] Associate Professor, Department of Civil Engineering, Virginia Polytechnic Institute and State University, Blacksburg, VA 24061, USA.

Introduction

Hydrogen bubble tracking techniques have been used in the past to interpret visual information in fluid mechanics or to alter pictures to highlight some interesting details. It has been extremely difficult to extract quantitative results by using these methods due to their algorithm deficiencies and hardware limitations. The recent advances in the high resolution cameras and the construction of dedicated computer software and hardware, however, allow for their use in the study of complicated fluid flow problems.

The current work adopts the hydrogen bubble technique to visualize the turbulent flow and therefore it aims at the development of a 2-D autotracking algorithm to obtain from images the bubbles position with time. The steps of the image development methodology that are deduced from this study are described next.

Experimental set-up

Preliminary experimental work has taken place in a laboratory flume for developing an automated method for recording the motion of hydrogen bubbles that are generated in flowing water. The device that generates the hydrogen bubbles consists of a fine platinum wire with diameter of 1/4 in. and length 2 ft, which constitutes the negative electrode of the electrical circuit, and from a generator. The wire is stretched in the transverse direction of the flow and is soldered to supports which are located 1 cm upstream of the measuring point (figure 1). By pulsing the voltage applied to the wire with a frequency of 20 Hz, bubble lines are produced (figure 2). On a single frame the shapes of the bubble lines indicate the integrated effect of the velocities experienced by the bubbles since their release from the wire. Furthermore, the position and timing of a particular type of flow structure can be determined by comparing line movements at successive frames.

The changes in the bubbles' position with time are recorded with a super VHS Panasonic (AG 76 model) camcorder. The filming process takes place from above the flume's bed by setting the camera in the "macro-focus" mode (figure 1). The "macro-focus" mode allows the focus to be within a range of 1 inch to 4 ft. Sufficient illumination of the control area is required to obtain images with small "noise". However, excessive amount of light can cause bubbles "saturation" and therefore it can lead to cases where the autotracking algorithm identifies 2 or more bubbles instead of a single bubble. In this work, three high - intensity tungsten lamps (DVY GE 650W, 120V quartz halogen lamps) are used as light sources. To minimize light reflectance the other side of the flume is covered with black poster paper (figure 1).

The obtained movie is imported into a Panasonic video cassette recorder (model AG-7750) for its timing process. The images that are created are filtered to remove low and high frequency "noises". The low image "noise" is caused by the uneven illumination during the experiment while the high frequency noise is due to the signal noise. By connecting the IRIS Indigo Board of the SGI with the Panasonic recorder the video tape is "grabbed" and then it is digitized to create a movie file. The frames are "grabbed" using the auto-increment choice at a rate of 15 frames per second. The storage of the files is done initially in FIT type of format. These files in turn are converted to binary ruster images (RIS) to be transferred to a Sun workstation for their image development.

Steps of the image analysis

The goal of the image analysis is met by computing the bubbles coordinates and by identifying them between two successive binary images. The imported images are analyzed using custom written procedure files in PV - wave command language. The first group of these files fulfills the quantitative part of the analysis while the second group provides the plots of different bubble pathlines. The following files have been created for the calculation of the bubbles coordinates: 1) the m_images.pro which opens the ruster images and converts them to bytarr arrays. The size of each of these arrays is 400 H x 300 V and they are saved as images.byt files. The "tvscl" command, which is included in the procedure code m_images.pro, scales the obtained bytarr arrays. Specifically, it finds the maximum and minimum values for each array and assigns them to (255) and (0) respectively. Furthermore, it draws the above arrays as black - white TV images; 2) the track.pro file is used to view the constructed above bytarr arrays. At the same time it creates files to restore the bubbles coordinates (x,y) at different time instants; and 3) the pick.pro procedure file that initiates the bubble tracking process. This process takes place here manually. Three windows are appeared simultaneously on the computers' screen. The first window informs the operator for the mouse buttons selections, the second window provides messages relatively to the progress of the image development, and the last window illustrates the image that is selected for analysis. With the left mouse button the bubbles controid is identified, with the middle button the image analysis is terminated, and with the right button the tracking process continues to the next frame. A flow chart of the constructed algorithm is shown in figure 3. To illustrate the bubbles' movement at different time instants the time is correlated with color. A color map bar is created which in turn is loaded to the graphics procedure file, zzz.pro, through the subroutine l_color.pro. The red color in this bar represents the bubbles maximum traveling time, which is equal to 1 sec, while the blue color indicates the initiation of the bubbles movement. From the imaging process it is deduced that most of the hydrogen bubbles reach the water surface in 1/3 and thus they can be typically traced in 3 consequtive frames. Therefore, the above methodology allows computation of the average bubbles velocity. For example, in figure 4 the pathlines of three bubbles involved in a vortex motion are illustrated. The average velocity of the second bubble that is shown in figure 4 is equal to 20 cm/sec. A multimedia presentation of the above results has been developed in this work with the use of the micromind director software. The presentation provides information on the steps of the conducted image analysis and offers several animations of bubbles movement in the 2-D space. The black and white copies included here obscure some of the features mentioned above, which are visible in the original pictures.

Conclusions

In this study, procedure files were created to compute the bubbles' position change with time. The current work enabled us, also, to determine the additional factors that are needed to be taken into consideration to make the code to run more efficiently. It is imperative that the image process be entirely automatic to avoid influencing the results with operator decisions. Therefore, the codes' automation could allow its use for the study of complicated fluid flow phenomena such as, the role of the turbulent episodes to particle entrainment.

Acknowledgment

The support for the present study by the U.S. Geological Survey (Grant. No. 14-08-0001-g2271) is gratefully acknowledged.

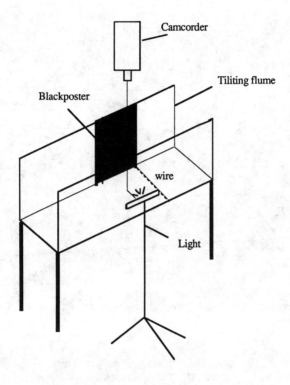

Figure 1. Experimental set-up for recording the motion of hydrogen bubbles

Figure 2. Color picture of bubble lines produced in the laboratory by pulsing the voltage applied to the wire.

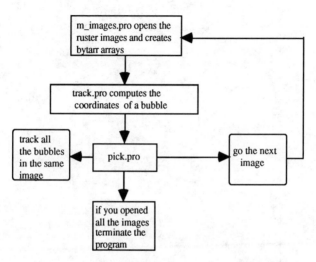

Figure 3. Flow chart of the tracking algorithm.

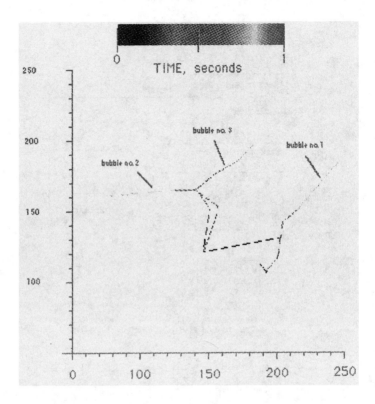

Figure 4. Pathlines of three bubbles in color.

HIGH-SPEED VIDEO ANALYSIS OF SEDIMENT-TURBULENCE INTERACTION

Yarko Niño[1], Fabián López[1], and Marcelo García[2], A.M. ASCE

Abstract

Results from the analysis of high-speed video recordings of particle motion and simultaneous flow visualizations in the near wall region of a turbulent open channel flow are presented. Analysis of the video images provided information about characteristics of particle motion and its interaction with coherent flow structures associated with the turbulent bursting process near the wall. The results presented herein demonstrate the valuable quantitative information that can be obtained by applying high-speed video imaging techniques to the study of sediment transport mechanics. Although the results are preliminary, they clearly indicate the fundamental role played by flow ejections on the particle entrainment mechanism. Further application of this technique would provide basic data, which could contribute to a better understanding of the physical processes involved in particle-turbulence interactions, and in particular, to clarify the intimate relation between bursting phenomena and sediment transport mechanics near the bed.

1. Introduction

One of the most important engineering applications of the theory of turbulence arises from the fact that it plays an essential role in transport phenomena. Either momentum, mass or heat transfer mechanisms are strongly related to turbulent processes. Turbulent diffusion of contaminants and heat transfer mechanisms have been investigated intensively, however the mechanics of sediment transport has yet to be thoroughly related to the knowledge of turbulent processes. Instead, the transport of sediment has been traditionally described by empirical or semiempirical formulations, usually having limited general validity.

---

[1] Research Assistant, [2] Assistant Professor. Department of Civil Engineering, University of Illinois at Urbana Champaign. 205 N. Mathews, Urbana, IL 61801. USA.

Progress in the understanding of the physics underlying sediment transport mechanisms is related to parallel improvements in the knowledge of turbulence dynamics in open channel flows. In particular, since the interaction between flow and natural sediment occurs mainly in the vicinity of the bed, detailed knowledge of the processes that govern the turbulence structure near the wall appears to be essential to advance a mechanistic approach to sediment transport phenomena.

It is currently well known that the streamwise velocity field in the near wall region of a channel flow is organized into alternating narrow streaks of high— and low—speed fluid, which are quite persistent in time. Intermittent, quasi—periodic events, consisting of outward ejections of low—speed fluid and inrushes of high—speed fluid towards the wall, are most responsible for the maintenance of turbulence in wall boundary layers. A number of studies have been conducted to investigate the implications of this so called "bursting" process in the mechanics of sediment transport. Several experimental techniques have been applied to visualize the flow field in the wall region of a boundary layer with the aim of inferring the mechanism by which sediment is entrained into suspension (Grass, 1974; Yung et al., 1988; Rashidi et al., 1990). However, to date, no precise description of the sediment—turbulence interaction that modulates the turbulence and is responsible for the motion of particles in the wall region exists. One of the main reasons for the poor understanding of the physics involved is the lack of sufficient data that provide the information and spatial—temporal resolution needed to estimate instantaneous values of variables like particle velocity and acceleration.

The main goal of this paper is to show the good performance of high—speed video techniques in providing valuable information for the study of sediment—turbulence interaction in boundary layer flows. In the experimental work reported herein, a high—speed video system, capable of recording up to 1000 frames per second, was employed to register the motion of sediment particles in the near wall region of an open channel flow. The high temporal resolution of the video system facilitated detailed observations of particle motion. The highly resolved particle trajectories resulting from the analysis of the video images provided unprecedented data from which kinematic variables, such as instantaneous particle velocities and accelerations, could be estimated. The results presented herein are preliminary, however they already reveal a clear and consistent picture of the role played by turbulent bursting on the entrainment of sediment into motion and suspension.

2. Experiments

Facilities

The experiments were carried out in an open channel, 18.6 m long, 0.297 m wide, and 0.279 m high. The slope of the channel was set to a value of 0.0009. The test section was located about 12 m downstream from the entrance, it was about 0.9 m long and had the right wall made of plexiglass to allow for visualization studies. A high—speed video recording system Kodak Ektapro TR

Motion Analyzer was used to record particle motion and flow visualizations. The system has the capability to record up to 1000 frames per second. A strobe−light synchronized with the video system was used in order to obtain sharp images, minimizing blur due to particle motion. Also, extension tubes for camera lenses were used for image magnification purposes. Video images were digitized into a personal computer using a frame grabber, and analyzed with the help of the National Institute of Health's Image public domain software.

Experimental Conditions

Experiments were carried out under uniform flow conditions, in a channel with smooth walls. Flow conditions corresponded to values of the Reynolds number ($Re = U h / v$, where U denotes the mean flow velocity, h denotes flow depth, and v denotes kinematic viscosity) in the range from about 5000 to 24000, with flow depths in the range from about 25 to 50 mm, which yielded relatively high values of the width to depth ratio, thus minimizing side wall effects. Five different particles were used in the experimental work, namely, glass beads with mean diameter, Ds, of 38 and 94 µm, and natural sand particles with Ds values of 112, 224 and 530 µm respectively. The wide range of particle diameters and Reynolds numbers covered in the experiments provided conditions to analyze particles moving within the viscous sublayer as well as in the transitional region.

Experimental Method

In each run, particles were released at a location far enough upstream from the test section in order for they to settle down toward the bottom of the channel before reaching the field of view of the camera. In some experiments, a solution of white clay and water was injected through an orifice in the channel bottom to act as a marker for flow structures developing at the wall. The dye discharge was controlled as to minimize disturbance of the flow, and to allow the tracer to displace attached to the wall before flow ejections lifted dye filaments away from it. Top and side views of particles moving in the near wall region of the open channel flow in study were recorded with the video system, using recording rates of 250 or 500 frames per second. Depending on the experimental conditions, this technique allowed recording particle motion at locations as close as about 5 wall units from the bed.

3. Method of analysis

Selected frames of the video recordings were digitized into a personal computer. Images were analyzed in order to obtain the position of particles in successive frames. Particle trajectories, velocities and accelerations were determined from this information, which allowed analysis of the main characteristics of the observed particle−turbulence interactions. For instance, top views allowed estimation of width and spanwise spacing of wall streaks, and mean and standard deviations of streamwise and spanwise velocities of particles moving along the streaks. From side views, trajectories, and instantaneous and mean values of vertical and streamwise velocities and accelerations of entrained particles could be estimated.

4. Results

Particle motion along wall streaks

Experimental results indicated that, in general, particles moving immersed within the viscous sublayer were sorted in the spanwise direction, such that they tended to group along lines corresponding to low−speed streaks of the flow. An example of this is shown in Fig. 1, which corresponds to a plan view taken for the experimental conditions Re= 12400 and Ds= 224 μm. From this figure, the mean spanwise spacing of the streaks expressed in wall units (i.e., made dimensionless with the length scale $v/U*$, where $U*$ denotes shear velocity), λ_+, is about 100 (see also Fig. 2), which agrees well with typical values reported for this variable.

Figure 1. Streaks in Viscous Sublayer − Re = 12400 − Ds = 224 μm

Observations of particle motion along the bed seem to indicate that particles are entrained into suspension from low−speed streaks, and deposited along high−speed streaks by sweep events, and rapidly displaced towards the low−speed streaks by transverse flows, which would be related to streamwise vorticity. An example of this situation is shown in Fig. 2. The instantaneous velocity vector map corresponding to particles moving along the bed, was obtained by using an automatic particle tracking algorithm. The results shown in Fig. 2 correspond to values Re=9900 and Ds=224 μm, and X_+, and Z_+ denote streamwise and spanwise coordinates expressed in wall units.

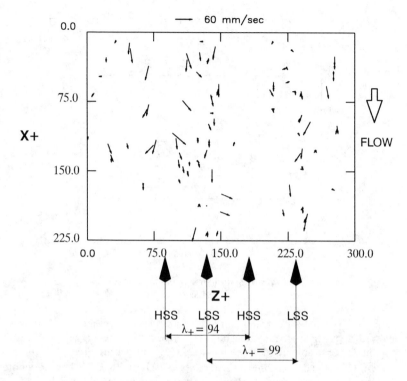

Fig. 2. Particle velocities along wall streaks. Reynolds number is about 9900, and mean particle size is 224 μm
HSS: High-speed streak ; LSS: Low-speed streak

Instantaneous values of the streamwise velocity of two particles, one moving along a high-speed streak and the other moving along a low-speed streak, made dimensionless with the fluid mean velocity at a distance Ds/2 from the bed estimated using the law of the wall, are shown in Fig. 3. These results correspond to values of Re=12400 and Ds=112 μm. The spacing between the streaks analyzed corresponded to 2.5 λ_+ = 230, which gives a value λ_+ = 92. Clearly, the velocity of the particle moving along the high-speed streak is, in general, considerably larger than the corresponding fluid mean velocity, whereas the velocity of the particle moving along the low-speed streak is lower than such value. The scatter of the velocity of the high-speed streak particle is also larger than that corresponding to the low-speed streak, which undoubtedly reflects the different level of turbulence associated with those streaks.

Threshold conditions for entrainment into suspension

In a series of experiments, particles were placed over the channel bottom and the wall shear stress was increased progressively, via flow discharge,

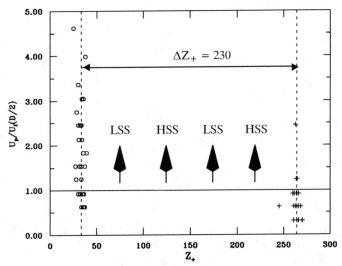

Figure 3. Particle Velocities of Sediment Moving in ○ High–Speed and + Low–Speed Streaks

until the threshold for entrainment into suspension was attained. The flow was allowed to reach a steady state before increasing the discharge. The results obtained after examination of the recorded video images are presented in Fig. 4, in the form of the dimensionless shear stress, $\tau_* = U_*^2 / (R\, g\, Ds)$, where R denotes the particle submerged specific density, versus a particle Reynolds number defined as $Re_* = U_* Ds / \nu$. Therein, clear circles correspond to flow conditions for which particles were not entrained, and black circles correspond to conditions for which particles were entrained by the flow. A curve corresponding to the estimated threshold for entrainment into suspension was traced. One conclusion that can be obtained from Fig. 4 is that, apparently, particles totally immersed in the viscous sublayer (values $Re_* < 5$) are indeed entrained into suspension from a smooth boundary, which is in opposition to the results reported by Sumer and Oguz (1978) and Yung et al (1988).

Particle entrainment into suspension

Some experiments were carried out to analyze particle entrainment into suspension, where side views of simultaneous particle and flow visualizations were recorded with the high–speed video system. An example of the results obtained, corresponding to Re=12400, and Ds=112 μm, is shown in Fig. 5. Therein a sequence of images of a particle being picked up by the flow, taken 0.012 sec apart, is shown. An inclined flow structure can also be observed in those images, which would correspond to a shear layer resulting from an ejection event as described by López et al. (1994). Their analysis suggested that a flow ejection

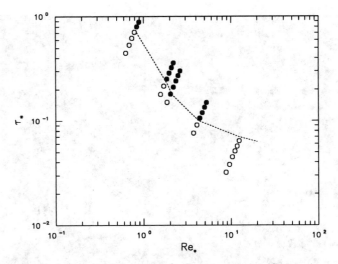

Figure 4. Threshold Condition for Entrainment into Suspension (associated with a positive vertical velocity fluctuation) occurs in the downstream vicinity of the shear layer, at a distance of about 120 to 160 wall units for values of Y_+ of about 40 to 55. The interaction between the particle and the flow structure, which indeed occurs at distance of about 100 wall units downstream from the structure as seen in Fig. 5, resulted in particle entrainment. Apparently, entrained particles respond to flow ejections through a mechanism that can be related either to drag or pressure gradients.

An example of observed trajectories of entrained particles is presented in Fig. 6, corresponding to the conditions Re=11300, Ds=112 μm. Therein, Y_+ denotes the wall normal coordinate expressed in wall units. The mean ejection angle with respect to the channel bottom, inferred from the observed particle trajectories, is about 20°. As seen in Fig. 6, ejected particles reached maximum heights in the range of 20 to 50 wall units, before starting to settle. A preliminary analysis of the particle trajectories seems to indicate that lift forces associated with particle entrainment are much larger than those expected to be exerted by the mean flow, which provide evidence that such particle trajectories would be driven by flow ejections associated with coherent structures developing near the wall.

Finally, some estimated values of instantaneous streamwise and vertical particle velocities during ejections and at locations very close to the wall are shown in Fig. 7 and 8 respectively, plotted as functions of Y_+. Figure 7 also illustrates the velocity distribution due to the law of the wall and the variation of the dimensionless standard deviation of streamwise flow velocity fluctuations (u_{rms}) with distance from the bottom. Figure 8 also depicts the variation of the dimensionless standard deviation of vertical flow velocity fluctuations (v_{rms}) with distance from the bed. In both figures velocities have been made dimensionless

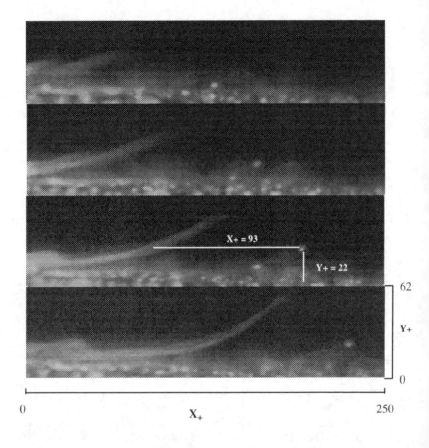

Figure 5. Sequence of entrainment process from top to bottom. Time between frames: 0.012 s. Diameter of Particles 112 μm.

with U_*, and experimental conditions correspond to Re=10500, Ds=224 μm. As observed in Fig. 8, the vertical particle velocities during ejection are as high as about 5 times the shear velocity, and about 9 times the mean local standard deviation of vertical flow velocities. On the other hand (Fig. 7), streamwise particle velocities during ejection are always lower than the values of the local mean flow velocity as given by the law of the wall, but higher than the local standard deviation of the streamwise velocity fluctuations. These results seem to indicate that particle entrainment is associated with strong flow ejection events, probably those associated with the turbulent bursting process.

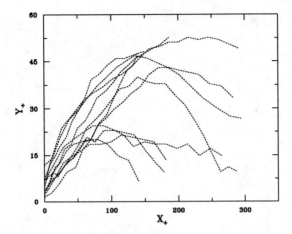

Figure 6. Trajectories of Entrained Particles — $Re = 11300$ $Ds = 112$ μm

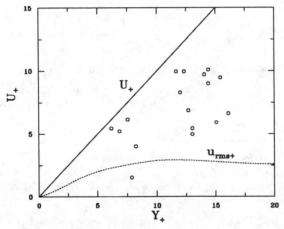

Figure 7. Streamwise Velocities of Entrained Particles
$Re = 10500$ $Ds = 224$ μm

5. Conclusions

Although the results presented herein are preliminary, they clearly show the valuable quantitative information that can be obtained by applying high-speed video imaging techniques to the study of sediment transport mechanics. It is believed that further application of this technique will provide more basic data, which could contribute to a better understanding of the physical processes involved in particle-turbulence interactions, and in particular to clarify

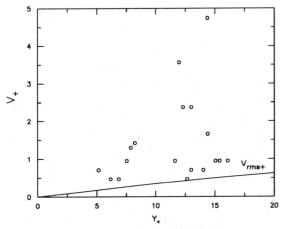

Figure 8. Vertical Velocities of Entrained Particles
Re = 10500 Ds = 224 μm

the intimate relation between the bursting phenomena and sediment transport mechanics near the bed.

6. Acknowledgments

The support of the Fluid, Hydraulic, and Particulate System Program of the National Science Foundation (Grant CTS-9210211) is gratefully acknowledged

7. References

LOPEZ F., NIÑO Y. and GARCIA, M.H., 1994, *Simultaneous Flow Visualization and Hot-Film Measurements*. ASCE Symposium on Fundamentals and Advancements in Hydraulic Measurements and Experimentation. Buffalo, NY.

GRASS, A.J., 1974, *Transport of Fine Sand on a Flat Bed: Turbulence and Suspension Mechanics*. Euromech 48. Inst. Hydrodynamic and Hydraulic Eng.. Tech. Univ Denmark. pp 33-34.

RASHIDI, M., HETSRONI, G. and BANERJEE, S., 1990, *Particle-Turbulence Interaction in a Boundary Layer*. Int. J. Multiphase Flow, vol 16, No 6, pp 935-949.

SUMER, B.M. and OGUZ, B., 1978, *Particle Motions Near the Bottom in Turbulent Flow in an Open Channel*. J. Fluid Mech. vol 86, pp 109-127.

YUNG, B.P.K., MERRY, J.D., and BOTT, T, 1989, *The role of Turbulent Bursts in Particle Re-Entrainment in Aqueous Systems*. Chemical Engineering Science. vol 44, No 4, pp 873-882

OPTICAL METHODS FOR SEDIMENT-LADEN FLOWS

R. N. Parthasarathy[1] and M. Muste[2]

ABSTRACT

A method of distinguishing between the signals from the liquid-seed particles and sediment particles, while using LDV in sediment-laden flows, is described. The validity of the method was tested in measurements made in a flume. It was also found that off-axis forward-scatter detection was more promising than backward-scatter detection for the measurement of particle velocities due to better signal-to-noise ratio. However, since signals from dust were also recorded in forward-scatter detection, an auxiliary system to identify signals from sediment particles was required.

INTRODUCTION

Sediment-laden flows form an important area of hydraulics and engineering. In most practical applications such as sediment-laden flows in rivers, reservoirs and canals, the presence of the sediment modifies the flow. Some issues that remain unresolved include the effect of sediment on the von Karman constant and turbulence of the liquid and the nature of sediment velocities. A better understanding of the coupling between the flow and sediment that is crucial in all sediment-transport processes requires experimental measurements under controlled conditions which, in turn need accurate methods of measurement of liquid and particle velocities, particle concentrations and sizes.

Measurements in sediment-laden flows

Early measurements in sediment-laden flows were attempted using hot wires; the data suffered from inaccuracies due to sediment collisions with the hot wires. The advent of laser-Doppler velocimetry (LDV) has enabled non-intrusive measurements of turbulent flows in a variety of configurations. Muller (1973) used LDV to measure mean velocities and longitudinal turbulent intensities in a flow with a movable bed. Later, Muller (1985) developed a method of simultaneous flow visualization and velocity measurement using a three-component LDV system to study the mixing layer behind dunes. However, it is not clear how Muller (1973, 1985) distinguished between the signals from the sediment particles and the seeding particles in the liquid.

---

[1] Assistant Professor, School of Aerospace and Mechanical Engineering, The University of Oklahoma, 235 Felgar Hall, 865 Asp Avenue, Norman, OK 73019
[2] Graduate Research Assistant, Iowa Institute of Hydraulic Research, The University of Iowa, Iowa City, IA 52242, Member ASCE

van Ingen (1981) used one-component LDV to make velocity measurements of both the fluid and sediment particles in a flume on the basis of signal amplitudes. A lot of data could not be assigned to the sediment or the fluid and, therefore, had to be discarded. This method was tedious and had a large uncertainty based on the threshold of the signal used to make the distinction. van Ingen (1981) recommended that an independent method of detecting the presence of sediment particles in the vicinity of the LDV measuring volume needed to be developed to improve the accuracy of the measurements. Lyn (1986) used a two-component LDV system to measure liquid velocities, and their temporal power spectra, in clear water flows and flows over sediment beds. Frequency isolation was employed to measure velocities in two directions simultaneously. Amplitude discrimination was again used to distinguish between the signals from the sediment particles and the fluid. It was found that for the sediment-laden flows, the shape of the mean velocity profile was unchanged from clear-water flows except near the bed. The upper half of the flow showed 5 to 10 percent increase in turbulence intensities due to the presence of the sediment.

Measurements in two-phase flows

In the early investigations of two-phase flows such as sprays, particle-laden and bubbly flows using LDV, simple amplitude discrimination was used to distinguish the signals due to dispersed elements from those arising from the carrier fluid. The usual reason cited for the validity of this method was that the dispersed-phase elements were at least two orders of magnitude larger than the small seeding particles that followed the motions of the carrier fluid. Modarress et al. (1984) recognized that such a method was inadequate for the accurate measurement of fluid velocities. The intensity of light scattered by the dispersed-phase elements would be large only if they were passing through the center of the LDV measuring volume. If the dispersed-phase elements were just grazing the measuring volume, the intensity of light scattered by them would be of the same order as the intensity of light scattered by the seeding particles in the carrier fluid. Thus, if only simple amplitude discrimination was used, the velocities of dispersed-phase elements grazing the LDV measuring volume would be mistaken to be fluid velocities, resulting in errors in the fluid-velocity measurements. This error would increase as the relative velocity between the dispersed phase and the fluid and the volume fraction of the dispersed phase increased.

Modarress et al. (1984) devised a method to avoid this error in their study of a turbulent jet of air carrying glass beads. They used a two-color LDV system based on the green and blue beams with the blue measuring volume enveloping the green measuring volume. To measure fluid velocities, the pedestal signal from the light scattered in the blue measuring volume was monitored together with the velocity signal. Thus, even when the glass beads grazed the green measuring volume, large pedestal signals would be obtained on the blue channel, and these measurements would be discarded.

Parthasarathy and Faeth (1987) simplified the method in their study of a water jet carrying glass beads. They used a single channel LDV (based on the green line of an Argon-ion laser) to measure velocities. To this system, they added a Helium-Neon laser, collection optics and a photomultiplier, and called this auxiliary system the 'discriminator system'. The measuring volume of the discriminator system enveloped the LDV measuring volume so that whenever a glass bead passed through the center of the LDV measuring volume or even grazed it, a strong signal from the light scattered on the discriminator system was recorded.

The output of the discriminator system was sampled simultaneously with the LDV output, and any velocity measurement accompanied by a pulse on the discriminator system was discarded. The concentrations of dispersed-phase elements could also be measured non-intrusively, by counting the number of pulses over a period of time (Parthasarathy and Faeth, 1987).

Objectives

It is clear that past methods of using LDV (employing simple amplitude discrimination) in the measurement of liquid velocities in sediment-laden flows were inaccurate and suffered from the ambiguity of not knowing clearly what was measured: velocities of sand particles or liquid velocities. The techniques used in studies of two-phase flows provide the background for the development of methods based on LDV that are applicable in sediment-laden flows. The objectives of this study were to develop a method to measure liquid velocities accurately when sediment is present in the flow, along the lines of Parthasarathy and Faeth (1987, 1990), and to study optical arrangements for the measurement of particle velocities.

MEASUREMENT OF LIQUID VELOCITIES

LDV System

The LDV system was based on a Helium - Neon laser (15 milliWatts) and was set up to measure two components of the velocity vector simultaneously. The system produced a three-beam arrangement with the three beams forming a triangle as shown in Figure 1. The common beam was polarized at 45° to the horizontal, while the other two beams were polarized in the vertical and horizontal directions respectively. The receiving optics was equipped with a polarization separator that separated the horizontally-polarized scattered light from the vertically-polarized. Thus, two components of velocities, along the two edges of the triangle from the common beam were measured. The common beam was frequency-shifted using a Bragg-cell arrangement so that reversed velocities could be measured.

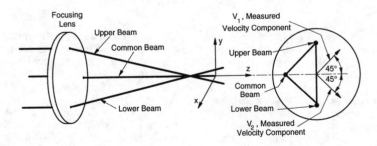

Figure 1. Schematic of the three-beam LDV arrangement

The LDV probe volume (the volume common to the three beams) was estimated to be 650 microns in diameter and 4 mm long. The resulting fringe spacing was 10.375 microns. The signals from the two photomultipliers were processed using two IFA 550 (TSI Inc.) signal processors. A time window could be specified to ensure that velocity measurements were made on the same particle. The operation of the entire system was verified by measuring the edge velocity of a disk which was rotated at a known constant speed. The difference between the velocities measured using the LDV system and the actual velocities was less than 1 percent. The IFA 550 signal processors were not equipped with any amplitude-control provision. Therefore, an auxiliary system was required to identify the source of the signal (dispersed phase or the carrier fluid).

Discriminator system
The discriminator system was similar to the one used by Parthasarathy and Faeth (1987, 1990). An auxiliary source of light, such as the additional Helium-Neon laser used by Parthasarathy and Faeth (1990) was not required as long as the measuring volume on the discriminator system was larger than and surrounded the LDV measuring volume. It was, therefore, decided to use the three beams in the LDV set up as the source of light for the discriminator system. The system essentially comprised of a light-collecting lens (600 mm in focal length), a focusing lens (100 mm in focal length), a color filter (632.8 nm) and a photomultiplier.

For the simultaneous recording of the velocities and the signal from the discriminator system, a Datalink Multichannel Interface (TSI Model DL100) was added to the system. The analog voltage from the discriminator photomultiplier served as the input to the system. The system was set such that the two Doppler frequencies together with the voltage present on the discriminator output were measured simultaneously and transferred digitally to the computer. The time interval between the validation of the signal and the transfer of the data to the computer was on the order of nanoseconds (as specified by TSI Inc.). This was verified by feeding a sine signal from a signal generator and a corresponding TTL signal that was triggered at the rising end of the sine signal as the input to the Datalink system. This system was, thus, more accurate than the system used by Parthasarathy and Faeth (1990) in which the digital frequency measurement was converted into an analog signal before sampling it again using an A-D converter.

Results
The LDV system was set up in the on-axis forward-scatter mode for measurement in a open recirculating flume with plexiglass walls. Natural dust in the water served as the source of scattered light for this measurement. The discriminator system was set in the backward-scatter mode (to collect the reflected light off the opaque sand particles) as illustrated in Figure 2. Accurate alignment of the two measuring volumes was critical to the success of this method. A preliminary alignment was done using a metallic wire, 1 mm in diameter, that was mounted on a traverse and served as the source of scattered light. Adjustments were made by observing (on an oscilloscope) the amplitudes of the Doppler bursts as the wire moved in the LDV measuring volume and the amplitudes of the voltage signal from the discriminator system and ensuring that they were triggered at the same time. The measuring volume on the discriminator system was found to be 6 mm long and 1 mm in diameter. Better spatial resolution could be achieved by moving the receiving optics off-axis; however, for the testing of operation of the system, the receiving optics was maintained on-axis.

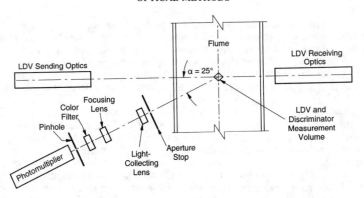

Figure 2. Arrangement of the LDV and discriminator systems

Sand particles were dropped individually in still water and in a uniform flow in the flume and LDV measurements were made. The data rates on the individual channels were on the order of 2 to 3 kHz while the coincident data rate varied between 1 and 1.5 kHz. In order to eliminate velocity bias in the computation of the statistics, the velocities were sampled at equal intervals of time and averaged, as recommended by Durst et al. (1981). The variation of streamwise (u) and vertical (v) velocities and the voltage on the discriminator system with time when a particle was present near the measuring volume is shown in Figure 3.

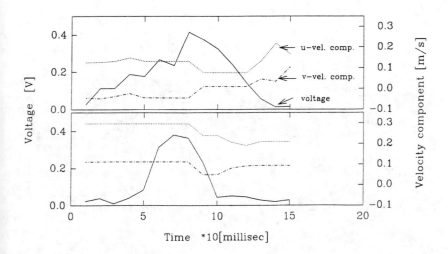

Figure 3. Representative liquid velocities and discriminator voltages

Since the particles had a vertically downward velocity component during these measurements, the presence of a downward velocity near the peak voltage suggests that the LDV measurement could have been made on the particle. The other instances, when the vertical velocity displays an upward trend near the peak voltage on the discriminator system, indicate particle passages outside the LDV measuring volume, but within the measuring volume of the discriminator system.

The validity of the present procedure is clearly demonstrated by these measurements. The streamwise and vertical velocities for a uniform open-channel flow were measured without particles and with particles dropped individually using the above procedure. In the latter calculations of the statistics, all velocity measurements associated with a voltage above the threshold on the discriminator system were eliminated and the statistics were computed with the remaining data. The mean and root mean square velocities for both cases agreed with each other within experimental uncertainties. This was expected since individual particles would not alter the velocities significantly. However, in a turbulent sediment-laden flow such a procedure needs to be adopted all the time, to eliminate errors in the measurement of liquid velocities.

MEASUREMENT OF PARTICLE VELOCITIES

When LDV is used to measure particle velocities, the measuring volume is enlarged due to the large size of the particles compared to the micron-sized dust that is used in liquid velocity measurements. In order to control the size of the measuring volume during measurement of velocities of large particles, off-axis detection is used. Two positions of the receiving optics were tested in this study: off-axis forward-scatter and off-axis backscatter, as illustrated in Figure 4.

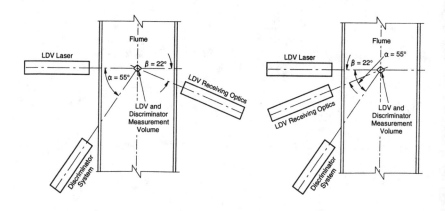

Figure 4. Forward- and back-scatter arrangements for particle velocity measurement

OPTICAL METHODS   229

In the forward scatter mode, the diffracted light off the sand particles was collected. Alignment was made as described in the section on measurement of liquid velocities. The gains on the two photomultipliers were decreased until no measurements of liquid velocities were made by the velocity-analysis system. It was not possible to completely eliminate signals from the dust and to be able to measure the velocities of sediment particles of the smallest size. It was, therefore, decided to use the voltage output from the discriminator system as an indicator of the source of the signal.

Typical Doppler bursts off sand particles (0.5-0.7 mm) and the associated signals on the discriminator system are shown in Figure 5. The nature of the voltage and velocities recorded in a typical data file containing particle velocity measurements, sampled at 10 millisecond intervals, is seen in Figure 6. The high voltages detected from the sand particles, 0.3 to 0.6 Volts, are seen in these measurements. The variation in the maximum recorded voltages is due to the size variation in the particles, and due to the passage of particles at different positions of the discriminator measuring volume. The measured velocities change occasionally, with the presence of the peak voltage in the discriminator system; sometimes they remain constant. This is primarily due to the constant sampling of velocities even when no new velocities are recorded.

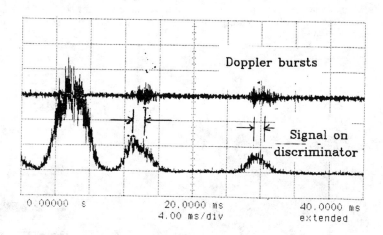

Figure 5. Typical Doppler bursts and voltages in the forward-scatter arrangement

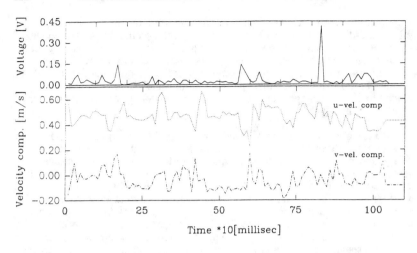

Figure 6. Sample particle velocities and voltages

In the off-axis backscatter mode, the photomultipliers could be operated at low gains without detection of signals from the dust, due to the strong signals from the reflected light off the sand particles. Reasonable signals were obtained, as shown in Figure 7.

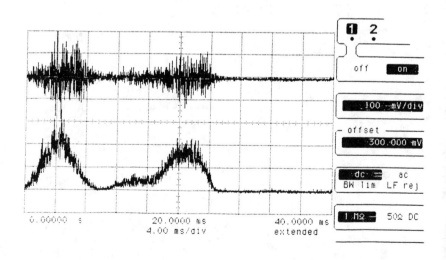

Figure 7. Typical Doppler bursts and voltages in the backscatter arrangement

It was found, however, that the cross-talk between the two channels of the LDV system could not be eliminated. This was due to the depolarization of the reflected light from the particles. Also, the signal to noise ratio was poor in the back-scatter arrangement even though the signals from the particles were large in amplitude. This could be due to the change in polarization of the reflected light from the large number of surfaces on the particle (since the particles were not smooth).

It is, therefore, recommended that the diffracted light from the sand particles be used in the measurement of particle velocities, with the receiving optics in the off-axis forward-scatter arrangement. The use of the discriminator system, in this case, is imperative to ensure that the recorded velocity measurements originate from the sand particles and not the dust in the liquid.

CONCLUSIONS

An accurate method of measuring liquid velocities in sediment-laden flows has been developed and tested. The method utilizes a conventional LDV system and a discriminator system consisting of light-collection optics and a photomultiplier. The system is arranged such that the volume viewed by the discriminator system envelops the LDV measuring volume to ensure that at all instances when sand particles pass through the center or graze the LDV measuring volume, strong voltage signals are recorded on the discriminator system. During the measurement of liquid velocities, these voltages are sampled simultaneously with the velocities; any measurement accompanied by a voltage above a specified threshold on the discriminator system is removed during the computation of the statistics. The validity of this method is demonstrated for the case of sand particles dropped in still water in a flume.

Two arrangements for the measurement of particle velocities were considered: forward scatter and backward scatter, both off-axis. In the forward scatter mode, the diffracted light off the sand particles was detected. In order to detect signals from even the smallest sand particles (0.4 mm), the gains on the photomultipliers on the LDV system needed to be set at values when signals from the dust in the liquid were also recorded. The accompanying signal from the discriminator system could, then, be used to ensure that only signals from sand particles are used in the computation of statistics. Strong signals from the reflected light off the sand particles were detected in the backscatter arrangement; no signals from the dust were recorded. However, since the reflected light got depolarized, a polarization-separation system could not be used to measure two velocity components simultaneously. Also, the back-scattered signals from sand particles was found to lack in signal to noise ratio.

The above results can be combined into the development of instrumentation capable of measuring liquid and sediment velocities, and sediment concentrations, in one sitting. The configuration essentially consists of LDV together with the discriminator system with the voltage from the discriminator sampled every time a velocity measurement is made. The velocities accompanying a voltage above a specified threshold correspond to particle velocities while the remaining data correspond to liquid velocities. The number of particle velocities recorded over the time of data collection is a measure of the local concentration of the particles. To get an absolute measurement of the concentration of particles, the measuring volume needs to be calibrated. The system needs to be tested in a practical flow containing suspended sediment. Such testing will reveal the maximum concentrations of sand particles at which reasonable signal-to-noise ratio can be achieved using this instrumentation.

ACKNOWLEDGMENTS
The financial assistance provided by the National Science Foundation, under Grant CTS 90-21149, is gratefully acknowledged. The authors are also thankful to Mr. Brennan Smith for assistance in the software development.

REFERENCES
Durst, F., Melling, A.H., and Whitelaw, J.H. (1981) *Principles and Practice of Laser-Doppler Anemometry*, Second Edition, Academic Press, London.

Lyn, D.A. (1986), "Turbulence and Turbulent Transport in Sediment-Laden Open-Channel Flows," Report No. KH-R-49, W.M. Keck Laboratory of Hydraulics and Water Resources, California Institute of Technology, Pasadena, California.

Modarress, D., Tan, H., and Elghobashi, S. (1984), "Two-Component LDA Measurements in a Two-Phase Turbulent Jet," *AIAA Journal*, Vol. 22, pp. 624-630.

Muller, A. (1973), "Turbulence Measurements over a Movable Bed with Sediment Transport by Laser-Anemometry," *Proceedings of the 15th Congress, International Association of Hydraulic Research*, Istanbul, Turkey, pp. 43-50.

Muller, A. (1985), "Simultaneous Visualization and Velocity Measurement," International Association of Hydraulic Research (IAHR) Symposium on Measuring Techniques in Hydraulic Research, Delft, The Netherlands.

Parthasarathy, R.N. and Faeth, G.M. (1987), "Structure of Particle-Laden Turbulent Water Jets in Still Water," *International Journal of Multiphase Flow*, Vol. 13, pp. 699-716.

Parthasarathy, R.N. and Faeth, G.M. (1990), "Turbulence Modulation in Homogeneous Dilute Particle-Laden Flows," *Journal of Fluid Mechanics*, Vol. 220, pp. 485-514.

van Ingen, C. (1981), "Observations in a Sediment-Laden Flow by Use of Laser-Doppler Velocimetry," PhD Thesis, California Institute of Technology, Pasadena, California.

Visual Investigation of Field Bed-Load Sampling

Moustafa T.K. Gaweesh[1]

Abstract

Bed-load sampling in alluvial streams is rather difficult, particularly when bed forms (ripples and dunes) are present. The most widely used method for bed-load sampling is the direct method by means of simple mechanical trap-type samplers. The basic principle of mechanical trap-type bed-load samplers is the interception of the sediment particles which are in transport close to the bed over a small incremental width of the channel bed. The accuracy of the bed-load transport measured by the use of a mechanical trap-type sampler depends on its sampling efficiency (instrumental design) and on the sampling location with respect to the bed form geometry (spatial variability of the physical process of bed-load transport).

A new bed-load sampler, called the Delft-Nile sampler, has been designed with unique characteristics which permit proper bed-load sampling. The sampler has been developed by Delft Hydraulics, the Netherlands, and the Hydraulics and Sediment Research Institute, Egypt. Field measurements were carried out to investigate a proper bed-load sampling procedure using a pioneering technique consisting of an under-water video camera setup to facilitate visual observations.

In this paper, the instrument design, its efficiency and the measuring procedure in field conditions are described. According to the visual observations of the under-water video camera recordings different types of bed-load sampling have been classified. The characteristics of these types are also described.

---

[1] Deputy Director, Hydraulics and Sediment Research Institute, Delta Barrage 13621, Egypt

Introduction

Bed-load transport is defined as the movement of bed material particles by sliding, rolling and jumping along the bed. In the lower transport regime, these types of motions appear in the form of mini ripples, with a height of about 0.01 to 0.05 m and a length of about 0.1 to 0.5 m, migrating over the upper side of larger dunes. In the upper transport regime, with plane bed conditions, the bed-load transport process shows a sheet flow behavior with a thickness of about 0.01 to 0.05 m. In practice, bed-load may best be defined as that part of the sediment load supported by frequent solid contact with the unmoving bed in a layer with a thickness of about 0.05 m rather than in the bulk of the flowing water.

The most widely used method to measure the bed-load transport is the direct method by means of simple mechanical trap-type bed-load samplers. The basic principle of mechanical trap-type samplers is the interception of the sediment particles which are in transport close to the bed over a small incremental width of the stream bed. Many forms of the trap-type sampler have been used with varying amount of success. The problems encountered with the trap-type sampler are the lowering and raising of the sampler to and from the stream bed and the efficiency of the sampler in collecting the bed-load particles.

The basic problem of bed-load sampling is that the sampler must be placed on the stream bed where bed forms, ripples and dunes, are usually present. Therefore, the flow pattern and bed-load movement in the vicinity of the sampler are altered to some extent. As a result, samplers do not catch the transported material of bed-load at the actual rate and must be calibrated to determine their trapping efficiencies under different flow conditions, transport rates and bed material particle sizes.

The bed-load transport measured by a mechanical sampler is dependent on its efficiency (instrumental design), on its location with respect to the bed form geometry (spatial variability of the physical process of bed-load transport) and on the near bed turbulence structure (temporal variability). The efficiency of the bed-load sampler depends on the hydraulic coefficient, the percentage of width of the sampler nozzle in contact with the stream bed (during sampling) and on sampling disturbances generated at the beginning and at the end of the sampling period.

To measure the bed-load transport, a new mechanical trap-type sampler (called the Delft Nile sampler; Van Rijn and Gaweesh, 1992) has been designed by Delft Hydraulics, the Netherlands, and the Hydraulics and Sediment Research Institute (HSRI), Egypt. The sampler has been extensively tested in a flume at HSRI (Egypt) using different sizes of bed material (sand) and flow velocities to determine its sampling efficiency. Field measurements in the Nile river has been carried out to investigate a proper sampling procedure and to study the performance of the new sampler. The field measurements were performed using an underwater video camera to observe the sampling process. According to the visual observations, five types of bed-load sampling can be distinguished.

Sampler Characteristics and Calibration

The Delft Nile sampler consists of a bed-load sampler and a suspended-load sampler attached to a supporting frame (see Figure 1). The sampler has a weight of about 60 kg. The emphasis is herein focused on the bed-load sampler. The new instrument has been designed with unique characteristics. The unique feature of the bed-load sampler is a movable nozzle (entrance width = 0.096 m, entrance height = 0.055 m, length = 0.085 m, rear width = 0.105 m, rear height = 0.06 m) connected to a bag. The bag consists of a nylon material (0.18 x 0.32 m^2) with a mesh size of 150 or 250 μm, depending on the size of the bed material. At the upper side of the bag a nylon patch (0.10 x 0.15 m^2) with a mesh size of 500 μm is present to reduce the blocking of the bag by fine silt particles as much as possible. The bottom side of the sampler nozzle has a sharp front edge and a forward slope of 1:10 to facilitate a perfect contact with the surface of the sand bed. The bed-load particles can easily enter the sampler nozzle and are then trapped in the rear side of the nozzle. The sampler is equipped with an underwater video camera to facilitate visual observation of the sampling process. More details are given by Van Rijn and Gaweesh (1992). The sampler was calibrated to determine its sampling efficiency factor under different flow conditions, transport rates, and bed material particle sizes (sand).

The sampling efficiency factor is herein defined as "the ratio of the bed-load transport measured by the sampler at a certain location during a certain period and the actual bed-load transport rate at the same location during the same period, if the sampler had not been there". The sampling efficiency factor of the Delft Nile sampler has been extensively investigated in a sand bed flume at HSRI, (Egypt). Different flow velocities and sand size ranges have been considered. The sampler has also been tested for different sampling periods of 1, 3, 5, 10 and 15 minutes. The maximum sampling period, in each test, was selected so as to provide samples that would not fill more than 50 % of the capacity of the sampler bag.

The sampling efficiency was determined by comparing the measured bed-load transport rates with the actual bed-load transport rates. More details are given by Gaweesh (1993). The individual bed-load transport rate, per unit width, measured by the sampler (in kg/s/m) is calculated as:

$$q_{b,m} = (G_s - G_o) / b\, t \tag{1}$$

in which:
G_s = dry mass of individual sand catch (kg)

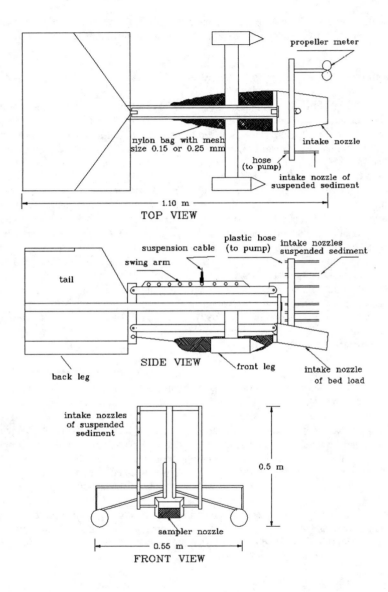

Figure 1 : The Delft Nile Sampler

G_o = dry mass of sand catch related to initial and scooping effect; determined during a "zero" sampling period (instantaneous sampling); the sampler was lowered to the bed and immediately raised after the nozzle had touched the bed (kg)
b = width of the sampler nozzle (m)
t = sampling period (s)

The sampling efficiency factor (α) of the Delft Nile sampler was determined as:

$$\alpha = \bar{q}_{b,m} / \bar{q}_b \qquad (2)$$

in which:

α = sampling efficiency factor of the Delft Nile sampler
$\bar{q}_{b,m}$ = mean bed-load transport rate per unit width measured by the sampler (kg/s/m)
$\bar{q}_b$ = actual time-averaged bed-load transport rate per unit width (kg/s/m)

The sampling efficiency factor, as determined from the flume tests, was found to be about 1.0 for bed material sizes (sand) with $D_{50} < 400$ μm and about 1.5 for finer bed material sizes.

Bed-Load Sampling

Laboratory and field observations of large variations in measured bed-load transport rates have been reported in the literature (e.g., Hubbell et al., 1985 and Carey, 1985). The variability of the bed-load transport rate at one location on a bed form is so large (factor 10 to 100, Carey, 1985) that the mean transport rate can only be determined accurately by taking a large number of samples. Different locations equally distributed along the bed-from length should be selected and many samples should be taken at each location. The bed-from length should be known from echosoundings. Quite often the sampling location is fixed because the sampler is operated from a bridge or a non-movable boat. In that case accurate determination of the mean transport rate requires sequential sampling over a period long enough for a (migrating) bed-form to pass the sampling location.

Typical sampling problems related to the variability of the physical processes of bed-load transport are:
- the sampling duration of each individual measurement.
- number of measurement locations along a bed form.
- number of measurement at each location.

The variability of the bed-load transport process related to the presence of bed forms at a certain station has been studied by performing measurements in the Nile river in Egypt.

The bed-load transport measurements were carried out in a cross-section which is located approximately 4 km upstream of the town of Bani Mazar (200 km south of Cairo), (Gaweesh, 1991 and 1992). At that location, the bed consists of sand with a median particle diameter of about 450 μm. Sand dunes with a mean height of about 0.2 m and a mean length of about 20 m were observed from echo-sounding recordings.

The Delft Nile sampler was used to collect bed-load samples. During the measurements an underwater video camera connected to a monitor was installed in front of the sampler nozzle to observe the sampling process. The measurements were performed at five locations distributed equally along the bed form length.

The complete measuring procedure at each location consisted of collecting the following data:
- ten instantaneous bed-load samples,
- eight to ten bed-load samples with sampling period of 3 to 5 minutes each (including video recordings),
- two velocity measurements at 0.15 m at 0.5 m above the bed and near the water surface,
- two bed material samples at the beginning and at the end of the measurements,

The completion of this measuring procedure took about 60 to 90 minutes, at each location, during which video recordings of the sampling process were observed. The basic data of these measurements are given by Gaweesh (1991 and 1992).

Visual Investigations

The video recordings clearly showed the presence of the initial effect. When the sampler nozzle touches the bed, sediment particles are stirred up and are then trapped into the sampler nozzle. On the average, the initial effect results in a relatively small catch (sample) of about 20 to 40 grams, which should be subtracted from the bed-load catches. To reduce the errors involved, the sampling period must be reasonably so long that the bed-load catch is relatively large compared to the initial catch ($G_s \gg G_o$). The scooping effect was never observed under conditions with mean flow velocities up to 1.3 m/s. According to the visual observations, five types of samplings can be distinguished; the characteristics of these five types are shown and described in Figure 2.

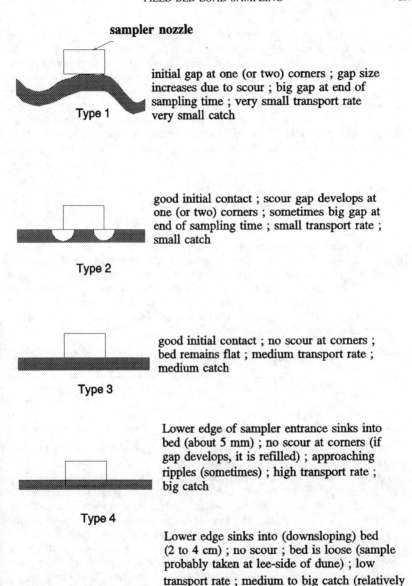

Figure 2 : Types of Bed-Laod Samplings (front view)

The type 1 and type 5 sampling are incorrect samplings related to the inevitable shortcomings of a mechanical sampler. type 1 sampling tend to underestimate the transport rate whereas type 5 sampling tend to overestimate the transport rate. Both types of samplings did occur occasionally in field conditions (percentage of occurrence of about 10 %), but they were hardly observed in flume conditions because the bed-form variability in a flume is less pronounced. The available visual information complicates the determination of the mean bed-load transport rate because not all samplings are correct. To study the relative errors involved, three methods were used to compute the mean bed-load transport rate, as follows:

Method A: all samples are used
Method B: all type 1 and type 5 samples are excluded
Method C: the 10 % biggest samples and the 10% smallest samples are excluded

The measured transport rates according to method A,B and C are given in Table 1. The maximum differences between the three methods are about 25 %. Method B is the most accurate method because all visual information are taken into account. The disadvantage of method B is that the use of a video camera is required which is not always available. Furthermore, the video camera cannot be used in silty conditions because of poor visibility in such conditions. Therefore, method C is preferred as a more general method.

Method	Bed-Load Transport Rate (kg/s/m)
A	26×10^{-3}
B	18×10^{-3}
C	21×10^{-3}

The individual bed-load transport rates measured at 5 locations (along the bed form) did show large cyclic patterns, which is most probably related to the migration of small-scale bed forms. The bed-load transport rates measured at only one location, did not show significant variations. Low transport rates can be observed near the trough region of a dune, while high transport rates can be observed near the dune crest. The number of samples necessary for a substantial mean of bed-load transport depends upon the accuracy required, the time scale of cyclic rate variations, and whether the sampler is fixed at one location or can be moved relative to the transport process. In the Nile river about 40 samples distributed over 5 locations, along a bed form, are necessary to obtain a proper mean of bed-load transport rate. The transport rate will be within $\pm$ 15 % of the actual rate (Gaweesh and Van Rijn, 1992). The sampling period of 3 to 5 minutes was found to be appropriate for the flow and sediment conditions of the Nile river, in Egypt.

Conclusions

Based on visual observations of the bed-load sampling process, five types of samplings were distinguished. The type 1 and type 5 samplings are incorrect samplings related to the shortcomings of the mechanical sampler. Both types of samplings did occur occasionally. A computation method was proposed to eliminate the effect of the incorrect samplings. The sampling period of 3 to 5 minutes was found to be appropriate for the flow and sediment conditions of the Nile river, in Egypt, to obtain sufficiently large samples. In the upper flow regime, smaller sampling periods may be used. The volume of the sample should not be more than 50 % of the volume of the sampler bag. The variability of the bed-load transport rate at different locations along a bed form (dune) is quite large. High transport values my occur near the bed-form crest and small transport values may occur near the trough. In the Nile river about 40 samples distributed equally over five locations are necessary to obtain a proper mean of bed-load transport rate, the transport rate will be within $\pm$ 15 % of the actual rate.

References

Carey, W.P., (1985). "Variability in Measured Bed-Load Transport Rates", Water Resources Bulletin, Vol 21, No. 1. USA.

Gaweesh, M.T.K., (1991, 1992). "Measurements of Sediment Transport in the Nile River at Bani-Mazar with the Delft Nile Sampler", HSRI, Delta Barrage, Egypt.

Gaweesh, M.T.K. and Van Rijn, L.C., (1992). "Laboratory and Field Investigation of a New Bed Load Sampler for Rivers", Proc. of Second Int. Conf. on Hydraulic and Environment of Coastal, Estuarine and River Waters, Vol. 2, Bradford, England.

Gaweesh, M.T.K., (1993). "Calibration of the Delft Nile Sampler at HSRI Flume", HSRI, Delta Barrage, Egypt.

Hubbell, D.W., Stevens, H.H., Skinner, J.V. and Beverage, J.P., (1985). "New Approach to Calibrating Bed-Load Samplers", Journal of Hydraulic Engineering, ASCE, Vol. III, No. 4.

Van Rijn, L.C. and Gaweesh, M.T.K., (1992). "A New Total Sediment Load Sampler", Journal of Hydraulic Engineering, ASCE, Vol. 188, No. 12.

Estimation of Mean Velocity for Flow Under Ice Cover

By Martin J. Teal[1], Member, and Robert Ettema[2], Member

Abstract: Methods of estimating mean velocity in a vertical from point velocity measurements are compared for an ice-covered channel. The comparison is based on velocity profiles obtained from a laboratory flume, a numerical model, and several rivers. The profiles are representative of flows subject to various combinations of bed and ice-cover conditions.

The numerically generated profiles use a two-parameter power law to model the vertical distribution of streamwise velocity in ice-covered flows. The velocity bias, defined as the percent error in estimating the average streamwise velocity by using a few point measurements instead of the entire vertical velocity profile, was calculated for several methods. The methods include those currently used by the United States Geological Survey and the Water Survey of Canada and proposed new methods.

The recommended method is the so-called two-point method in which velocity measurements at 0.2 and 0.8 of channel depth are averaged to obtain an estimate of the mean velocity. The accuracy of the estimate will be enhanced if a coefficient of 0.98 is applied to the two-point method.

Introduction

Ice cover formation complicates gauging of flow in streams and rivers, causing the flow records for many North American streams to contain either gaps in winter-flow information or winter-flow information of suspect accuracy. The consequent flow record deficiencies and inaccuracies attributable to ice-cover formation have long troubled organizations engaged in routine monitoring of stream and river flow. The deficiencies and inaccuracies cloud understanding of watershed behavior and hamper sound water resources planning with respect to rivers and streams subject to frigid winters. The U.S. Geological Survey, for example, has to contend with ice-cover effects for over half of its gauging stations (Melcher and Walker, 1990). Ice effects are more common for Canadian gauging stations (Pelletier, 1988a).

For a given discharge and channel slope, cover formation increases flow resistance and thereby flow depth. Consequently, a stage-discharge relationship developed for open water flow may no longer be accurate for ice-covered flow. The presence of the additional boundary almost doubles the wetted perimeter for wide channels; the increased flow depth leads to the indicated discharge from an open water rating curve being greater than the actual discharge with ice cover present.

---

[1]Project Engineer, WEST Consultants, Inc., 2111 Palomar Airport Rd., Ste. 180, Carlsbad CA 92009
[2]Professor, Iowa Institute of Hydraulic Research, Iowa City IA 52242

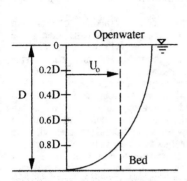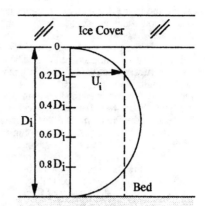

Figure 1. Velocity profiles for openwater and ice-covered channels.

The practice of many organizations that routinely monitor stream flow is to use the same measurement method(s) whether an ice cover is present or not. As the cover presence alters velocity profiles (see Figure 1), this practice likely produces inaccurate estimates of mean flow velocity, and thereby inaccurate measurement of ice-affected discharge. The purpose of the present study is to evaluate the accuracy of point-velocity methods of estimating mean velocity in a vertical, and to suggest possible improvements of those methods.

Estimation of Mean Velocity From Point-Velocity Measurements

The mean discharge at a stream cross section is commonly estimated using the velocity-area approach, which entails adding the mean discharges from a series of subsections; i.e.,

$$Q = \sum_{i=1}^{m} b_i d_i \bar{u}_i \qquad (1)$$

where Q is the total calculated discharge in the cross section; b_i is the width of subsection i; d_i is the depth of the vertical i; $\bar{u}_i$ is the mean velocity in the vertical i; and m is the number of verticals. It is customary to make a minimum of 20 to 25 verticals per cross section.

As current meters measure velocity at a point, the vertical distribution of streamwise velocity is determined on the basis of point velocity measurements. The mean velocity for a subsection is determined from the vertical distribution of velocity. To minimize the time, cost, and discomfort of working under difficult weather conditions, the mean velocity is usually estimated by measuring only a few points and using a known relation between velocities at those points and the mean velocity in the vertical. The following point methods for estimating the mean velocity with ice cover present are analyzed herein:

1. Two-point.
2. Six-tenths-depth.
3. Three-point.
4. Five-tenths-depth.
5. Proposed USGS 0.4 & 0.8 depth.
6. Proposed four-tenths depth.

where the term "depth" used in these methods is in accordance with the usual USGS convention whereby depth positions are measured from the surface (or ice-water interface) downwards.

In the two-point method, observations are made at 0.2 and 0.8 of the depth below the surface in each vertical. The average of these two velocities is taken as the mean vertical velocity. For the six-tenths depth method, the velocity measured at 0.6 depth, multiplied by a coefficient of 0.92 is taken as the mean vertical velocity. This method is favored over the two-point method under several conditions: (1) whenever the depth is between 0.3 and 1.5 feet for the Price pygmy meter, or between 1.5 and 2.5 feet for the Price type AA (or A) meter, (2) when the two point method cannot be used due to slush, debris or a sounding weight preventing measurements at 0.2 or 0.8 of depth, or (3) when the stage is changing rapidly and a measurement must be made quickly. The three-point method combines the two-point and 0.6-depth method. The average of the velocities computed by those two methods is taken as the mean velocity.

Pelletier (1988a) reports that in making discharge measurements under ice cover conditions, the Water Survey of Canada uses the two-point method for effective depths equal to or greater than 0.75 meter, and the five-tenths method (a single measurement at 0.5 depth multiplied by a coefficient of 0.88) for effective depths less than 0.75 m.

Two additional methods are not currently being used in the field but are included in this paper. The USGS is studying a method whereby two measurements at 0.4 and 0.8 depth, multiplied by coefficients of 0.32 and 0.68, respectively (determined by regression analysis), are added to estimate the mean velocity. Also, Alford and Carmack (1988) proposed a method whereby the mean vertical velocity is approximated by the velocity measurement at 0.4 depth multiplied by a coefficient of 0.86.

Accuracy of Measurements and Methods

Several authors, including Rantz et al. (1982), Carter and Anderson (1963), and Pelletier (1988b), have presented studies on the accuracy of current meter measurements in general and specific methods of estimating mean velocity from point sampling of velocity in particular. Some factors they indicated as affecting the accuracy of a discharge measurement are: equipment, characteristics of the measurement section, spacing and number of observation verticals, rapidly changing stage, measurement of depth, velocity pulsation, ice in the measuring section, and wind. Estimates of error for open channel flow at the 95% confidence level (two standard deviations) for the two-point method range from five to six percent under ideal conditions. Rouse (1950), without citing a confidence level, reported that the error obtained by using the two-point method instead of a full velocity profile to obtain mean velocity is under two percent. He also reported that a single measurement at 0.6 depth will yield errors up to 5% for open water flows. All the above results considered streams under open water conditions. Few studies of the accuracy of discharge measurements under winter ice conditions are in existence. Lau (1982), however, found that for ice-covered flows the error in using the two-point method is also under 2% for most conditions.

Two-parameter Power Law

A two-parameter power law (or "two-power law") was used to generate vertical distribution of streamwise velocity. This type of function describes the entire flow with a single continuous curve, can be linearized for numerical computations and yields an unambiguous value of D_b, the depth of maximum velocity. Tsai (1991) proposed a form of the two-power law based on physical reasoning as

$$u = K_0 \left(\frac{z}{D}\right)^{\frac{1}{m_b}} \left(1 - \frac{z}{D}\right)^{\frac{1}{m_i}} \tag{2}$$

where K_0 is a constant for a given flow rate and m_b and m_i are flow resistance parameters related to the bed and ice cover respectively. The expressions to the right of the constant K_0 in Equation (2) make up the integrand of the beta function. Note that the m parameters are not true roughness parameters, as each is dependent on the surface conditions of the other boundary. Also, as m_i approaches infinity, Equation (2) approaches the normal power law for open water flow, with K_0 becoming u_{max}.

The mean value of the velocity distribution described by Equation (2) is

$$\begin{aligned} U &= K_0 \int_0^1 \left(\frac{z}{D}\right)^{\frac{1}{m_b}} \left(1 - \frac{z}{D}\right)^{\frac{1}{m_i}} d\left(\frac{z}{D}\right) \\ &= K_0 \frac{\Gamma(1 + 1/m_b)\Gamma(1 + 1/m_i)}{\Gamma(2 + 1/m_b + 1/m_i)} \\ &= K_0 \beta(1 + 1/m_b, 1 + 1/m_i) \\ &= K_0 K_1 \end{aligned} \tag{3}$$

in which $\Gamma(n)$ = a gamma function, i.e., $\int_0^\infty e^{-x} x^{n-1} dx$; and, $\beta(a,b)$ = the beta function equivalent to $\frac{\Gamma(a)\Gamma(b)}{\Gamma(a+b)}$.

Comparison of Power Law with Field Data

Velocity profiles from ice covered steams, in the form of point velocity measurements, were compared with profiles generated using the two-power law (Teal and Ettema, 1993). The close match between the measured data points and the two-power profile lends credibility to the two-power expression as a descriptor of the velocity distribution.

A similar comparison was performed in the same report using measured velocity data from flume experiments. The results were the same; i.e., the two-power expression matched the measured data very well.

Velocity Bias

Numerical experiments were carried out using the two-power law to determine the bias of point-velocity methods previously outlined. The velocity bias is defined as

$$\varepsilon = (U_{meas} - U)/U \tag{4}$$

where U_{meas} is the average velocity obtained by one of the methods mentioned previously, and U is the true average velocity obtained by using the complete velocity profile. Using the two-power law, ε can be expressed as a function of m_b and m_i.

The most common method for estimating U, the two-point method, consists of averaging the flow velocities measured at 0.2 and 0.8 of the total stream depth to obtain the mean stream velocity:

$$U_{2pt} = \frac{u_{0.2} + u_{0.8}}{2} \tag{5}$$

The velocity bias resulting from the estimated mean velocity is

$$\varepsilon = \frac{U_{2pt} - U}{U} = \frac{U_{2pt}}{U} - 1 \tag{6}$$

Rewritten as a function of m_b and m_i and using Equations (2), (3), and (5), Equation (6) becomes:

$$\varepsilon = \frac{0.8^{1/m_b} 0.2^{1/m_i} + 0.2^{1/m_b} 0.8^{1/m_i}}{2\beta(1 + 1/m_b, 1 + 1/m_i)} - 1 \tag{7}$$

A contour plot of ε for combinations of m_b and m_i is presented as Figure 2.

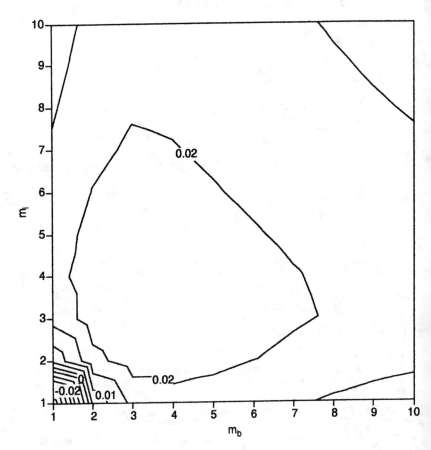

Figure 2. Velocity bias for the two-point method.

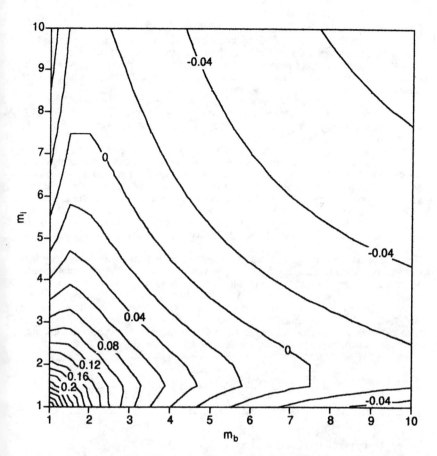

Figure 3. Velocity bias for a single measurement at 0.5 depth multiplied by a coefficient of 0.88.

In addition, values of ε were calculated for other methods of computing mean flow velocity from combinations of point velocity measurements. Expressions similar to Equation (7) were obtained for each of the other methods for the velocity bias as a function of m_b and m_i. The fraction of the effective depth at which these measurements are taken is expressed as the distance downwards from the bottom of the ice cover divided by the total effective depth, standard USGS practice. Note that this definition varies from that of z which is taken as the distance from the channel bottom up. Hence for the one point measurement method at 0.6 depth, z is actually 0.4 of the depth.

Sample contour plots showing the velocity bias as a function of m_b and m_i for two of the mean velocity schemes are presented as Figures 2 and 3.

Comparison of Methods

An overall evaluation of the accuracy of the six methods can be obtained by comparing the average velocity bias of each method for a range of values of m_b and m_i that is representative of velocity profiles in ice-covered rivers and streams. In addition to the average velocity bias over the range, the distribution of values and the maximum and minimum bias values should also be considered.

Analyses conducted for this study show that values of m_b and m_i are bracketed by the range 1.5 to 8.5. The average values of velocity bias determined for each method over this range are presented in Table 1, which also indicates the maximum and minimum bias values for each method.

Measurement of velocity at 0.5 depth with use of a coefficient of 0.88 gives the lowest average velocity bias. However, the distribution of velocity bias over the range shows that this method yields bias values ranging from 19 percent to -5 percent. Measurement of velocities at 0.2 and 0.8 of the flow depth (the two-point method) yields the second lowest average velocity bias, and has a narrowly distributed range of velocity bias, never surpassing 2.5 percent within the given range of roughness parameters. This bias is roughly the same as the expected measurement error of approximately 2 percent. The other methods give higher average biases than do these two.

Comparison of Figures 2 and 3 further reveals the difference in velocity bias distribution for the two leading methods. The velocity bias from the two-point method (Figure 2) is approximately two percent over the entire range of m_b and m_i values considered. Velocity measurement at 0.5 of depth (Figure 3) yields widely varying velocity bias depending on the bed and ice-cover roughness parameters. Therefore, the two-point method is recommended as being the most accurate method overall. Also note that while the velocity bias for the two-point method is always positive, indicating overestimation of the mean velocity, the velocity bias for the 0.5 depth method may be either positive or negative.

Table 1. Comparison of Measurement Methods.

Method	Average Bias (%)	Velocity Bias Range (%)
0.5 depth[*]	-0.31	$19.05 \geq \epsilon \geq -5.83$
Two-Point	1.97	$2.43 \geq \epsilon \geq 0.47$
0.6 depth[†]	3.21	$21.12 \geq \epsilon \geq -8.73$
"486"	-3.52	$13.20 \geq \epsilon \geq -14.68$
0.4&0.8 depth[‡]	5.24	$25.85 \geq \epsilon \geq -16.38$
Three-Point	7.08	$16.06 \geq \epsilon \geq 0.36$

[*] With coefficient of 0.88
[†] With coefficient of 0.92
[‡] With coefficients of 0.32 and 0.68

Use of a Coefficient with the Two-point Method

To correct a consistent positive bias of about 2 percent when using the two-point method, it is appropriate to adjust the calculated mean velocity by use of a coefficient of 0.98. The velocity bias is calculated in the manner previously outlined and yields an average velocity bias for the given range of m_b and m_i of -0.0705%. The velocity bias range is $0.39 \geq \epsilon \geq -1.54$.

The much improved accuracy resulting from this refinement suggests that the two-point method with a coefficient of 0.98 should be the preferred measurement method. As the two percent bias is a systematic offset from the true mean velocity, not merely a fluctuation about the true mean, use of a coefficient of 0.98 with the two-point method should be encouraged.

One-point Methods

The two-point method is recommended for use when the stream depth is greater than 1.5 to 2.5 feet (depending on current meter type), velocities are not extremely high, and the stage is not rapidly changing. If such conditions prevail, a one-point method must be used. Comparison of the three one-point methods included in this study does not elicit a preferred method for the range of roughnesses presented previously. Although the 0.5 depth method yields the overall lowest average bias, the range of velocity bias values for all three methods are comparable.

Conclusions

This study confirms the accuracy of the two-point method for estimating mean velocity in a stream section vertical. The accuracy of this method, which entails estimating mean velocity from the average of velocities measured at 0.2 and 0.8 of flow depth, is reflected in the consistently low bias (bias defined as the percent error between the estimated and actual mean velocities) for a wide range of bed and ice cover roughnesses.

Although this method should then be preferred, there are cases (such as in very shallow streams) where a single-point method should be used. No single point method from the three compared herein showed an advantage over the others for the range of roughnesses considered.

In addition, it was shown that using a coefficient of 0.98 with the two-point method will further reduce the velocity bias. Therefore, its use is recommended.

The above conclusions are based on agreement of measured velocity profiles with those modeled using the two-power law. This study has shown that the two-power law may be applied with confidence to modeling real velocity profiles. The goodness of fit to both laboratory and field data, as expressed in the R values, was excellent.

Acknowledgments

Support for this work was provided by the United States Geological Survey, Grant Number 14-08-0001-G2066, and the National Science Foundation, Grant Number CTS90-02697.

APPENDIX. REFERENCES

Alford, M.E., and E.C. Carmack, 1988, "Observations on Ice Cover and Streamflow in the Yukon River Near Whitehorse During 1985/86," NHRI Paper No. 40, IWD Scientific Series No. 162, National Hydrology Research Institute, Saskatoon, Saskatchewan.

Carter, R.W., and I.E. Anderson, 1963, "Accuracy of Current Meter Measurements," *Journal of Hydraulic Engineering*, ASCE, Vol. 89, No. HY4.

Lau, Y.L., 1982, "Velocity Distributions Under Floating Ice Covers," *Canadian Journal of Civil Engineering*, Vol. 9, No. 1, pp. 76-83.

Melcher, N.B., and J.F. Walker, 1990, "Evaluation of Selected Methods for Determining Streamflow During Periods of Ice Effect," USGS Water-Supply Paper 2378.

Pelletier, P.M., 1988a, "Techniques Used by Water Survey of Canada for Measurement and Computation of Streamflow Under Ice Conditions," Proceedings of the 5th Workshop on Hydraulics of River Ice/Ice Jams, Winnipeg, Manitoba.

Pelletier, P.M., 1988b, "Uncertainties in the single determination of river discharge: a literature review," *Canadian Journal of Civil Engineering*, Vol. 15, No. 5 pp. 834-850.

Rantz, S.E., et al., 1982, "Measurement and Computation of Streamflow," 2 volumes, USGS Water-Supply Paper 2175.

Rouse, H., 1950, *Engineering Hydraulics*, John Wiley and Sons, New York, NY.

Teal, M.J., and R. Ettema, 1993, "Estimation of Mean Velocity for Ice-Affected Stream Flow," IIHR Report No. 390, Iowa City, Iowa.

Tsai, W. F., 1991, "A Study of Ice-Covered Bend Flow," Ph.D. thesis, The University of Iowa, Iowa City, Iowa.

Effects of Simulated Ice on the Performance of Price Type-AA Current Meter Rotors

Janice M. Fulford[1]

Abstract

Slush ice readily adheres to the standard metal rotor of the winter Price type-AA current meter and affects the ability of the meter to measure the flow velocity accurately. Tests conducted at the U.S. Geological Survey Hydraulics Laboratory at Stennis Space Center, Mississippi, attempt to quantify the effects of slush ice on the performance of standard Price type-AA meter metal rotors. Test data obtained for rotors filled with simulated slush are compared with data for solid-cup polymer and standard hollow-cup metal rotors. Partial filling of the cups only marginally affects rotor performance at velocities greater than 15.24 centimeters per second. However, when cups are filled or over-filled with simulated slush, rotor performance is noticeably affected. Errors associated with slush over-filling and filling of cups are also significant when flows are angled vertically.

Introduction

During extended periods of below-freezing air temperatures, stream discharge measurements may be made in slush or frazil ice conditions. When slush or frazil ice is present, ice can adhere to the inside of standard metal winter Price type-AA meter cups and alter the mass and hydrodynamic shape of the rotor. These changes in mass and hydrodynamic shape affect the meter's ability to measure the velocity accurately and can prevent the rotor from rotating. In similar conditions, however, the solid polymer rotor accumulates much smaller amounts of slush ice.

Because the solid polymer cup resists slush accumulations, stream gagers have been interested in its use for winter measurements. However, previous test results (Fulford, 1990) indicated that the solid-cup polymer rotor is not as accurate as the metal-cup rotor in vertically angled flows. These previous tests did not compare the effects of slush accumulations on rotor performance. This paper attempts to quantify

[1]Hydrologist, U.S. Geological Survey, WRD, Stennis Space Center, MS 39529

the effects of slush ice on the performance of standard Price type-AA meter metal rotor cups on the basis of tests conducted at the U.S. Geological Survey hydraulics laboratory at Stennis Space Center. Tests include linear response and vertically angled flow response of the slush-affected rotors.

Simulated Ice in Test Rotors

Because a constant configuration and amount of slush ice on a meter rotor cannot be maintained in a refrigerated flume, slush ice was simulated using various types of materials. Slush ice was simulated in six metal rotors that were removed from regular use because of minor surface dents. Three types of materials, Room Temperature Vulcanizer (RTV) sealant made of silicone rubber, cellulose sponge, and foam sponge, were secured inside the standard metal rotor cups to simulate various accumulations of slush ice. The test rotors and meters are shown in figure 1.

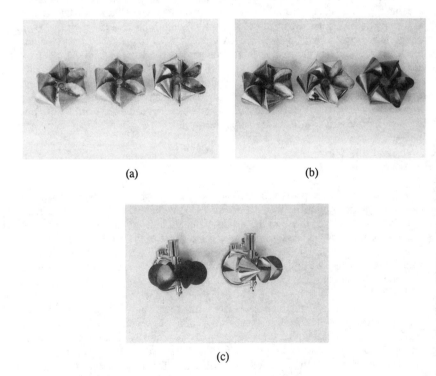

Figure 1. Photographs of tested "slush" affected rotors from left to right: a) RTV over-filled, full-filled, and half-filled; b) full-filled cellulose, foam over-filled, and foam half-filled; and c) winter Price type-AA with polymer and standard metal rotor.

The cups of three rotors were, half-filled, filled, and over-filled with RTV. The cups of two rotors were half-filled and over-filled with foam sponge. The cups of the remaining rotor were filled with cellulose sponge. Both types of sponges were glued to the cups using small amounts of RTV. Two winter Price type-AA yokes were used to house the "slush" filled rotors for testing. Additionally, a standard metal winter Price type-AA rotor without "slush" and a solid-cup polymer rotor were tested for comparison.

Ideally, the material used to fill the rotor cups should duplicate the density of ice or slush. This would best duplicate the effect of additional mass due to slush ice on the bearing pressure of the meter pivot and on the inertia of the rotor. The submerged and dry weights of the rotors tested are listed in table 1. The foam and cellulose sponges used in the rotors were completely wetted before tests and submerged weighing. The full cellulose rotor and the half-filled RTV rotor most closely duplicate the submerged weight of the standard metal rotor. The two rotors filled with foam and the over-filled RTV rotor have submerged weights that are smaller than the standard metal rotor. The remaining "slush" filled rotors have larger submerged weights.

Table 1. Dry and submerged weights of tested rotors

			METER ROTORS					
	Polymer	Standard metal	Half-filled foam	Full-filled cellulose	Over-filled foam	Half-filled RTV	Full-filled RTV	Over-filled RTV
Dry weight, in grams	198	150	156	170	179	207	292	340
Submerged weight, in grams	57	128	122	128	113	128	136	113

Linear Response of "Slush" Affected Rotors

The linear response of the "slush" affected rotors was determined by towing winter Price type-AA meters equipped with the rotors at speeds used by the U.S. Geological Survey to determine rating equations for standard Price type-AA current meters (Kaehrle and Bowie, 1988) and at one additional low speed. The various "slush" filled, solid-cup polymer, and standard-metal rotors were towed in the tow-tank facility at 4.57, 7.62, 15.24, 22.86, 31.38, 42.52, 67.37, 141.75, and 226.80 centimeters per second (cm/s). The distance traveled by the tow carriage, the number of revolutions of the meter, and the elapsed time were measured simultaneously for each towing speed. Two measurements were made at each tow speed. For each rotor tested, the data were fitted by two linear equations using linear regression. The breakpoint for polymer and standard metal rotor equations was

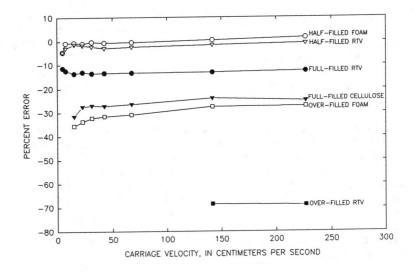

Figure 2. Percent errors for rotors filled with simulated slush.

constrained to be 1 revolution per second (r/s). This is the standard breakpoint for USGS winter Price type-AA meters. The data for the various "slush" filled rotors were fitted by the two lines with the smallest standard error. The regression coefficients, the breakpoint between the two equations, and the standard error determined for each rotor are listed in table 2. The slopes and intercepts (regression coefficients) determined for the "slush" filled rotors are generally larger than those determined for the standard metal rotor, because the "slush" accumulation alters the rotor mass and hydrodynamic shape. The rotor over-filled with RTV did not rotate except at 152.40 and 243.84 cm/s, therefore no linear regression was performed on the data collected for that rotor. The velocity required to start rotation of "slush" filled rotors is as large or larger than that required by the standard rotor.

"Slush" accumulations change the linear response of the meter. Flow velocity is under registered when the rating equation determined for the meter unaffected by "slush" is used in "slush" conditions. To determine the error in velocity measurement caused by "slush" accumulation, a "slush" affected velocity was computed using the revolutions per second measured by the "slush" filled rotors at each test velocity in the rating equation determined for the empty metal cup. Percent error was computed as 100 times the difference between the "slush" affected velocity and the cart velocity divided by the cart velocity. The effect of various amounts of "slush" accumulation on measurement error is shown in figure 2. The half-filled foam rotor had insignificant errors that averaged -0.7% over the test velocities. The

Table 2. Regression coefficients determined for the "slush" filled rotors and for the polymer and standard metal rotor
[cm/r, centimeters per revolution; r/s revolution per second; cm/s centimeters per second]

Rotor type	Slope cm/r	Intercept cm/s	Breakpoint r/s	Slope cm/r	Intercept cm/s	Standard error cm/s
Standard metal	68.458	0.671	1.00	69.891	0.762	0.219
Polymer	77.358	0.671	1.00	77.541	0.518	0.195
Half-filled foam	68.641	0.762	2.701	67.605	3.597	0.229
Full-filled foam	66.690	2.286	0.579	92.415	5.029	0.643
Full-filled RTV	79.248	0.671	0.584	79.644	0.457	0.189
Full-filled cellulose	91.501	1.920	1.84	92.202	0.640	1.018
Half-filled RTV	70.226	0.671	2.716	69.616	2.316	0.332
Over-filled RTV			not determined			

half-filled RTV rotor had errors that were somewhat larger in magnitude than those for the half-filled foam rotor, averaging -2.3%. The remaining rotors had the following average errors: -27.0% for the full-filled foam, -13.0% for the full-filled RTV, -31.3% for the over-filled foam and -68.7% for the over-filled RTV. In general, the magnitude of the errors increased as "slush" accumulations increased. All "slush" filled rotors, except for the half-filled foam rotor, under registered the flow velocity, resulting in negative errors.

The increased mass and changes in hydrodynamic shape due to the "slush" also increased the velocity necessary to start rotation of the rotor. Rotors that were filled did not rotate at test velocities less than 15.24 cm/s, and the over-filled RTV rotor did not rotate at test velocities less than 152.40 cm/s. Partial filling of the cups affected the measured velocities only marginally. However, when the cups are filled or overfilled, the velocities are under registered by the rotor and the stall velocity of the meter increases. Because the polymer solid-cup rotor does not readily accumulate slush, it is assumed that its linear response would be unchanged by slush.

Vertical Angle Flow Response in "Slush"

Some of the "slush" filled rotors were tested for their response to vertically angled flows. This test, a measure of how accurately a meter measures the appropriate vector component of the flow, is also known as the cosine response, because an ideal meter would register the cosine component of an angled flow. When the flow is angled in the vertical plane, the solid-cup polymer rotor is known to under register velocity in comparison to the standard metal cup rotor unaffected by slush (Fulford, 1990). It was unknown whether metal rotors affected by slush would have larger under registration of the velocity than that observed in the solid-cup polymer rotor.

For vertical-angle response testing, a half-filled and a full-filled rotor were selected that most closely matched the submerged weight of the standard metal rotor. The over-filled foam rotor was also selected for angle testing, because the over-filled RTV rotor did not turn at low speeds. The three selected "slush" filled rotors—half-filled foam, full-filled cellulose, and over-filled foam—were tested at vertical-flow angles ranging from 90° to -90° in increments of 10° and at ±5°. Flows directed downward onto the meters are positive angles and flows directed upward onto the meters are negative angles. This test was conducted in the submerged jet tank of the hydraulics laboratory. Jet velocity is determined by timed volumetric measurement and the area of the jet orifice. Meter angular velocity in r/s is determined by counting and timing the revolutions. Because only the meter and not the actual flow could be angled, the meters were positioned with the axis perpendicular to the force of gravity. Positioning the axis perpendicular to gravity ensured a consistent loading of the meter bearings throughout the test.

Jet velocities for the vertical-angle response testing were 31.36 cm/s for the half-foam rotor and 76.20 cm/s for the full-filled cellulose and over-filled foam rotors. Similar to the linear response test, a "slush" affected velocity was computed using revolutions per second measured by the "slush" filled rotors at each test velocity in the rating equation determined for the empty metal cup. Percent error was computed by comparing the ideal meter response at a jet velocity (jet velocity times cosine of the flow angle) to the "slush" affected velocity. In figure 3, percent error computed for the "slush" filled rotors is shown plotted with the percent error for the polymer rotor at 30.48 cm/s and 91.40 cm/s and for the standard rotor at 30.48 cm/s. The percent errors for the polymer rotors were obtained from previous testing and were computed assuming that the solid-cup polymer rotors are unaffected by slush accumulations. As in the previous test, the error is dependent on the amount of "slush" in the cups. The largest errors are for the full-filled cellulose and over-filled foam rotors. The half-filled foam rotor behaves similar to the standard metal rotor without "slush". For small vertical angles, the polymer rotor has errors of less than 12% in magnitude that are considerably smaller than the almost 50% errors of the full-cellulose and over-filled foam rotors. For angles greater than 70° in magnitude, the polymer rotor did not rotate at 30.48 cm/s.

If it is assumed that the polymer rotor does not accumulate slush, the errors for the polymer rotor at angles between ±60° are less than the error for either "slush" full-filled or overfilled rotor. The errors found for "slush" full-filled or over-filled rotors are larger than the vertically-angled flow errors for the polymer rotor between ±10°. The linear response errors for "slush" affected rotors averaged from -0.7% to -2.3% for the half-filled rotors, -13.0% to -27.0% for the filled rotors, and from -31.3% to -68.7% for the over-filled rotors. In comparison, the vertically-

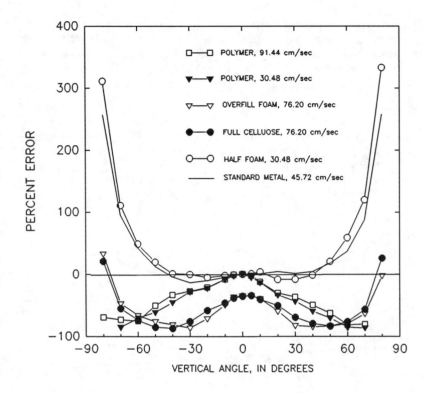

Figure 3. Cosine response error for "slush" filled rotors and for polymer rotors.

angled flow errors for the polymer rotor between ±10° varied from -8.3% to -13.0% for flow speeds of 30.48 and 91.44 cm/s. Thus for all tested flows with vertical angles less than ±10°, the meter equipped with the polymer rotor should have less under registration of velocity than the "slush" full-filled or over-filled rotors.

Summary

Partial filling of the rotor's cups by "slush" only marginally affects rotor performance of the winter Price type-AA meter. However, when rotor cups are full-filled or over-filled by "slush", rotor performance is noticeably affected. Solid-cup polymer rotors are resistant to slush ice accumulation but do not perform as well as metal cups in vertically angled flows. However, the velocity errors associated with "slush" over-filled and full-filled cups is significantly larger than the error for a polymer rotor with a $\pm 5°$ vertical flow angle. Therefore, for slush ice conditions, the solid-cup polymer rotor should give more accurate velocity measurements than a standard metal rotor in the winter Price type-AA meter.

References

Fulford, J.M. (1990). "Effects of turbulence on Price AA meter rotors." Proceedings of 1990 ASCE Hydraulic Engineering Conference, July 30-August 3, p.909-914

Kaehrle, William R. and Bowie, James E. (1988)."Calibration of Water-Velocity Meters." Proceedings of 1988 ASCE Hydraulic Engineering Conference, August 8-12, p.60-65

Innovative Instrumentation for a Physical Model
of River Ice Transport

Johannes Larsen,[1] M. ASCE, Jon E. Zufelt,[2] M. ASCE,
and Randy D. Crissman,[3] M. ASCE

ABSTRACT

Large physical models often require the use of innovative instrumentation and measurement techniques. Hydraulic models that attempt to simulate the effects of stationary ice covers put constraints on the types of instrumentation that can be used to successfully measure water velocities and stages. Ice transport models compound these measurement problems, since the motion of the ice is one of the primary variables of interest. The challenges of measuring the thickness, velocity, and concentration of moving ice in the field are significant. The challenges are only marginally easier in a physical model. These obstacles were addressed in a physical model study of the ice transport processes in the Grass Island Pool of the Upper Niagara River, which forms part of the border between the State of New York (USA) and the Province of Ontario (Canada).

INTRODUCTION

The Grass Island Pool (GIP) is a wide reach of the upper Niagara River immediately upstream of Niagara Falls, from which diversions are made for hydropower production (Fig. 1). The GIP has three major inflow channels; the Tonawanda or East Channel, the Chippawa or West Channel, and the Little West Channel. The distribution of water flow among the three channels depends on the relative channel resistance and the distribution of weed or ice accumulations within the river. Under average open

---
[1]Vice President, Alden Research Laboratory, Inc., 30 Shrewsbury St., Holden, MA 01520
[2]Research Hydraulic Engineer, USACRREL, 72 Lyme Road, Hanover, NH 03755-1290
[3]Senior Hydraulic Engineer, New York Power Authority, P. O. Box 277, Niagara Falls, NY 14302

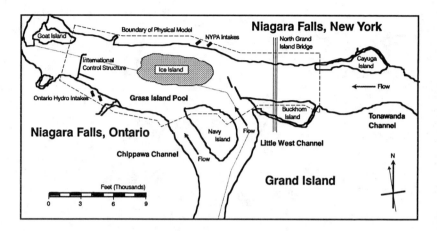

Fig. 1: Map Showing the Physical Features of the Grass Island Pool and the Boundaries of the Physical Model.

water flow conditions, the flow split is about 42, 38, and 20% of the total flow into the Tonawanda, Chippawa, and Little West Channels, respectively.

The GIP is bounded at its downstream end by the International Control Structure (ICS). The ICS extends about two-thirds of the distance across the pool from the Canadian shore and is used to maintain water levels in the GIP to within limits specified by the International Niagara Board of Control and meet other requirements of the 1950 Treaty Between the U.S. and Canada Concerning the Uses of the Waters of the Niagara River.

The New York Power Authority's (NYPA) Niagara Power Project intakes are located on the U.S. shore, and the Ontario Hydro Sir Adam Beck (SAB) intakes are located on the Canadian shore just upstream of the ICS. Two dikes at the downstream end of Buckhorn Island serve to direct the Tonawanda Channel flow toward the NYPA intakes. The 1950 Treaty specifies that 1,416 cms (50,000 cfs), during non-tourist periods, and 2,832 cms (100,000 cfs), during tourist periods, should flow over Niagara Falls. The remainder of the river flow is shared equally by the NYPA and Ontario Hydro to produce hydropower.

Portions of the GIP were excavated to depths of 2 to 3 m (6 to 10 ft) in the vicinity of the power intakes when the generating stations were built to improve the ice transport capacity through the GIP. The depths in the remainder of the GIP are generally about the same as in the ice escape channels, except in some areas near the banks of the river. However, there is a large shallow area in the center of the pool that averages less than 0.6 m (2 ft) in depth at normal operating levels in the GIP.

Early winter storms having prevailing southwest winds can cause wind setups of the surface of Lake Erie at the entrance to the Niagara River before a stable ice cover forms on the lake. These setups can significantly increase the flow into the river and force massive amounts of unconsolidated lake ice into the river. When the ice reaches the GIP, a large island of grounded ice typically forms in the shallow areas in the center of the pool, reducing the ice transport capacity through the pool. The ice island and other areas of ice accumulation along the banks of the GIP can restrict the flow of water and ice past the NYPA intakes. The ice escape channel that was excavated along the U.S. shore was intended to facilitate the movement of ice past the NYPA intakes. During severe ice runs, ice stoppages may occur in the ice escape channel at the NYPA intakes, which can require reductions in power diversion flows to provide additional water to transport ice through the GIP. This mitigating action reduces power generation during severe lake ice runs or ice stoppages. Icebreakers are used by both the NYPA and Ontario Hydro to assist in reducing the potential of ice stoppages, to aid in maintaining ice movement through the GIP, and to clear ice when it stops moving.

The NYPA initiated a comprehensive study to investigate ice jamming on the Upper Niagara River in 1991 (Crissman et al., 1992). The Grass Island Pool Physical Model was constructed as one component of this study to investigate the ice transport and jamming processes within the pool. Other modeling objectives included the identification of variables that contribute to ice jamming in the GIP, and operational and structural mitigation measures that might increase the ice transport capacity through the GIP.

PHYSICAL MODEL DESIGN

Model Construction

The boundaries of the model of the Grass Island Pool and short reaches of the three branches of the river leading into the pool are shown in Figure 1. The geometric scales selected for the model were 1/120 in the horizontal and 1/50 in the vertical, yielding a distortion ratio of 2.4. A distortion ratio of less than three was required to minimize the distortion effects on the ice accumulation processes in the model. Templates were constructed from 2-foot-contour-interval field data and were spaced 110 m (360 ft) apart (prototype) in the model basin. Vertical placement was controlled using a laser beam level and electronic beam locators, yielding a placement accuracy of 0.15 m in the prototype (0.5 ft). Topography was reproduced up to the 173-m (565-ft)[4] contour. Crushed stone was placed between the templates to within 1.5 m (5 ft) of the prototype finished grade, and a weak cement and sand mix was screeded to the top of the templates to provide a fixed stable bed.

The hydraulic structures were constructed from engineering drawings of the prototype structures. The NYPA intakes were fabricated from Lucite to facilitate visualization of flow and ice accumulation patterns. The geometry of the intakes was accurately reproduced at the distorted scale to ensure that any effect of the geometry on approach flow patterns to the intake would be reproduced. The SAB intakes were simulated using a scaled area distribution over the horizontal length of the intakes but

---

[4]International Great Lakes Datum of 1955.

without the prototype's intake vanes. The vanes were not considered important for the planned tests. The ICS was constructed to the distorted scale using plastic materials to ensure structural stability with time. The Buckhorn Dikes were constructed in situ using reinforced concrete placed on top of the topography.

The model was designed following Froude similitude criteria. To ensure that the flow patterns were correctly reproduced, large-scale roughness elements were placed in the model to reproduce longitudinal and transverse water surface profiles measured in the prototype. Field measured velocity and flowline data were obtained by releasing floats in the prototype and recording their position with time using a global positioning system (GPS) with onboard timing and recording. These data were subsequently used in guiding the placement of the model roughness elements.

Model Ice Material

Many types of materials have been used in the past to model ice, including paraffin, wax, wood, and plastics (Wuebben, 1994). The quantities of model ice needed for this model study dictated that a plastic material be used. For ice transport models, a material is required whose shape, specific gravity, and accumulation characteristics are close to those of natural ice. Both polypropylene and polyethylene have specific gravities very close to natural ice and are commercially available in large quantities. These materials are commonly available in pellets having a 1-to 3-mm diameter and are used as a raw material for injection molding. However, these uniformly graded pellets have an angle of internal friction that is appreciably less than that of natural ice. As a result, physical models using these pellets produce thicker accumulations of ice and make it difficult to reproduce ice jamming in wide sections without the use of some artificial surface restriction.

Experiments were undertaken to crush molded polypropylene parts in a special crusher to produce irregularly shaped pieces with a wider size distribution. The resulting product proved to be far superior in reproducing the prototype ice behavior and would formed ice bridges and ice jams without the addition of artificial surface obstructions. The angle of internal friction of this crushed polypropylene material was very close to that of natural ice. The D_{50} of the material simulates ice pieces about 0.3 m (1 ft) thick, which is reasonable for ice entering the upper Niagara River during early winter lake ice run events. Approximately 15 metric tons (16.5 short tons) of molded polypropylene parts were crushed for use in the model study.

Ice and Flow Handling Systems

The amount of model ice that was handled in the most severe ice transport conditions simulated in the physical model was about 13 metric tons (14.3 short tons). Furthermore, the model ice needed to be introduced into each of the three channels in a predetermined and repeatable spatial and temporal schedule. A flow system was designed to allow the ice to be introduced into the supply lines for the river flow and transported back to the supply point with the return water flow from the downstream

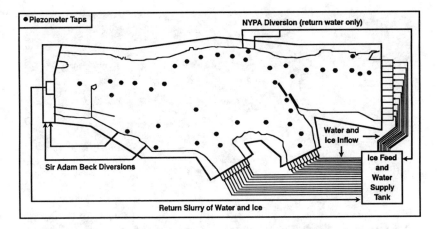

Fig. 2: Water and Ice Circulation System and Locations of Piezometer Taps for the Grass Island Pool Physical Model.

end of the model. Upon return, the ice material was separated from the water flow by gravity, and reintroduced into the river flow at the appropriate rate. This allowed the model to be operated continuously without manually handling the ice material (Fig. 2).

The water flow system consisted of a large tank with three weirs whose lengths were proportional to the flow split between the three channels flowing into the GIP. The three flows were divided into separate compartments, each containing ten variable-width weirs that further separated the flow into ten pipes leading to the appropriate river channel. Each of the ten pipes provided flow to one tenth of the channel width in an amount required to replicate the field-measured flow distribution in each channel. The flows from the downstream boundary of the model and from the power diversions were pumped back into the distribution tank, which allowed the model to be operated as a closed water system.

The model ice material was metered into each channel supply flow immediately downstream of the ten adjustable-width weirs using ten revolving scoops (Fig. 3) that could be adjusted both in speed and scooping depth. This allowed both the ice distribution across each channel and the total ice discharge to be varied with time. The ice that was transported through the model was collected at the downstream boundary, where a variable-speed propeller pump was used to return the water/ice slurry back to the distribution tank.

MODEL INSTRUMENTATION

Hydraulic Measurements

A key hydraulic variable measured in tests conducted in the model was the water level distribution, since it influences water and ice conveyance as well as the

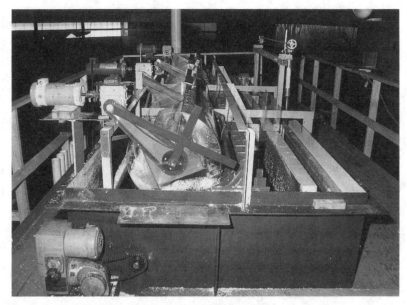

Fig. 3: Ice Feed Tank with Revolving Ice Scoops.

jamming processes. The effects of the ice accumulations on the water levels were also important. A total of 40 piezometer taps were installed in the model as shown in Fig. 2. Each piezometer tap was connected by plastic tubing to a centrally located valve array, which allowed each piezometer to be switched to a differential pressure transducer and compared to a reference pressure. A dedicated micro-computer was used to control the switching, settling, measuring, timing, and unit conversion for the piezometers. Measurement of the pressure transducer was accomplished by a 12-bit A/D converter. The measured water levels were reported to a central computer, which controlled the overall model coordination, data display, and data storage.

Key information that was recorded on the central computer was also transmitted to an alpha-numeric display board that alerted the model operator to conditions requiring changes in flow rates or ICS gate settings. This allowed the model to be operated in a quasi-unsteady manner without using complex computer-controlled valves, pumps, and gates. Water flow rates were measured using calibrated venturi meters, and water velocities were measured using electro-magnetic and miniature propeller meters.

Ice Measurements

To compare the performance of measures aimed at improving ice conveyance through the GIP, it was necessary to measure ice velocities and thicknesses. These measurements were used to estimate ice discharges passing selected cross sections of the GIP and to estimate the volumes of ice accumulation within reaches of interest. Sets of instruments were placed on support beams across a section to measure ice thicknesses and velocities at discrete locations. These discrete measurements were then integrated to estimate ice discharges.

Ice speeds were measured using paddle wheels suspended above the ice surface such that the paddle tips were in contact with the moving ice (Fig.4). Multi-contact magnetic counters in the paddle wheels were used to count and sum the number of contacts between readings. These data, when properly calibrated, were used to accurately measure ice speeds. The paddle wheels were mounted on a vertical axis support connected to a 360-degree potentiometer, which was monitored to provide measurements of the direction of ice movement. A vane connected to the paddle wheel and in contact with the ice downstream of the paddle wheel tracked the direction of ice movement and kept the paddle wheel properly oriented.

A novel probe was developed to automatically measure ice thickness. It consisted of an optical transmitter and receiver located on opposite legs of a two-pronged fork. The transmitting frequency was selected to be impeded by the polypropylene ice material such that the ice thickness could be calculated by measuring the time that the receiver was shadowed during a constant velocity penetration of the ice cover. All the

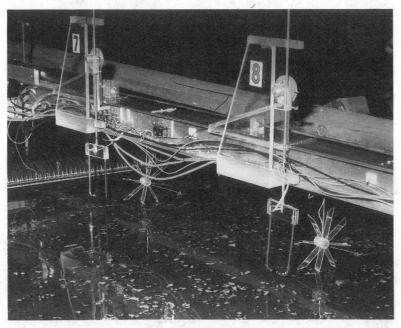

Fig. 4: Photograph of Instrumentation Used to Measure Ice Speeds, Directions and Thicknesses in the Physical Model.

ice thickness probes on the support beam simultaneously passed through the ice cover when a horizontal motorized piston connected to a series of ramps was actuated. This caused the thickness probes to be lowered and raised through the ice cover. The probe motion and data sampling were controlled by the central computer, while the ice penetration time was measured by the probe circuit. Figure 4 shows two ice velocity and thickness measuring stations on a support beam. Figure 5 is an example of ice speeds and thicknesses recorded during a test conducted in the model at two different measurement stations.

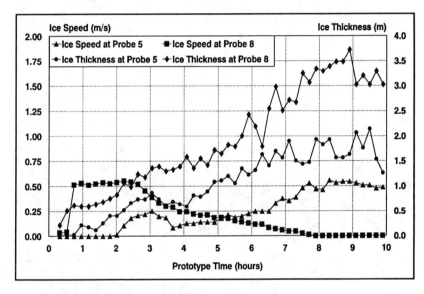

Fig. 5: Example of Measurements From a Test Conducted in the Physical Model.

CONCLUSIONS

The construction and instrumentation of the physical model of the GIP overcame a number of obstacles. The problems of model ice material handling in a large physical model were overcome by creating a slurry of ice and water, which was pumped to a head tank from the downstream boundary of the model. The subsequent separation of the ice and water in the head tank (via buoyancy) and the introduction of ice into the water flow into the model before it reached the upstream end of the model was accomplished with a novel ice handling system of rotating scoops. This provided an easy way to set the proper ice distribution across each channel as well as the magnitude and timing of the ice discharge into the model. Measuring ice speeds and directions at specific locations in the model using paddle wheels suspended above the ice cover proved to be both reliable and repeatable. One of the most innovative instruments was an optical ice thickness probe. This probe allowed ice thicknesses to be measured in a moving ice cover without affecting the ice movement and the ice accumulation processes occurring in the model. The system simultaneously measured ice thicknesses and velocities at several locations across up to two sections of the model that could be used to estimate ice discharges.

REFERENCES

Crissman, R.D., Ettema, R., Gerard, R.L., and Andres, D. (1992) "A Plan for Studying Ice Jamming on the Upper Niagara River," Proceedings of the IAHR 11th Symposium on Ice, Banff, Alberta, Canada, Aug., 1992, pp. 527-538.

Wuebben, J.L. (1994) "Physical Modeling of River Ice Jams" In: River Ice Jams, S. Beltaos, ed., prepared by the NRCC Working Group on Ice Jams, Ottawa, Canada.

Developing Air Concentration and Velocity Probes for Measuring Highly-Aerated, High-Velocity Flow

Kathleen H. Frizell, David G. Ehler, and Brent W. Mefford[1]

Abstract

The U.S. Bureau of Reclamation's stepped spillway overtopping protection research program required velocity and air concentration profile data be obtained during testing in a large, 15.2-m-high outdoor flume. Both velocity and air concentration profile measurements were needed to evaluate energy dissipation, bulking, and model/prototype correlation with a smaller indoor model.

Measuring highly-aerated, high-velocity (12-17 m/s) flow is not practical using available instruments. Although air concentration measurement devices exist, most are custom made by researchers. As a result, two instruments were developed and calibrated by Reclamation to gather prototype data in our overtopping facility. The instrument development, calibration, and verification will be discussed.

Introduction

The instruments were developed to measure in highly-turbulent, highly-aerated, high-velocity flow. The instruments were developed for obtaining prototype scale velocity and air concentration measurements to verify and extend laboratory data on overtopping embankment dam protection schemes.

Both instruments were developed by Reclamation to quantify energy dissipation in flow over rough surfaces on steep slopes. Tests were performed in a large, outdoor flume facility located at Colorado State University in Fort Collins, Colorado (fig. 1). The 2:1 sloping facility is 15.2-m-high by 1.52-m-wide, and is capable of passing unit discharges of 2.9 $m^3/s/m$, thus theoretical velocities, of up to 17 m/s.

---

[1] Respectively, Hydraulic Engineer, Electronic Engineer, and Head, Hydraulic Structures Section, U.S. Bureau of Reclamation, Hydraulics Branch, Denver, Colorado 80225

Figure 1. - Near-prototype test facility where instruments were used.

Quantifying energy dissipation requires measurement of flow velocity and air concentration profiles along the slope of the flume. No standard measuring devices were identified that could be used to measure flow velocity or air concentration for the two-phase flow and high-velocity test conditions. In addition, shallow flow depths limited the size of the instruments.

Velocity profile measurements required a very sturdy instrument capable of measuring point velocities in air-water flow. A back flushing pitot tube concept, used earlier in Reclamation's laboratory to determine velocities in slurry flow, was utilized. The components of the velocity probe will be discussed.

An air bubble detector developed by Wood, 1978, and later improved by Bachmaier, 1988, was chosen to measure air concentration. The principle for the air bubble detector is based upon the difference between the electrical resistivity of air and water. The detector electronics and physical probe assembly will be discussed.

Both probes required laboratory calibration to provide relationships to known values or baseline information and prove reliability. The calibration apparatus, procedures, and results are discussed. The instruments were tested and calibrated in Reclamation's hydraulic laboratory prior to their use on the prototype facility. Verification of the instrument's measurements will be presented.

Velocity probe

The development of a velocity probe focused on using something technically simple and stout. This plan was devised after attempting to use several "off the shelf" propeller-type meters. A pitot-static tube designed for mounting on the fuselage of an airplane was attached to the end of a 2-m-long, triangular-shaped shaft. The pitot tube and structural shaft provided good strength for the measuring device.

Flow from a constant head source is fed to both the static pressure ports and the total head port of the pitot tube. This flow provides continuous back flushing of the pitot tube during operation. Back flushing is necessary to maintain a fluid of known density (or in this case solid water) within the pitot tube and connecting tubing. The constant head supply pressure used for back flushing must be larger than the maximum expected velocity head to be measured in the flow. The back pressure is set with a pressure regulating gage. The back flushing flow rate to each port is controlled by low-flow-rate rotameters. Through laboratory testing, a back flushing flow rate of 3.79 l/hr in air was chosen. This flow rate was the minimum flow rate that could be passed and still produce a continuous flow out of the pitot tube ports in air. Back flushing flow rates of 7.57 and 11.36 l/hr also produced good results; however, the lower flow rate provided better low end sensitivity of the instrument.

A differential pressure cell measures the pressure difference between the static and total head ports. The voltage output from the differential pressure cell is sent to an HP3457A integrating voltmeter to determine the average voltage. Using the cell calibration the voltage is converted to pressure head.

The back flushing flow rate is set in air and a baseline reading, or zero offset, is measured from the differential pressure cell. The baseline reading is used to shift the measured data to a common zero velocity that is independent of the back flushing flow rate. This zero reading in air eliminates differences between initial settings of the back flushing flows through the rotameters. Figure 2 shows a schematic of the velocity probe components. After recording the pressure zero, the pitot tube is ready to measure velocity in flowing water.

The velocity from the measured pressure of the pitot tube is determined by:

$$V = \sqrt{\frac{2(p_t - p_s)}{\rho}}$$

where: V = velocity (m/s), p_t = total pressure head (Pa), p_s = static pressure head (Pa), and ρ = fluid density (kg/m^3).

This equation shows that pitot tube velocity measurement is dependent on fluid density. In the air entrained flow of the tests, the density of the two-phase fluid varies through the flow depth with air concentration. Therefore, the pitot tube must be calibrated as a function of air concentration and the air concentration of the flow at the point of the velocity measurement must also be determined.

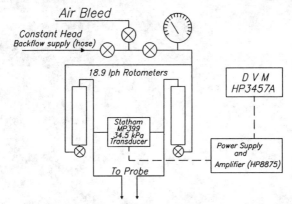

Figure 2. - Schematic of the components used for the velocity probe.

Velocity probe calibration

The velocity probe was calibrated for air concentrations of 0 to about 70 percent by volume. The same calibration setup was used for both the velocity probe and the air concentration probe (Bachmaier, 1988). The calibration apparatus consists of an inverted "U" shaped water pipe system with the water and air volumes entering the system measured with rotameters (fig. 3). Water is supplied to the pipe system on the up leg and an air line with pressure regulator and rotameter is connected to the water pipe on the down leg of the system.

For the velocity probe calibration, a 10-cm-long tip of smaller diameter pipe is attached to the end of the piping system to provide backpressure and obtain a relatively smooth uniform jet of air/water mixture leaving the pipe. Pressure in the tip is monitored using a piezometer tap located two tip diameters upstream of the end. The pressure in the tip is needed to determine the tip cross-sectional area occupied by air as opposed to water. It is desirable to have near atmospheric pressure in the tip to prevent rapid volumetric changes in air concentration at the tip exit. The pitot tube is placed within one tip diameter of the end and centered within the free jet. Water velocity impinging on the total pressure port of the pitot tube is assumed to be equal to the water velocity internal to the calibration apparatus tip. Water velocity is determined as the water discharge divided by the area of the tip minus the area of the entrained air. Air and water discharges are adjusted to achieve a wide range of air/water concentrations and velocities impinging on the pitot tube.

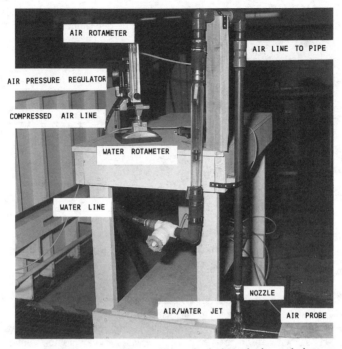

Figure 3. - Test apparatus used to calibrate both the velocity and air concentration instruments.

The calibration produced a family of curves of pressure differential versus velocity for each percent air concentration tested (fig. 4). Good agreement with the theoretically predicted coefficients for air/water density was achieved for 0 and 40 percent air; however, with the higher air concentrations of 58 and 68 percent, the coefficients from the calibrations are significantly higher. The reason for this discrepancy is not known at this time. As may be seen, the data scatter is limited and it was also repeatable. These curves were used with the air concentration data from the large scale flume to correct the measured velocity profiles for the influence of varying density throughout the flow depth. This primary calibration was checked in water (0 percent air concentration) against velocities measured using a laser doppler velocimeter.

Air concentration measurements

An air concentration probe, which acts as an air bubble detector, was based upon previous instrumentation reported by Ian Wood of the University of Canterbury, Christchurch, New Zealand, in 1988. The air concentration measurement is based

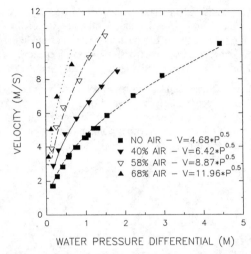

Figure 4. - Calibration curves for the velocity probe.

on the difference in electrical resistivity between air and water. The air probe consists of two concentric conductors encased in a protective support. The tip of an insulated platinum wire is encased in a stainless steel sleeve. The probe polarity is periodically reversed to prevent plating, gassing, and degradation of the probe conductors due to electrolytic action. An air bubble passing across the probe tip interrupts the current flowing between the two conductors. This current is amplified and conditioned to produce a step output. Integrating the resulting signal over time gives the probability of encountering air in the air/water mixture.

The air concentration probe discussion will be separated into the three components that comprise the instrument: the physical probe, the electronics, and the output or data collection equipment. Finally, the calibration results will be presented.

Air probe

The function of the air probe is to provide the conductor or air bubble sensor. The concentric conductors are a 0.2-mm platinum wire set in a 1.59-mm-outside-diameter stainless steel tube 50.8-mm long with waterproof epoxy. The probe sensor and the wire leads are then encased in a sturdy, streamlined support to prevent damage under the high-velocity testing conditions and to carry the wires out to the electronics. The critical part of the probe construction is preventing water from reaching and shorting out the conductors.

The probe sensor is epoxied into a brass cone-shaped tip with a threaded base. The two leads from the platinum wire and stainless tube are soldered to the center and inside shield of a long triax cable. The outside shield is grounded at the electronics end. Every wire connection is protected from water seepage by placing each in shrink tubing and filling the tubing with a silicone compound before shrinking.

After making the wire connections, the brass probe tip is mounted on a threaded 190-mm pipe welded perpendicularly to one end of a 127-mm steel rod. The entire cavity inside the pipe is filled with silicone potting compound to prevent water seepage. The coax cable is brought out to a point above the water surface through copper tubing fitted at the downstream end of the pipe (fig. 5).

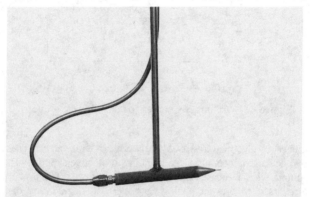

Figure 5. - Photograph of the air concentration probe assembly.

Air bubble detector electronics

The probe electronics for the original design from Wood were built initially, but gave inconsistent and unstable results. The electronics were redesigned using bipolar logic and updated devices. This eliminated the mix of transistor-transistor logic (TTL) and bipolar logic conversions between the two types of logic. The replacement integrated circuits (IC's) were all current production devices to assure continuing availability.

The air concentration probe detects the change between water and air by using the conductivity of water to change the gain of an operational amplifier (opamp). Relatively high gain is used to create an on-off signal that corresponds to the water-air environment of the probe rather than an output proportional to the actual conductivity of the water.

Measuring conductivity in water with single polarity excitation of the probe causes gassing on the probe and a corresponding drift toward higher air concentration. This problem is resolved by periodically switching the probe polarity. The opamp output is switched along with the excitation of the probe to maintain a single polarity output. A common clock signal is used to excite the probe and switch the inputs to an amplifier. The switch also reverses the inputs to maintain a single polarity output from the probe.

This block diagram describes the basic components of the system:

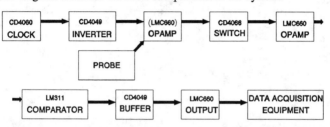

Probe output

Two probe outputs are provided. An analog output using a resistor capacitor filter, provides a running average between 0 and 5 volts over time. The second, is a digital on (5 volts or air) off (0 volts or water) signal that follows the air/water mixture that the probe is sensing. This output can be recorded using a computer or calculated directly using an integrating voltmeter over a fixed period of time.

Air concentration instrument calibration

The air probe electronics are always balanced in water with no air. The probe output is used as input to a Tektronics 454 oscilloscope for balancing and monitoring the quality of the signal. The same apparatus used to calibrate the velocity probe was used to calibrate the air concentration probe (fig. 3) for varying air/water volumes. The air concentration probe tip was located flush with the end of the pipe system nozzle. The probe voltage output was recorded with a Systron+Donner 1033 integrating digital voltmeter.

The air concentration is determined as a ratio of the air volume over the air volume plus the water volume measured in the pipe system. The voltage output from the probe is 0 volts when in water and 500 volts in air. The air/water mixture measured by the probe is a percent of the air voltage (500). Figure 6 shows the final calibration with air concentration versus percent air measured by the probe.

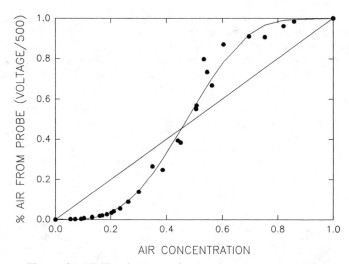

Figure 6. - Calibration curve for the air concentration probe.

When the probe is entirely in water then the probe reads zero. When the probe is entirely in air the probe reads one. If the probe were entirely linear, then this curve would show 50 percent air from the air probe when the measured air and water volumes were identical. When comparing this calibration curve to a linear relationship this curve shows an underprediction of air concentration at small air volumes and an overprediction of air concentration at high air volumes. This calibration, in spite of not being linear, was consistent and repeatable and was used to determine the air concentration in the near-prototype facility. The detector electronics and/or the calibration system could perhaps be improved to linearize the results.

Field operation and problem solving

The field test program began with the air concentration probe fully operational and confidence in the laboratory calibration. As usual with field applications, several problems arose. The primary problems were water leaking into the probe causing it to always indicate water, and the low water conductivity at the field site compared to the laboratory. The difference in water conductivity prevented balancing of the probe at the field site. To solve the problem of water intrusion into the probe it was rebuilt and sealed as discussed. The gain (feedback resistance) for the probe was increased threefold to provide sufficient gain for the detector in the low conductivity water. The higher gain, however, caused a non-symmetric output and oscillation of one of the opamps. The TL074 opamp originally used was replaced with a National LMC660 opamp. The LMC660 still showed a tendency to self oscillate on one polarity of the probe. Therefore, capacitors were installed to eliminate the noise (oscillation), but the non-symmetrical output still existed. Therefore, at high gains, only one polarity of the antiplating period could be used and data were taken only on one side of the antiplating period. These data were used in the final analysis of the project, and work is continuing to attempt to stabilize the electronics.

The LMC660 was replaced with a National LF444CN. This eliminated the noise, but produced an objectional offset. This device was again replaced, this time with a PMI OP470. The OP470 reduced the offset by 10. The noise problem has been reduced considerably, but the output is still not symmetric between the two polarities. One cycle of the output and a longer antiplating period will be used until the symmetry problem is resolved for low conductivity water. Work will continue to correct the symmetry problem and improve the ease of operation.

Conclusions

Both instruments performed very well. Continuity was checked to verify the data gathered using the instruments at various locations down the slope. To check continuity, the profiles obtained on the flume centerline were assumed to be representative of the entire flume cross section. The depth at 90 percent air concentration was assumed to be the water surface. The computed discharges, using the average water velocity times the depth times the flume width, were compared with the known water discharges to the flume. The computed discharges were, on the average, within 8 percent of the known discharges. The largest differences were at the low flow rates where the flows tumbled down the slope. Given the highly turbulent nature of the flow and the assumption that the flow was uniform across the flume width, this is excellent agreement. This simple calculation verified the excellent performance of the developed instruments.

Both instruments will be used in the same facility during the summer of 1994 to measure velocity and air concentration in flow over riprap.

Appendix

Bachmaier, Guido, "Setup, Calibration and Use of a Measuring Probe for Determination of the Air Concentration in a Spillway," Translated from German Ph.D Thesis for the Institute for Hydromechanics of Karlsruhe University, Karlsurhe, Germany, March 1988.

Cain, Paul and Ian Wood, "Measurements within Self-aerated Flow on a Large Spillway," Research Report 78-18, Ph. D Thesis, Department of Civil Engineering, University of Canterbury, Christchurch, New Zealand, April 1978.

VOID FRACTION MEASUREMENT TECHNIQUES FOR GAS-LIQUID BUBBLY FLOWS IN CLOSED CONDUITS; A LITERATURE REVIEW

Mahmood Naghash[1]

ABSTRACT

This paper provides a review of void fraction measuring techniques for gas-liquid bubbly flows in closed conduits with emphasis on the suitability of these methods for air-water bubbly flows in reaeration applications. The techniques addressed include Electrical Conductance Probes, Radiation Attenuation Techniques, Optical Probes, and Thermal Anemometers. Each technique is described and the principles of measurement and applications are provided.

1.0 INTRODUCTION

The complexity of gas-liquid flow phenomena, the need for verification of hypotheses used in development of two-phase flow mathematical models, and the diversity in industrial applications of gas-liquid flows, combined with the advancement of electronic equipment, have resulted in the availability of a large number of measurement techniques to determine the detailed distribution of various gas-liquid flow parameters. The flow parameters of interest may include the local distribution of gas and liquid phases, pressure fluctuations, phase density functions, the bubble velocity and its spectrum, the fluctuating liquid velocity, bubble size distribution, and the bubble interfacial area. Among all these parameters, the primary property of gas-liquid flows to be measured is the local volumetric fraction of the free gas (i.e. the void fraction).

The measurement techniques for determining the local void fraction are numerous. The most common techniques can be categorized as: *Electrical Conductance Probes,* which are designed based on the significant difference between electrical conductivity of the gas and the liquid; *Radiation Attenuation Techniques,* which are based on the absorption characteristic of different phases under a constant intensity beam; *Optical Probes,* which are based on the

---

1   Senior Hydraulic Engineer, Bechtel Corporation, Geotechnical and
     Hydraulic Engineering Services, Gaithersburg, Maryland 20878-5356

refractive index of the surrounding medium, and *Thermal Anemometers,* which are based on the difference between the heat exchange capacities of air and water from a hot wire.

2.0 GENERAL CONSIDERATIONS

For analytical modeling of a gas-liquid flow system accurately, detailed information on the internal flow structure is required. This internal structure can be described by two geometrical parameters, namely the void fraction and the interfacial area. Both of these parameters are particularly important in determining the rate of mass, momentum, and energy transfer between phases or components [Wallis, 1969]. While this paper concentrates on the methods for void fraction measurement, some of the techniques described here can be used to measure both parameters simultaneously.

2.1 Void Fraction Definition and Gas Liquid Signal

The void fraction (α) represents the volumetric fraction for the gas phase at the measuring point. The volume average void fraction in a gas-liquid mixture is defined as:

$$\alpha = \frac{\text{volume of gas in mixture}}{\text{total volume of mixture}}$$

For a finite control volume, the above relation can be approximated by

$$\alpha = \frac{A_g}{A_g + A_\ell} \tag{1}$$

where A_g and A_ℓ are cross-sectional areas of gas and liquid phases, respectively. All the measuring techniques described in this paper (except for attenuation techniques) provide an analog signal proportional to relation (1) and are suitable for processing. The signal from the sensor contains two basic types of information, namely (i) identification of phase (gas and liquid) and (ii) residence time of the phase. For $X_k(x,t)$ representing the k-phase density signal, $X_k(x,t)$ is defined as 1 if x is in phase k at time t, and as zero otherwise (Figure 1) [Cartellier and Achard, 1991].

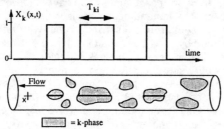

Figure 1 k-Phase Density Function in a Two-Phase Flow [Cartellier et al., 1991]

All experimental determinations of X_k are obtained from a point sensor probe. The probe emits a two-state signal indicating which phase surrounds the sensing part of the probe by taking advantage of a specific phase-dependent physical phenomena occurring at that point. The measured signal, caused by a known amount of injected air in the conduit, is the basis for calibration of the technique.

2.2 Statistical Properties of Bubbly flows

The inherent unsteadiness aspect of multi-phase flows makes them suitable for statistical studies, [Ishigai et al., 1965]. Bubbly flows in particular have been shown to be stochastic [Jones and Zuber, 1975]. A useful technique for studying void fraction fluctuations of bubbly flows is the use of the probability distribution function (PDF). The PDF allows characterization of mean square fluctuations and the mean signal which is the average local void fraction. For a statistically stationary bubbly flow, the successive ensemble averages of void-time records become relatively constant; this represents the average void fraction. Figure 2 shows the use of this technique in a low-void-fraction bubbly air-water flow in a horizontal pipeline [Naghash, 1989].

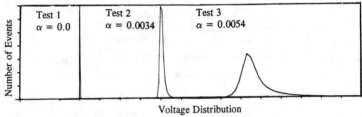

Figure 2 PDF of Fluctuating Void Fraction [Naghash, 1989]

3.0 ELECTRICAL CONDUCTANCE PROBES

Direct contact conductance/resistivity probes that utilize on the significant difference in electrical conductivity of gas and liquid, have been used extensively for void fraction measurement [Delhaye and Jones, 1975]. The technique is based on the instantaneous measurement of a local electrical parameter (such as voltage) around a sensor (probe) with an exposed surface placed in the gas-liquid flow. As the circuit is opened or closed (depending on whether the sensor is in contact with the gas or liquid), the voltage drop across the probe fluctuates between two reference voltages. From the timing of the shift in the voltage, the time when the gas-liquid interface passes the probe can be recorded [Revankar and Ishii, 1993]. These types of probes are classified in the literature as *Impedance Probes, Conductance Probes, Resistivity Probes, or Capacitance Probes* depending on the electrical circuitry design and the measured electrical parameter. Typical configurations of electrical probes are shown in Figure 3.

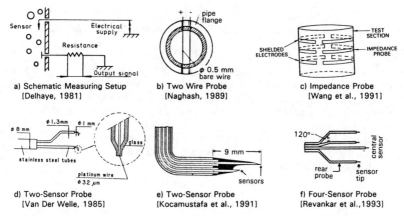

Figure 3 Typical Conductance/Resistivity Probes and Measuring Setup

The signals from two-sensor and four-sensor probes can also provide the bubble velocity and interfacial area information in addition to the void fraction data.

3.1 Measurement Principles

The conductivity of a fluid depends directly upon the ion density of the fluid, the surface area of the electrodes, and inversely upon the distance between the electrodes. For a geometrically fixed probe configuration, the dispersion of voids in the liquid results in a decrease either in the ion density or the effective electrode surface area. Such a decrease results in a variation of the probe's electrical conductance from its initial value, which then becomes a means for void fraction measurement [Lafferty and Hammitt, 1967].

Using theoretical terms, the above arguments are represented for an impedance probe [Halliday and Resnick, 1978; Brown et al., 1978; Andreussi et al., 1988]. The conductance (G) between two electrodes immersed in an electrolyte (G=1/R where R is the resistance) is proportional to the ratio of electrode effective sensing area (A) and the distance between electrodes (ℓ) times the conductivity (σ):

$$G = \sigma A/\ell$$

The capacitance (C) between electrodes is proportional to the ratio of the electrode effective sensing area (A) and the distance between electrodes (ℓ) times the dielectric constant (ϵ):

$$C = \epsilon A/\ell$$

Consequently, the conductance becomes a simple function of the capacitance between identical electrodes:

$$G = (\sigma/\epsilon) C \qquad (2)$$

The impedance (Z) for a series circuit can be calculated [Ma et al., 1991] as

$$Z = 1/(G+j2\pi fC) \qquad (3)$$

where f is the frequency of the applied signal. With applying a sufficiently high frequency AC voltage signal for suppressing undesirable electrode polarization and capacitance effects, the measured electrical impedance across the electrodes becomes essentially resistive. For a fixed probe configuration, the difference between the impedance of the pure liquid flow and gas-liquid flow can be directly related to the void fraction. The complete electrical circuit of an impedance probe and its application can be found in Ma et al. [1991] and Wang et al. [1991], respectively.

3.2 Applications

The number of applications of electrical conductance/resistance probes to gas-liquid bubbly flows is large, see reviews in Hewitt, 1978; Banerjee and Lahey, 1981; and Cartellier and Achard, 1991. For ring type electrodes, Chang et al. [1984] have provided theoretical equations for capacitance calculations for various two-phase flow patterns including bubbly flows. Using a double-sensor resistivity probe, Van Der Welle [1985] measured void fraction, bubble velocity, and bubble size for an air-water bubbly flow in a horizontal pipe section. This resistivity probe provided good accuracy for void fraction measurement but only fair measurement accuracy for bubble size determination. Modifying this probe, Kocamustafaogullari and Wang [1991] measured void fraction, interfacial area concentration, mean bubble diameter, and bubble interface velocity in air-water bubbly flows for void fractions from 4.3 to 22.5%. They found that the local bubble velocity and the mean bubble diameter generally increase with the air flow rate.

The major advantages of conductance probes include good sensitivity, ease of installation with multiple units, and low cost of equipment. The temperature and conductivity of the liquid must remain constant during the calibration and test experiments.

4.0 RADIATION ATTENUATION TECHNIQUES

Radiation attenuation techniques are non intrusive where they allow the measurement of void fraction without disturbing the flow. These techniques differ depending on the attenuation of the strength of photons (x- or γ-rays), electrons (β-ray), or neutrons [Teyssedout et al., 1992]:

(i) x-ray systems where a source of electromagnetic radiation in the range of 25-60 ke V is provided by an x-ray vacuum tube;

(ii) γ-ray systems where the source is a radio-isotope which emits photons with energies usually between 40 and 600 ke V;
(iii) β-ray systems which obtain a stream of electrons from radio-isotopes with energies up to 10 Me V; and
(iv) neutron systems where neutrons are supplied by a source in the range up to about 2 Me V.

The basic difference between these techniques lies in the differences in the attenuation laws and the hardware required to accomplish the measurement.

4.1 Measurement Principles

The basic principle for void fraction measurement using radiation attenuation techniques is as follows [Thiyagarajan et al, 1991]. When a collimated narrow beam of monoenergetic γ-rays (as an example) is passed through some matter of pathlength χ, the attenuated beam intensity I is given by

$$I = I_0 \exp(-\mu\chi) \tag{4}$$

where I_0 is the intensity of the incident beam and μ is the linear attenuation coefficient which is a characteristic of the material. Using relation (4) in a two-phase flow medium, with the respective linear attenuation coefficient and pathlength of each phase, the following relation for void fraction α can be obtained:

$$\alpha = \log(I/I_\ell)/\log(I_g/I_\ell) \tag{5}$$

where I is the attenuated γ-ray intensity of the two-phase flow medium and I_g and I_ℓ are the attenuated intensities when the gas or liquid phase alone is present throughout the probe volume. Since the measured void fraction is related to three intensities in relation (5), the statistical function of counts affects the result. Hence, point differentials are not adequate in themselves to completely describe the system behavior. For a cross sectional average void fraction, the PDF should be obtained by a method of successive averaging until stationary results are obtained. A typical arrangement for measurement of void fraction using attenuation techniques is shown in Figure 4.

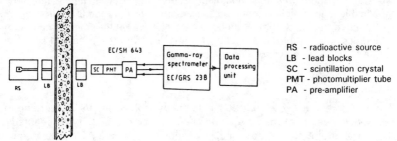

Figure 4 Typical Radiation Attenuation Measuring Setup [Thiyagarajan et al., 1991]

The radioactive source is placed in a lead container, producing a narrow beam of γ-rays. The source is shielded in all other directions. Between the source and the flow system, lead blocks with a central hole are used to obtain a collimated narrow beam. Similar lead blocks between the detector and the flow system are used to prevent scattered secondary radiation from reaching the detector. The detector unit is connected to a γ-ray spectrometer which has an attached data processing unit for analyzing the data.

4.2 Applications

Radiation attenuation techniques are widely used in nuclear engineering research due to their non-intrusive characteristics. Using a fast response, linearized x-ray void measurement system for air-water flow in a rectangular channel, Jones and Zuber [1975] have demonstrated that the PDF of void fraction fluctuations is a reliable technique for flow pattern recognition and quantitative void fraction measurement. Teyssedout et al. [1992] used a γ-ray absorption technique for measuring the void fraction of a steam-water two-phase flow under high-temperature and high-pressure conditions for simulating the flow regime of a boiling water reactor (BWR). Through use of an automated data acquisition system, they measured a wide range of void fractions with reasonable accuracy. Morooka et al. [1989] have used a x-ray CT scanner to measure the void fraction of a steam-water two-phase flow in a simulated vertical rod bundle of a BWR fuel assembly.

Although radiation attenuation techniques provide good accuracy in void fraction measurements and are non-intrusive, they have limited use, especially in hydraulic applications. Their cost, the complexity of the data acquisition system required, safety, potential for site license, and the fact that they provide a single point measurement, make them unsuitable for air-water reaeration research.

5.0 OPTICAL PROBES

An optical probe is sensitive to the change in the refractive index of the surrounding medium and is thus responsive to the interfacial passages which is applicable to local void fraction measurement. This method is applicable to non-conductive fluids. A typical probe consists of an optical fiber protected inside a stainless steel tube. The end part protrudes from the tube and is cut on an angle to act as the sensor (Figures 5.a and 6).

5.1 Measurement Principles

Light of core index n_f is emitted inside the optical probe, and, according to the refractive index n_k of the surrounding medium, some intensity is reflected back from the probe tip. The principle is normally defined by considering the sensitive tip to be completely immersed in a single phase, and by using the refractive law (Figure 5.b) [Cartellier and Achard, 1991]:

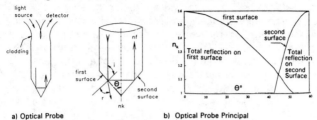

a) Optical Probe   b) Optical Probe Principal

Figure 5 Optical Probe Details and Principle [Cartellier et al., 1991]

$$n_f \sin i = n_k \sin r. \qquad (6)$$

The total reflection occurs for i larger than the critical angle i_c which is given by:

$$i_c = \arc \sin(n_k/n_f) \quad \text{for } n_k > n_f \qquad (7)$$

For air water two-phase flows, optimal detection is predicted for half tip angle (θ) of 43 to 51 degrees (Figure 5.b).

A typical optical method setup for measuring the void fraction is shown in Figure 6 [Minemura et al., 1993]. The measuring system consists of a semiconductor laser beam, coupler, photodiode, fiber connector, and the probe.

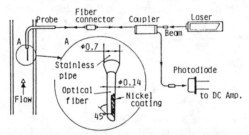

Figure 6 A typical Optical Probe Measuring Setup [Minemura et al., 1993]

When the inserted probe in the pipe is surrounded by the gas, the supplied laser beam is reflected and develops a voltage output to the photodiode for processing. The local void fraction is defined as the ratio of the time the gas phase was present over the probe to the total duration of the experiment.

5.2 Applications

Minemura et al. [1993] have used an optical probe to measure the local void fraction for an air-water bubbly flow in a straight channel rotating horizontally around an axis. Their experiments were intended to simulate the bubbly flow patterns in turbo-machinery. Cartellier [1992] has made simultaneous measurement of void fraction and bubble velocity in a vertical standing pipe for an air-water two-phase flow using an optical probe similar to that shown in Figure 5.a. Using signal processing techniques, he has determined the void fraction measurement is sensitive to the probe type, orientation of the probe, and the interface curvature. Using optical fibers and photo-sensitive devices, Haruyama [1987] has measured the void fraction of a two-phase helium bubbly flow. An advantage of optical probes is that the hydrodynamic and optical phenomena are uncoupled. Their disadvantages include single point measurement and higher cost of equipment compared to the electrical probes.

6.0 THERMAL ANEMOMETERS

This technique has been primarily developed as an extension of hot-wire or hot-film anemometry to two-phase flows. Thermal anemometers can simultaneously provide information on local void fraction in addition to the velocity and turbulence intensity measurements.

6.1 Measurement Principles

The distinct heat transfer capacity of gas and liquid is the basis for development of two-phase heat exchange methods. From analysis of the spikes of the signal due to impingement of a gas bubble on a hot wire, diameter histograms can be determined which become the basis for void fraction measurement [Cartellier and Achard, 1991]. Details of hot-wire anemometry technique can be found elsewhere [Bradshaw, 1975].

6.2 Applications

Although the thermal anemometers may seem attractive due to their capability of measuring few flow parameter simultaneously, they deliver signals which are difficult to use. The interpretation of such signals is still a subject for active research. The interaction between the probe and the pierced interfaces can misshape the signal causing difficulties in signal processing [Delhaye, 1981; Cartellier and Achard, 1991]. The probes are very sensitive to fouling and their calibration problem is well known to single phase anemometry.

7.0 SUMMARY AND DISCUSSION

This paper provides a review of void fraction measurement techniques for gas-liquid bubbly flows in closed conduits with emphasis on the suitability of these methods for air-water bubbly flows in reaeration applications. From the methods described, the electrical conductance/resistivity probes are considered to be the most suitable. The major advantages of these probes include good sensitivity, ease of installation with multiple units, and low cost of equipment.

Electrical wire sensors are a common instrument for hydraulic engineers used in measuring unsteady water levels and their capability for measuring the void fraction adds to their advantage. Signal processing has been addressed very briefly; however, such information is contained in most of the literature cited. Signal processing of two-sensor and four-sensor probes can provide additional information such as interfacial area, phase density functions, and bubble velocity.

8.0 REFERENCES

Andreussi, P.; Di Donfrancesco, A.; Messia, M.; 1988, "An impedance method for the measurement of liquid hold-up in two-phase flow", International Journal of Multiphase Flow, Vol. 14, No. 6, pp. 777-785.

Banerjee, S.; Lahey, R. T. Jr; 1981, "Advances in two-phase flow instrumentation", In Advances in Nuclear Science and Technology, Vol. 13 ,Edited by J. Lewins and M. Becker, pp. 227-414, Plenum Press, London.

Bradshaw, P.; 1975, An Introduction to Turbulence and its Measurement, Pergamon Press Inc., Elmsford, New York.

Brown, R.C.; Andreussi, P.; Zanelli, S.; 1978,"The use of wire probes for the measurement of liquid film thickness in annular gas-liquid flows", Can. J. Chem. Engng, Vol. 56,pp. 754-757.

Cartellier, A.; 1992, "Simultaneous void fraction measurement, bubble velocity, and size estimate using a single optical probe in gas-liquid two-phase flows", Review of Scientific Instruments, Vol. 63, No. 11, November, pp. 5442-5453.

Cartellier, A.; Achard, J.L.; 1991, "Local Phase Detection Probes in Fluid/Fluid Two-Phase Flows", Review of Sci. Instruments, Vol. 62, No. 2, pp. 279-303.

Chang, J. S.; Girard, R.; Raman, R.; Tran, F. B. P.; 1984, "Measurement of void fraction in vertical gas-liquid two-phase flow by ring type capacitance transducers", Mass Flow Measurement, Eds. Hedrick and Reimer, Vol. FED-17, ASME press, New York, pp. 93-99.

Delhaye, J. M.; 1981, "Instrumentation", A chapter in Two-Phase Flow and heat Transfer in the Power and Process Industries, A.E. Bergeles, J.G. Collier, J.M. Delhaye, G.F. Hewitt, and F. Mayinger authors, Hemisphere Publishing Company, New York.

Delhaye, J. M.; Jones, O. C.; 1975, "Measurement Techniques for Transient and Statistical Studies of Two-Phase, Gas-Liquid Flows", ASME paper 75-HT-10.

Halliday, D.; Resnick, R.; 1978, Physics, Parts 1&2, 3rd Ed., John Wiley & Sons, New York.

Haruyama, T.; 1987, "Optical method for measurement of quality and flow patterns in helium two-phase flow", Cryogenics, Vol. 27, August, pp. 450-453.

Hewitt, G.F.; 1978, Measurement of Two-Phase Flow Parameters, Academic Press,London, UK.

Ishigai, S.; Yamane, M.; Roko, K.; 1965, "Measurement of the component flows in a vertical two-phase flow by making use of the pressure fluctuations (Part 1)", Bulletin of JSME, Vol. 8, No. 31, pp. 375-382.

Jones. O. C.; Zuber, N.; 1975, "The Interrelation Between Void Fraction Fluctuations and Flow Patterns in Two-Phase Flow", Int. J. of Multiphase Flow, Vol. 2, pp. 273-306.

Kocamustafaogullari, G.; Wang, Z.; 1991, "An Experimental Study on Local Interfacial Parameters in a Horizontal Bubbly Two-Phase Flow", Int. J. of Multiphase Flow, Vol. 17, No. 5, pp. 553-572.

Lafferty, J. F.; Hammitt, F. G.; 1967, "A conductivity probe for measuring local void fractions in two-phase flow", Nuclear Applications, Vol. 3, May, pp. 317-323.

Ma, Y-P; Chung, N-M; Pei, B-S; Lin, W-K; Hsu, Y-Y; 1991, "Two simplified methods to determine void fractions for two-phase flow", Nuclear Technology, Vol. 94, April, pp. 124-133.

Minemura, K.; Uchiyama, T.; Ishikawa, T.; 1993, "Experimental investigations on bubbly flows in a straight channel rotated around an axis perpendicular to the channel", International Journal of Multiphase Flow, Vol. 19, No. 3, pp. 439-450.

Morooka, S.; Ishizuka, T.; Iizuka, M.; Yoshimura, K.; 1989. "Experimental study on void fraction in a simulated BWR fuel assembly (evaluation of cross-sectional averaged void fraction)", Nuclear Engineering and Design, Vol. 114, pp. 91-98.

Naghash, M.; 1989, "Steady Turbulent Gas Desorption in a Horizontal Pipeline", Ph.D. Dissertation, School of Civil Engineering, Georgia Institute of Technology, Atlanta, GA.

Revankar, S.T.; Ishii, M.; 1993, "Theory and measurement of local interfacial area using a four sensor probe in two-phase flow", International Journal of Heat and Mass Transfer, Vol. 36, No. 12, pp. 2997-3007.

Teyssedout, A.; Aube, F.; Champagne, P.; 1992, "Void fraction measurement system for high temperature flows", Measurement Science & Technology, Vol. 3, pp. 485-494.

Thiyagarajan, T. K.; Dixit, N. S.; Satyamurthy, P.; Venkatramani, N.; Rohatgi, V. K.; 1991, " Gamma-ray attenuation method for void fraction measurement in fluctuating two-phase liquid metal flows", Measurement Science & Technology, Vol. 2, pp. 69-74.

Van Der Welle, R.; 1985, "Void Fraction, Bubble Velocity and Bubble Size in Two-Phase Flow", International Journal of Multiphase Flow, Vol. 11, No. 3, pp. 317-345.

Wallis, G. B.; 1969, One Dimensional Two-Phase Flow, McGraw-Hill, New York.

Wang, Y. W.; Pei, B. S.; Lin, W. K.; 1991, " Verification of using a single void fraction sensor to identify two-phase flow patterns", Nuclear Technology, Vol. 95, July, pp. 87-94.

ADDITIONAL REFERENCES

Chang, J. S.; Morala, E.C.; 1990, "Determination of two-phase interfacial areas by an ultrasonic technique", Nuclear Engineering and Design, Vol. 122, pp. 143-156.

Coney, M. W. E.; 1973, "The Theory and Application of Conductance Probes for the Measurement of Liquid Film Thickness in Two Phase Flow", J. of Physics E: Scientific Instr., Vol. 6, pp. 903-910.

Kang, H. C.; Kim, M. H.; 1992, "The development of a flush-wire probe and calibration method for measuring liquid film thickness", Int. J. of Multiphase Flow, Vol. 18, No. 3, pp. 423-437.

Kataoka, I.; Ishii, M.; Serizawa, A.; 1986, "Local formulation and measurements of interfacial area concentration in two-phase flow", Int. J. of Multiphase Flow, Vol. 12, No. 4, pp. 505-529.

Klausner, J. F.; Zeng, L. Z.; Bernhard, D. M.; 1992, "Development of a film thickness probe using capacitance for asymmetrical two-phase flow with heat addition", Review of Scientific Instruments, Vol. 63, No. 5, May, pp. 3147-3152.

Sekoguchi, K.; Inoue, K.; Imasaka, T.; 1987, "Void signal analysis and gas-liquid two-phase flow regime determination by a statistical pattern recognition method", JSME International Journal, Vol. 30, No. 266, pp. 1266-1273.

Serizawa, A.; Kataoka, I.; Michyoshi, I.; 1975, "Turbulent Structure of Air-Water Bubbly Flow - I. Measuring Techniques", Int. J. of Multiphase Flow, Vol. 2, pp. 221-233.

Sheng, Y.Y.; Irons, G.A.; 1991, "A Combined Laser Doppler Anemometry and Electrical Probe Diagnostic for Bubbly Two-Phase Flow", Int. J. of Multi. Flow, Vol. 17, No. 5, pp. 585-598.

Tsochatzidis, N. A.; Karapantsios, T. D.; Kostoglou, M. V.; Karabelas, A. J.; 1992, "A conductance probe for measuring liquid fraction in pipes and packed beds", International Journal of Multiphase Flow, Vol. 18, No. 5, pp. 653-667.

Wang, Y. W.; Pei, B. S.; King, C. H.; Lee, S. C.; 1990, "Identification of two-phase flow patterns in a nuclear reactor by high-frequency contribution fraction", Nuclear Technology, Vol. 89, February, pp. 217-226.

Xie, C.G.; Stott, A.L.; Plaskowski, A.; Beck, M.S.; 1990, "Design of capacitance electrodes for concentration measurement of two-phase flow", Measurement Sci. & Tech., Vol. 1, pp. 65-78.

Measuring Air Concentration in Flowing Air-Water Mixtures

By: Boualem Hadjerioua[1], Tony A. Rizk[2] Member, ASCE, Emmett M. Laursen[3] Member, ASCE, and Margaret S. Petersen[4] Honorary Member, ASCE.

Abstract

This paper describes the use of an electrical conductivity probe to determine aeration of a plunge pool by an overfall nappe. The probe was calibrated using an apparatus designed to provide known concentrations of an air-water mixture with controlled bubble sizes. The calibrated probe was then used to measure air concentration in a full-scale physical model of an aerating weir with different nappe impinging velocities, unit discharges, and tailwater depths. Calibration curves and measured representative plunge pool air concentrations are presented.

Introduction

As a result of air entrainment by a plunging liquid nappe, bubbles are dispersed below the liquid surface in a pool to some maximum depth related to the flow pattern of the submerged jet. The process is known as self-aeration.

The earliest reported investigations of air concentrations in self-aerated flow used mechanical samplers. Principal disadvantages of mechanical samplers are:

---

[1] Graduate Student Intern, Tennessee Valley Authority, Eng. Laboratory, Norris, Tennessee.

[2] Tennessee Valley Authority, Engineering Laboratory, Norris, Tennessee.

[3] Emeritus Professor, Dept. of Civil Eng., and Eng. Mech., University of Arizona, Tucson, Arizona.

[4] Emeritus Associate Professor, Dept. of Civil Eng., and Eng. Mech., University of Arizona, Tucson, Arizona.

(1) physical dimensions of the device limit the range in which accurate readings can be obtained (the air chamber fills in a very short time at high air concentrations and does not fill uniformly at low air concentrations), (2) the time required to obtain the correct aspiration setting and to balance the system is about 30 minutes for every point, and (3) it is not possible to study short-time variations in air concentration.

Lamb and Killen (1950) introduced the first electrical probe for measuring air concentration in air-water mixtures on a very steep chute. It was based on the premise that the electrical conductivity of a suspension of particles in a fluid varies with the relative amount of suspended material. The method basically consists of measuring the difference in conductivity between a mixture of air and water and that of water alone. A pair of electrodes mounted on a mechanical strut is combined with electrical circuitry in a manner so that air-concentration measurements can be made in a small volume surrounding the electrodes.

Experimental Investigation

The probe used in this investigation was fabricated in the TVA Laboratory several years ago, following the concept suggested by Lamb and Killen (1950), but had not been calibrated. Size and spacing of the electrodes were selected to ensure a uniform field between electrodes and to provide a large enough gap for representative bubbles to pass between so that conductivity across the gap would not shift over the full scale (measuring only air) during passage of a single bubble. The electrodes are small enough to measure a very limited region of flow and cause minimum disturbance, but have sufficient rigidity to prevent excessive vibration.

A setup was designed and constructed to calibrate the probe before it was used to measure air concentration in the plunge pool of an impinging nappe.

Probe Calibration

The calibration apparatus is shown schematically in Figure 1. The setup included a Plexiglas pipe 0.125 meter in diameter, with perforated plates (screens) just downstream of the air injection to assist in distributing the bubbles uniformly. Two small pumps used to supply water to the system were regulated by a manual valve to provide the desired flowrates. Air was supplied by a compressor, regulated by a manual valve to control the air flowrates.

A constant water discharge of 50 gallons per minute (gpm) was used throughout the calibration procedure. Air was injected into water flowing in the Plexiglas pipe through two perforated tubes, each time at a different volume fraction of the water flow rate. Holes in the tubes were of various diameters; 2, 4, and 6 mm. For each run 20 readings were taken at different locations across the pipe and averaged to minimize measurement errors in the air concentrations.

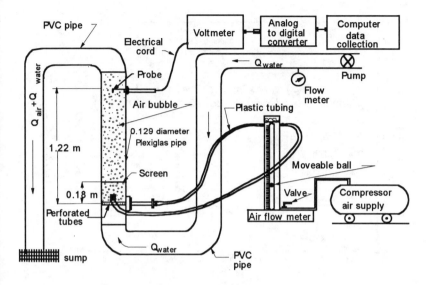

Figure 1. Probe Calibration Setup

Results of the Probe Calibration

As shown in Figure 2, the effects of injecting air through different-sized holes in the injection tubes was not significant. As the bubbles travelled through the perforated plates and through the water column they broke up and/or coalesced. As expected, calibration of the probe gave consistent readings for the entire set of runs. A least squares cubic regression equation gave the best fit for the experimental results, and was used to compute plunge pool air concentration contours.

Plunge-pool Experimental Setup

The plunge pool testing apparatus consisted of a 1-m wide, 12-m long flume, sharp-crested weir; adjustable tailgate; and the probe, Figure 3.

The flume was equipped with windows 2.74-m long downstream of the weir. A grid was marked on the windows for visually determining the locations of the hydraulic characteristics. The tailgate was adjusted manually to control the tailwater elevation required for each experiment.

A sharp-crested weir, 2.13-m high, was used. To avoid depletion of air beneath the lower surface of the nappe, two 1.3-cm diameter holes were used to provide a continuous air supply below the nappe. Water discharge was measured with an electromagnetic flow meter.

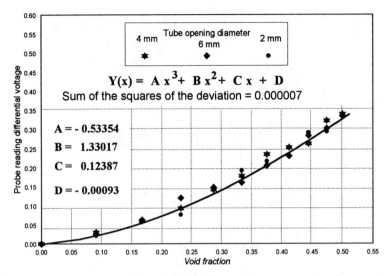

Figure 2 Probe Calibration Curve

Testing Procedure

Objectives of testing in the plunge pool were to observe the behavior of entrained air bubbles and dynamics of the plunge pool, and to measure the distribution of air concentration.

The ranges of discharge and drop height tested were based on the fact that the optimum rate of aeration by a straight weir is achieved with a unit discharge of about 0.065 m^2/s with a drop height of about 1.20 m (Nakasone, 1987).

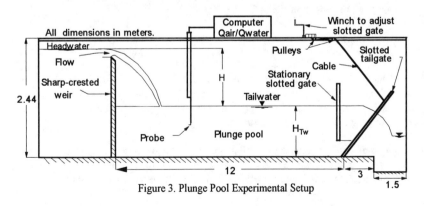

Figure 3. Plunge Pool Experimental Setup

Bubble Characteristics

High-speed photography was used to determine average bubble size in the plunge pool. The mean diameter, Figure 4, represents an equivalent spherical shape having the same volume as the ellipsoidal bubble. Numerous visual observations and measurements of bubble sizes in the plunge pool were made (Hadjerioua et al., 1994). Large bubbles, occurring at the location of the impinging nappe, escape through the water surface very quickly. The smallest bubbles are near the bottom of the submerged jet where shear forces are maximum. Away from that zone, the bubbles either coalesce or break-up and reach a state of equilibrium. The high turbulence intensity and shear stress created by the bubbles themselves tend to promote bubble break-up.

At the maximum depth of penetration, buoyancy forces overcome the momentum imparted to the bubbles by the jet. The bubbles decrease in size as they move deeper due to shear forces (pressure forces are not significant for small depths). Bubbles tend to segregate by size, with larger bubbles rising and escaping more quickly than smaller bubbles.

The structure of the "bubbly region" is very complex, and understanding the mixing processes is essential for determining rate of oxygenation. The flow pattern in the pool is similar to that caused by a submerged jet in an infinite fluid, and mixing is probably enhanced as a result of bubble dynamics in the pool. With increasing velocity of the impinging nappe, the depth of bubble penetration increases. Since the pool is of finite size, a circulation pattern is created. One of the most important aspects of the oxygen transfer process is the size and composition of the submerged aerated region.

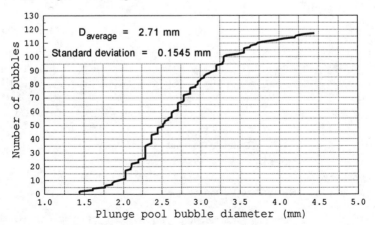

Figure 4. Plunge Pool Bubble Size Distribution

Plunge Pool Air Concentration Contours

Air concentration contours throughout the plunge pool were determined for each test based on probe measurements. Each data point is the average of 20 concentration measurements (one reading per second) at each location. Typical results are shown in Figure 5 for a drop height of 1.067 m and unit discharge of 0.05 m²/s.

The amount of air dragged into the pool, per unit length of the crest, was computed from

$$q_{air} = \frac{q_{water} \; \alpha_{max}}{1 - \alpha_{max}}$$

where q_{air} and q_{water} are the air and water unit discharge, respectively, and α_{max} is maximum void fraction in the plunge pool which occurs near the point of nappe impingement. The probe is very sensitive, and the readings are inacurate if measurements are taken too close to the surface. Therefore, the α_{max} measured is conservative, and the amount of air entrained by the nappe is underestimated to some extent.

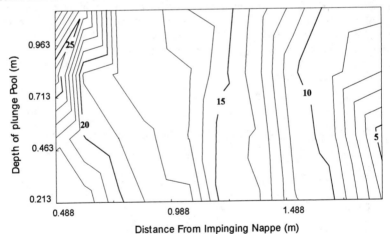

Figure 5. Plunge Pool Air Concentration Contours in Percent

Advantages and Disadvantages of the Probe

The principal advantages of the probe are ease and simplicity of use and accuracy of results.

The main disadvantage is the initial time required to reach equilibrium. Once the probe is placed in the water, it takes many hours to reach stability. Therefore, the probe must remain in the water until all measurements are completed. Also, because the probe is very sensitive, readings are inaccurate when it is too close to the metal bottom of the flume or to the water surface.

Conclusions and Recommendations

The electrical conductivity probe is a reliable instrument for measuring air concentrations in a self-aerated plunge pool.

Data collected in this study show good agreement with previous experimental data collected by others (Ervine and Elsawy, 1975). For future use, such a probe should be calibrated for different water velocities and for buoyancy effects (water flowing downward and bubbles rising in the tube).

Acknowledgements

The authors would like to thank the following people at the TVA Engineering Laboratory for their support and assistance: M. Mobley and G. Hauser for providing funding for the design of the calibration setup and testing; H. Pearson for design of the probe and help with instrumentation, and P. Hopping for his helpful suggestions for the design of the calibration setup.

APPENDIX I. REFERENCES

Ervine, D. A., and E. M. Elsawy (1975). "The Effect of a Falling Nappe on River Aeration," Proc. 16th Congress I.A.H.R., Sao Paulo, Brazil, Vol. 3, pp 390-397.

Hadjerioua, B., E. M. Laursen, M. S. Petersen, and T. Rizk (1994). "Air Entrainment and Bubble Behavior in Plunge Pools," ASCE National Conference on Hydraulic Engineering, Buffalo, New York.

Lamb P. O., and M. J. Killen (1950). "An Electrical Method for Measuring Air Concentration in Flowing Air-Water Mixtures," St. Anthony Falls Hydraulic Laboratory, University of Minnesota.

Nakasone, H. (1987). "Study of Aeration at Weirs and Cascades", ASCE Journal of Environmental Engineering, Vol. 113, No. 1.

APPLICATION OF A NEEDLE PROBE IN MEASURING LOCAL PARAMETERS IN AIR-WATER FLOW

A.R.ZARRATI[1] and J.D.HARDWICK[2]

Abstract

In this paper, development of a double needle probe for measuring local parameters in air-water mixtures is explained. This probe is designed to measure flow velocities up to 10 m/s and bubbles as small as 1 mm. Careful programming made it possible to collect up to 2 minutes of data with a speed of 52 KHz from 2 channel.

Introduction

The measurement of bubble velocity, bubble size and air concentration is essential for studies of air-water mixtures. It is known that conventional flow metering instruments are not practicable in two-phase flows with markedly different densities. It appears from the literature that a resistivity probe similar to that used by Cain and Wood (1981) is the most promising because it is relatively easy to use and widely applicable (Herringe and Davis, 1974). This probe was redesigned and further developed in the present study.

Principles of the Resistivity Probe (Needle Probe)

The resistance of water is one thousand times lower than that of air bubbles. Therefore, when a probe consisting of two conductors with a potential difference across them and separated by a small distance, is placed in a flowing air-water mixture the resistance of the probe will increase as it pierces an air bubble and the

---

[1] Department of Civil Engineering, AmirKabir University of Technology, Hafez Ave, Tehran, Iran.

[2] Department of Civil Engineering, Imperial College of Science Technology and Medicine, London, U.K.

voltage drops. The voltage will remain at this low level until the bubble has been swept beyond the probe when it will then rise to its former value (Figure 1). Through analysis of the probe signals air concentration, bubble velocity and size may be determined as follows:

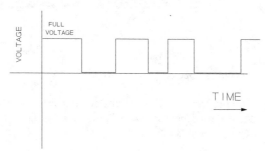

Figure 1- The typical output signal from a probe

Air concentration may be predicted by adding the gas contact time over a total sampling time. If the tips of two resistivity probes are aligned in the direction of the flow then the bubble velocity can be inferred from the time that an air-water interface takes to travel between the first and second probe tips.

Ideally, the signals from the two tips would be identical but separated by a time delay. In practice they differ because the upstream probe disturbs the flow to some extent and consequently the *probable* time delay is found by cross-correlating the two signals.

As a bubble approaches the probe, it is pierced by the probe tip and the electrical circuit will remain disconnected until the bubble passes. The time of disconnection is representative of the bubble size, and provided the velocity is known, the bubble size can be estimated (Neal and Benkoff, 1963).

Design of the Probe

The dimensions of the tips of the resistivity probes and their spacing depend on the flow velocity, bubble sizes to be detected and the data acquisition system. In this work, the probe was designed for a velocity up to 10 m/s and bubble sizes down to about 1 mm. It was reasoned that if the probe is to pierce a bubble, its tip should be at least 5 times smaller than the bubble diameter. A 0.14 mm insulated wire in an alloy of copper and steel was used as the probe tip and located within a surgical needle of 0.25 mm outside diameter (Figure 2).

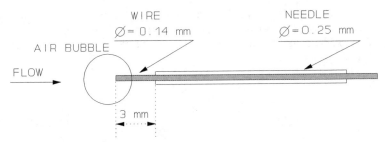

Figure 2- Needle probe dimensions

The distance between the upstream and downstream needles depends on the flow velocity and data acquisition speed. A lower sampling speed rate will demand a greater distance between the probes.

Data Acquisition System

The analog signals from the probe progressed through an analog-to-digital convertor, (A/D), to the computer storage. The speed of data collection must be sufficiently high to provide a reliable representation of the analog signals. This probe was designed for up to 10 m/s flow velocity and bubbles as small as 1 mm, and consequently the corresponding signal of such a bubble would have 0.1 ms width. To record such a wave, the frequency of data collection should be at least 20 KHz. For the present work, a 26 KHz frequency was used for data collection.

Time of Data Collection

The analysis of data was based on the statistical treatment of probe signals and was valid only if the sampling time was sufficiently long. It is not possible to be precise about a minimum sampling time because it will depend on the characteristics of the flow in each experiment. Serizawa et al (1975) examined a bubbly flow in a vertical pipe with a velocity of the order of 1 m/s and suggested a sampling time of 1 to 3 minutes to obtain reliable results. The main difficulty for any researcher in this field is the limited capacity of a computer Random Access Memory (RAM). Chanson (1988) used a 64 kbyte computer to collect data from two channels (one for each probe). With a sampling frequency of 10 KHz in each channel he was restricted to a very short sampling time of 1.2 seconds. In this study a 80386 personal computer with 640 kbytes RAM capacity was used. Even with this computer the number of data samples which could be stored for each measurement was limited to 400,000. With a 26 KHz frequency of data collection and two channels, the total sampling time was limited to 7.7 seconds. This was considered unacceptably short.

The output signal from the probe varies only between two values when the probe is either in air or in water. The A/D used in this work allocated 16 bits (2 bytes) for any number, with the more significant bit (MSB) reserved for the sign. Since the numbers to be stored were only varying between two figures it was considered unnecessary to devote 16 bits for each number. Therefore, 1 was assigned to the maximum voltage and 0 to the lower level voltage. When the 15 such bits had been stored, the completed bytes were sent to the computer memory to be stored as a single figure. In this way the effective storage space was increased 15 fold.

Calibration of the Needle Probe

To calibrate the needle probe for velocity measurement air was injected through a tapping in the bed of a flow channel such that only the central region of the stream was aerated. The needle probe was located in this region and aligned with the general trajectory of the air bubbles. The local velocity was also measured with a total head probe located just outside the boundary of the plume. Many sets of data were analyzed over a velocity range of 1.5 m/s to 6 m/s and discrepancies between the two methods were rarely found to exceed 5%.

Further tests were conducted to check the probe. In these tests air was injected from a slot at the channel bed while velocity and air concentration were measured simultaneously at various depths in the deaeration zone downstream of the slot. The flow discharge was then calculated from the following equation:

$$q_w = \int_0^H (1-c) \, U dy$$

where H is the flow depth, c is volumetric air concentration and U is the flow velocity. The flow discharge calculated in this way was compared with the measurements and the discrepancy was less than 6%.

Conclusion

The use of a double needle probe with a high speed sampling arrangement and careful programming made it possible to store a large quantity of data. These data could later be analyzed to provide more information about air-water mixtures than has so far been available with more conventional measuring systems.

Acknowledgements

The authors wish to express their gratitude to Dr. Mohammad Movahedi and Dr. Ahmad Kardan for their helpful advice throughout this study.

Appendix I. References

Cain P, and Wood I R, (1981), "Instrumentation for Aerated Flow on Spillways", ASCE, HY11, pp. 1407-1423.

Chanson H, (1988), "A Model Study of Aerator Performance", Thesis submitted to the university of Canterbury for the degree of PhD (Civil Engineering).

Herringe R A, and Davis M R, (1974), "Detection of Instantaneous Phase Change in Gas-Liquid Mixtures", Journal of Physics E:Scientific Instruments, Vol 7, pp. 807-812.

Neal, A. and Benkoff, S.A., (1963), "A High Resolution Resistivity Probe for Determination of Local Void Properties in Gas-Liquid Flow", J. of A.I.Ch.E., pp. 490-494.

Serizawa A, Kataoka I, Michiyoshi I, (1975), "Turbulence Structure of Air-Water Bubbly Flow-I, Measuring Technique", International Journal of Mutiphase Flow, vol 2, pp. 221-233.

A New Basic Principle for a New Series of Hydraulic Measurements. Erosion by Abrasion, Corrosion, Cavitation and Sediment Concentration

Lucien Chincholle[1]

Abstract - The DECAVER is a novel apparatus able to make new measurements in various domains where an erosion is produced in an aqueous medium. It gives the erosion rate and the accumulated erosion. This device finds using in hydraulic machinery and in sediment transport, both in laboratory and in field. One of the main advantages of the sensor is that its size can be adjusted to any place. The time response characteristics indicate that it's capable to make instantaneous measurements quite accurately.

Keywords: measurements, abrasion, cavitation, sediment transport

Introduction

Generally, a novel instrument is consecutive to a displayed physical phenomenon that is little or not at all known. It involves a new sensor. Our work, relative to bubble mechanics in liquids has been developed in two stages. First, in 1966, we display which we call "Rocket Effect" that points out very considerable local energies and pressures, able to create a big erosion [1]. Then in 1982, we show some properties of the passive layer on metals [2]. A novel measuring device had resulted from this. It's named DECAVER (DEtection CAVitation and ERosion). It concerns different hydraulic fields. Now, we have only to use it, on various experimental devices, to obtain interesting results. We have particularly investigated the erosion scope (mechanical abrasion, electrochemical corrosion, cavitation) and sediment transport.
The aim of experimental results that we present, is not to study particular subjects but to show possibilities of the DECAVER in various domains.

---

[1] Emeritus Professor, Laboratoire de Génie Electrique de Paris, ESE, Plateau du Moulon, 91190 Gif sur Yvette, France

1. - Device

1.1. - Principle

The DECAVER is a novel electrochemical detector based on a particular property of the passive layer [3]. Metals are covered, particularly in an aqueous medium, with a oxide layer characterized by a semiconductor aspect. When this layer is destroyed in any erosion, it's instantaneously reconstructed. This process sets electrons free. The device collects the corresponding current. As this one is proportional to the erosion rate, it can be used to measure the physical phenomenon of the erosion. We measure a corrosion current.

Each time an erosion is produced on the DECAVER sensor, whether by mechanical abrasion or corrosion or cavitation, the device gives an electric signal. After a specific calibration, we deduce the erosion rate, the total erosion, the sediment concentration or the sediment transport rate.

1.2. - Device calibration [3]

The calibration consists to establish a correlation between the electric signal and the mass loss rate of the sensor that is obtained by weightings. This operation asks to work in strict same experimental conditions.

As the signal intensity depends on the surface value of a particle, we have to know the corresponding removed material that is not oxidized. This one varies owing to the different working techniques. A specific calibration will be necessary to make able DECAVER measurements.

Calibration curves have been drawn for any type of erosion.

Electrochemical corrosion - This calibration is ruled by the Faraday Law.

Mechanical abrasion by a grinding stone - A calibration was established in our laboratory for given experimental conditions: a sensor of 1 mm diameter, standard metal, given pressure and stone wheel velocity.

Cavitation erosion - The calibration working concerns both a *cavitation channel* and a *vibratory device*. With the cavitation channel, for example, it results a specific curve that presents two parts when it is placed in the erosion condition (Figure 1). The first one is a straight line corresponding to an erosion without loss of material (incubation time). Only the surface is distorted. The reconstitution of the passive layer that is destroyed in some places, involves an electricity quantity. Then, the other line corresponds to a current that is proportional to the erosion rate. A relation exists between the current signal and the particle mean diameter.

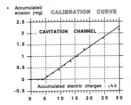

Figure 1 - Calibration curve on a cavitation canal

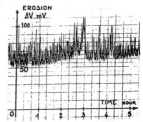

Figure 2 - Sand erosion on a hydraulic turbine

1.3. - Measuring instrument

1.3.1. - *Industrial device.*

Two laboratory prototypes, built nine years ago, work satisfactorily. Today, industrial devices are available. Characteristics of the device output signal are interesting.

- The device gives the current value. This one (i µA), having a velocity dimension (Coulomb/second), corresponds to an erosion rate value (mg/s or mm/year).
- These values being instantaneous, we shall know the erosion rate at any moment. It's interesting to record these parameter variations. So, concerning the erosion, we can see the evolution of any physical phenomenon. Perhaps, it shall result a best understanding of it.
- By integration of the electric current signal, we obtain a quantity of electricity (It) equal to: $\int_{t0}^{t1} i \, dt$. This relation shall give the corresponding value of the total erosion that is whether the accumulated erosion or the mean penetration depth when we know eroded surface value. On the device, a counter assumes this task.
- The device is very sensitive. We can easily detect about 0.01 µA, because we have demonstrated that the current issued from the corrosion region is of a current-source type [2].

1.3.2.- *Sensor setting up*

Three electric wires connect the DECAVER to the equipment. For the auxiliary and reference electrodes, there is no problem. Concerning the sensor, the knowledge of certain information is necessary. The sensor is level with the surface to be observed; so, there is no flowing disturbance. It is made with a standard metal or any metal. Its surface depends on the objective, either a wire about one millimeter in diameter or a stuck sheet of any surface, or the piece itself on condition that it's electrically insulated from the ground. Several sensors can be placed on the surface and successively connected to the DECAVER. An important point is the probe activation (incubation time) when the cavitation erosion rate is weak or when the measurement is required immediately and for the first time.

1.3.3.- *Erosion Scale* [3]

To simplify the use of the DECAVER and to generalize its calibration, we also propose a Cavitation Erosion Scale, similar to the Richter Scale for seisms. It permits us to establish a correspondence between any erosion and a number from 1 to 10. At the same time, the erosion relative to each material is pointed out. At a glance, all informations on erosion rate, choice of material, life span and technical visit periods are given. For example, by using titanium as a basis, an erosive power 6, read on the DECAVER, corresponds both to 1 µA/cm^2 current density and to an erosion of 1 mm per year. For different materials (metal, ceramic, concrete, etc.) we can make a comparison by way of two successive tests between titanium and the considered material with the same experimental conditions. So, stainless steel offers a resistance five times greater than titanium.

2. - Erosion in Hydraulic Machinery

2.1. - Different Types of Erosion

2.1.1. - Erosion definition
In this paper, we only consider the erosion on solid materials, in an aqueous medium. For us, the erosion is the sum of mechanical abrasion, electrochemical corrosion, cavitation erosion and synergistic effect between them. The last one is often very consequent. We have developed many tests in various fields such as cavitation canal, hydraulic machinery, water jets and corrosion. Every one knows the corrosion that is ruled by the Faraday Law. We have do not forget that the corrosion effect takes place in each metal erosion and that the device measures a corrosion current.

2.1.2. Mechanical erosion
Many experiments have been conducted by us, particularly with a grindstone, jet, sand or with erosive sponge balls inside a condenser tube. This study is complicated owing to many parameters.
Abrasion by a grindstone - The grindstone works as a gouge. Only the surface of the torn out metal ribbon is oxidized. The total removed material depends on grindstone and material characteristics.
Erosion by sand - Owing to its velocity, the erosion by sand resembles to the precedent one or it can be due to shocks that divide material into ejected particles.
Erosion by jet with or without sand - In this case the erosion is of same type or of a cavitation one.
2.1.3. Erosion by cavitation - The material is principally removed by imploding cavities' shocks.

2.2. - Experimental Results

2.2.1. - Experimental facilities
Cavitation canal - Hydraulic turbine - Only we present some results relative to a cavitation canal and a hydraulic turbine (2 MW). The cavitation canal is the principal device that permits us to develop the cavitation erosion study by using the DECAVER. The cavitation is created by a cylinder placed in the flow along the cross section of the pipe. It evolutes when various parameters vary: pressure (P), drop in head (ΔP), velocity (V) or flow rate (Q), temperature (T), dissolved gases, etc. . Variations in erosion and various parameters are recorded while only one parameter varies. As the cylinder is also the sensor it results that the output signal corresponds to the total erosion on it. During the pressure evolution, for instance from 6 bars to 3 bars, curves present a maximum erosion and two erosion thresholds (Figure 3).
Erosive jet tests. In the case of jet devices we have experimented with successively clear water, water with sand and high velocity water. Using clear water we have worked up to 450 bars. The experimental device consists of a jet in front of a target. Parameters are the jet length and pressure. Tests were developed at the Institut Français du Pétrole (450 bars - 10 seconds for one test). In this case we have a cavitation erosion.

2.2.2. - *Experimental results. Abrasion and Cavitation*

On the cavitation canal, we have obtained a series of 230 runs giving 230 curves. They are *Cavitation erosion characteristic curves*. These curves present a great importance because they permit to know the erosion rate continually and to choice the good working point for a hydraulic machine (Figure 3).

By considering successively the three following parameters: P, T and ΔP that is proportional to V^2, we have deduced the erosion variations versus each one and as a function of $P/\Delta P$, $(P-p_v)/0.5\rho V^2$, and σ_c.

$$\sigma_c = \frac{P}{\Delta P} + \frac{1}{T} - .00565\ T$$

This relation corresponds to the maximum erosion.

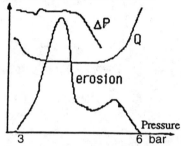

Figure 3 - Erosion characteristic curve when the pressure decreases from 6 to 3 bars

- For hydraulic turbine testings, the operation mode is easy. The wicket-gates open, the wheel starts and slowly, goes up to the nominal power. The DECAVER gives variations in erosion rate. There is a maximum erosion rate at 55 % of nominal power. Results are similar to the cavitation channel ones. Experiments on a hydraulic turbine have been carried out during a period where the water was very dirty. On figure 2, it can be seen, without cavitation erosion, an abrasion erosion due to the sand in the water, specially during quarter of an hour periods when there are sudden showers.

2.3. - Remarks

2.3.1. - *Stochastic analysis presentation of experimental results* [4]

- <u>Fundamental and secondary parameters</u> - Only considering maximum erosion points (230 runs) which are well defined, we deduced some specific relations conducing to erosion laws. For instance, the ratio $P/\Delta P$ stays constant for a fixed temperature. More, the role of the temperature was shown. It can be said that *fundamental parameters* are temperature, pressure and drop in head. They characterize one working point.

Secondary parameters define the erosion intensity for a given working point; there are the pressure, the gas content, etc.

2.3.2. - Cavitation Erosion Coefficient: $\sigma_C = P/\Delta P + 1/T - 0.00565T$
For us the more important interest of this notion is to have a specific point. It can be used by researchers to characterize all experimental results in cavitation erosion on a hydraulic device. This type of relation permits also to engineers to define the instantaneous erosion rate in any working point.

2.3.3. - *Flow Erosive Intensity*
The notion of the F.E.I. (Flow Erosive Intensity) seems to us very useful. It shows that the erosion depends first and foremost on the flowing properties. This important value can be measured instantaneously by the DECAVER. We note that this definition and this measure concern any erosion. Only the device detects the total erosion independently of the type of erosion.

2.3.4. - *Applications in hydraulic machinery - Jets*
Using precedent results and tests relative to erosion by ultrasounds, we have deduced several facts:
. the DECAVER calibration permits to foresee cavitation erosion and to choice the most convenient material,
. the Erosion Scale graph gives directly the Flow Erosive Intensity by an elementary reading of the DECAVER,
. from the measured current, by calculation, we obtain particular informations relative to ejected particles: number, mean diameter, also the thickness of the passive layer.

3. - Sediment Concentration Measuring

The DECAVER can measure the instantaneous sediment transport rate and concentrations. Now, various mechanical, optical, acoustical and impact measuring instruments are used to study sediment transport and concentration but with certain difficulties. Though our work is in progress, we shall try to present the DECAVER as a new measurement technique in this domain. The concentration is defined as the ratio:

$$C\% = \frac{\text{sediment volume}}{\text{total volume}}$$

3.1. Measuring device
3.1.1. *Fundamental aspect*
We have seen that the sensor of the device was based on the ability of metal to create a passive layer on its surface when it's placed in an aqueous mdium. When this layer is destroyed by mechanical abrasion due to sediment, an associated current can be measured. This one depends on particles that strike the sensor with a certain velocity. So, this sensor can be used to measure the concentration and the instantaneous sediment transport rates. Only, we have to proceed to a calibration. Considering sediments in water, evidently particles must strike the sensor to erode it if their kinetic energy is sufficient. It appears that we shall have various parameters to consider:
- The kinetic energy is $W = 1/2 \, mV^2$.
- The particle impact velocity is not always the same that the water one, particularly when particles strike the sensor.

- The mass of particle flux depends on the specific mass but also of the concentration and of the velocity. So, we can write the kinetic energy under relation:

$$W = k\, 1/2\, \rho\, C\, V^3$$

- The coefficient k characterizes liquid and sediment particularities such as particle diameter, or nature of their surface (hardness, erosive intensity).

3.1.2. Cavitation device

It's a simple loop including a pipe (8 mm diameter, 212 cm^3 volume), a motor-pump, a sensor and the reference electrode. A manometer gives the drop in head between two points of the hydraulic circuit and permits to deduce the flow velocity at any moment. The sensor is placed along the cross section of the pipe. The sediment, introduced in the circuit, moves inside it and strikes the sensor. The electric signal, goes from the probe to the device.

3.1.3. Operating process

First we choice one sediment for example Rugos 2000 and we vary the velocity from 0 to 3.3 m/s. We have made 300 runs corresponding to 5 diameters, 5 concentrations and 12 velocities.

3.2. - Experimental results

3.2.1. - Laboratory results. Calibration

A first attempt calibration was begun at the Hydrosystems Laboratory of the University of Illinois (Professor Garcia). The main objective of these experiments was to assess the capabilities of the sensor to measure instantaneous sediment transport rate and concentrations with 0.5 mm sand diameter.

Now in our laboratory, we conduct testings in water flow with different types of sand.

We dispose of the following values:
- For the sand: diameter (from 0.09 to 0.50 mm), density (2.65).
- For the flow: velocity from 0 to 3.3 m/s.
- For tests: the concentration being constant for one run, the velocity takes 12 values in the range 0 - 3.3 m/s. The concentration is successively 0.1%, 0.25%, 0.50%, 1% and 2.5 %.

Three series of tests are developed:
- Case of sediment with 0.50 mm diameter (USA silica sand)
- Case of sediment with 0.09 mm diameter (Loire sand)
- Case of a new sediment with 0.15-0.25-0.35-0.45-0.60 mm diameter (Rugos 2000). This last work is in progress.

Remarks

° Many experimental results have been obtained. Only we present a few of them to give an idea on this work.

° In Figure 4, the current intensity (microamperes), measured by the device is shown. From this graph, it is clear that for a constant flow velocity, variations in sediment concentration are easily detected by the sensor. At the same time, for a constant sediment concentration, a strong dependence of the sensor output on flow velocity and on diameter, can be observed.

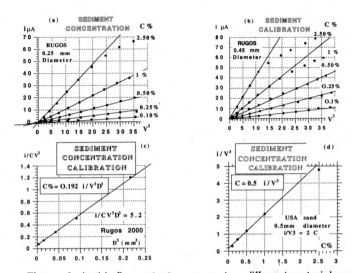

Figure 4 - In this figure, tests concern two different materials:
- RUGOS sand 0.25 (a) and 0.45 (b) mm diameter. Calibration curve (c).
- USA sand 0.50. Calibration curve (d).

○ It's important to mention that the range of flow velocities and sediment concenrations covered in the laboratory experiments was not limited by the capabilities of the device but rather by the particular characteristics of the calibration set up.

3.2.2. - *Field results*

The field experiments were conducted at several sites to detect the nature of sediment transport events, in different natural environments (Figure 5).
○ Ibiza (Islas Baleares, Spanish)
○ Loire River at Orleans (France)
○ Lake Michigan at Waukegan (USA)

Measurements taken at the beach of Cala Charco (Ibiza) are shown. The sensor (1 mm diameter and 80 mm length) was placed near the bottom at different distances from the shore that is covered by sand with sizes ranging from coarse sand to fine gravel. The intensity of the onshore / offshore sediment transport clearly decreases with the distance from the shore. Also we see offshore observations taken at the Ibiza-Formentera Pass. The flow depth was 4.5 m. The same sensor was placed at different vertical locations to detect the variations of the sediment transport with distance from the bottom. The current therein was transporting sediment in suspension up to a height of about 2.3 m above the bed.

Measurements conducted in the Loire River at Joffre Bridge, show specially the difference between suspended and carried sediments. So, measurements made from the pier at Waukegan, Illinois, display the reliability of measures.

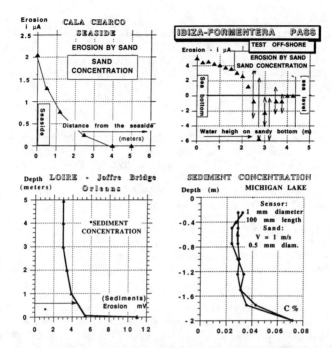

Figure 4 - Four types of tests have been made in various fields to show the possibilities of the DECAVER. We have worked in sea (onshore and offshore), in river and in Lake Michigan.

3.2.3. - Remarks
At the moment, we think that the DECAVER cannot measure scour concentrations with below 70 mm diameter or with very low erosive intensity.

3.3. - Applications

3.3.1. - Device possibilities
The basic principle of the DECAVER gives to it many advantages such as very little response time characteristics, form of the probe, etc. As we have pointed out the current-source nature of the corrosion phenomenon, it results an important sensitivity. So, we detect weak concentrations (0.1% or less). For that, we increase the sensitivity of the recorder or the velocity of the flow. The sensors can be made in any form and any surface according to working conditions.

3.3.2. - Laboratory
Due to the device sensitivity, it's using seems useful in laboratory both for fundamental research on sediments (erosion, concentration) and for tests to prepare measurements in field.

3.3.3. - *Field using*

In field, working conditions can be very different concerning the place or the time. If there is a particular erosion, the measuring is ever qualitatively possible. To have a quantitative value of the concentration, several parameters have to be known such as the erosive intensity of the particle surface, its diameter and density. If these values are unknown, a specific calibration shall be made in laboratory; it's easy and rapid.

Conclusion

By its nature, the DECAVER affords to make any measurement each time any erosion occurs. It opens a big range of using that we have not entirely exploited. Our stochastic measurements relative to the cavitation erosion only aim to show device possibilities. Experimental results and facts pointed out evince potentialities of this new tool working in real time. Various domains such as hydraulic machinery, ultrasonic devices, jets, mud and debris flow, also all random eroding processes are concerned.

More generally, regarding all erosions in an aqueous medium, we have seen that the first step should be to measure the Flow Erosive Intensity that is independent of the used material. At the same time, for standard metal, the instantaneous erosion rate can be known. The device assumes this measurement.

Then, we take notice that it's possible to record erosion characteristic curves of a hydraulic machine in relation to any parameter and to deduce cavitation erosion laws or others characteristic properties. Direct measurements of instantaneous sediment transport rates, either as bedload or suspended load in natural environment can be made with this novel sensor.

The use of the DECAVER permits progress in hydraulic machinery for researchers, planners and operating staff.

References

[1] CHINCHOLLE L. and COELHO R.: Rocket Effect and Electrohydrodynamics phenomena, in liquids dielectrics, First International Symposium on Electrohydrodynamics, M.I.T., Cambridge, Mass.,1969.
    CHINCHOLLE L.: Thesis, Faculté des Sciences, Paris, 1967.
    CHINCHOLLE L.: Bubbles and the Rocket Effect, Journal of Applied Physics, Vol.41, N°11, 4532-4538
    CHINCHOLLE L.: New views of cavitation erosion by using an instantaneous erosion measurement device (DECAVER), 1st ASME-JSME Fluids Engineering Conference, Portland, Oregon, 1991.
[2] CHINCHOLLE, L.: Répartition du courant à la surface d'une électrode plongée dans un milieu aqueux, lorsque cette surface est partiellement abrasée", Compte rendu Académie des Sciences Electrochimie, T.296, 1983.
[3] CHINCHOLLE L.: Measure of the Flow Erosive Intensity on a specific surface, in an aqueous medium, by abrasion, corrosion or cavitation erosion, Flucome' 91, Sheffield, September 1988.
[4] CHINCHOLLE L.: Stochastic measurements of cavitation, Sixth IAHR Intern. Symposium on Stochastic Hydraulics, Taipei, Taïwan, R.O.C. 1992.

Anomalous Measurements in a Compound Duct

David G. Rhodes[1] and Donald W. Knight[2], Member, ASCE

Abstract

An anomalous velocity distribution on the "flood plain" of a compound duct is reported and discussed. The tentative conclusion is that the phenomenon is to be explained by the co-existence of laminar and turbulent flow regimes on the flood plain, with a transitional zone between them.

1 Introduction

The complex interaction between main channel and flood plain flows has a profound effect upon the hydraulic performance of two-stage channels, and its significance is indicated by the considerable amount of previous research work undertaken. Most of the experimental effort has been devoted to laboratory compound open channels, however there are a few examples of the use of closed compound ducts. In the latter case a zone of interaction, analogous to that in open channel flow, is engineered in a conduit of the shape illustrated in Figure 1.

Compound duct work carried out by the present authors consisted of the measurement of time averaged velocity and boundary shear stress distributions, by Pitot-static tube and Preston tube respectively, in a closed duct 13.1 m long and 1231.5 mm wide. The roof panel of the duct was adjustable in height and measurements were carried out for relative depths in the range $0.05 \leq \frac{H-h}{H} \leq 0.67$ (refer to Figure 1). Experiments were conducted for three values of the wall angle, $\theta = 90°$, $\theta = 45°$ and $\theta = 26.6°$, with two step heights $h = 20$ mm and $h = 40$ mm. The ranges of relative width corresponding to these step heights

---

[1]Lecturer, School of Mechanical, Materials and Civil Engineering, Cranfield University, Shrivenham, Swindon, SN6 8LA, UK.
[2]Reader, School of Civil Engineering, The University of Birmingham, Edgbaston, Birmingham, B15 2TT, UK.

were $1.96 \leq \frac{B}{b} \leq 2.16$ and $1.41 \leq \frac{B}{b} \leq 1.65$ respectively. The working fluid was air and the Reynolds number based upon cross-section mean velocity and hydraulic diameter varied in the range $6.12 \times 10^4 \leq \text{Re} \leq 10.30 \times 10^4$ (Rhodes et al (1991) and Rhodes (1991)).

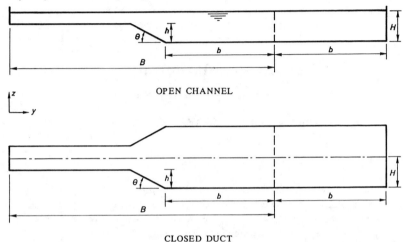

Figure 1: *Equivalent duct and open channel cross-sectional geometries*

In designing the duct cross-section and prescribing the range of relative depths, it was intended that fully developed turbulent flow would be achieved throughout the measurement section. However in order to satisfy the other objective of near two-dimensional flow on the flood plain and in the main channel, comparatively shallow flood plain flows resulted at the lower relative depths. The effect upon flood plain Reynolds numbers is significant to what follows.

During the course of the experimental work, some anomalous results were observed in the velocity distributions, and these form the subject of this paper. However, the time constraints of the project precluded an adequately detailed investigation, and therefore any conclusions drawn here are tentative and subject to further experimental work being undertaken by the first author at the time of writing.

2 Results

When the time averaged primary velocity is depth averaged across a fully developed turbulent flow in a compound duct, the lateral distribution of depth mean velocity on the flood plain assumes the general form illustrated in Figure 2, with a wall angle $\theta = 90°$. (Henceforth we use the term "flood plain" to refer to the shallow part of the duct of semi-depth $(H - h)$, and "main channel" for the deep part of semi-depth H.)

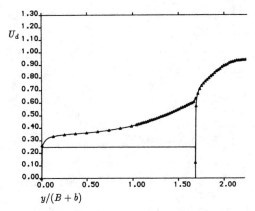

Figure 2: *General form of depth mean velocity distribution on the flood plain*

Proceeding from the outer wall there is at first a region of steep lateral velocity gradient due to wall shear. Provided that the flood plain aspect ratio is wide enough, the gradient gradually subsides to a near zero value in a region of approximately two-dimensional flow, and then steepens again in the region of strong lateral shear created by the interaction of the slower moving flow on the flood plain with the faster moving flow in the main channel. Thus the lateral gradient of depth mean velocity is generally monotonic from the outer wall to the main channel–flood plain interface. There is a well known exception to this in the immediate vicinity of the interface, where secondary currents directed outwards from the salient corner and convecting low momentum fluid may depress the local velocity field. As a result, the lateral gradient in depth mean velocity changes sign, and the distribution exhibits a maximum value. This phenomenon is most easily observed at high relative depths.

Contrary to the general form of depth mean velocity distribution previously described, it was found that at one particular relative depth, $(H-h)/H = 0.1$, a very clear dip in the distribution on the flood plain was observed. This occurred for all three wall angles.

Subsequently it was discovered that the pressure transducer, connected to the Pitot-static tube, had reached its lower limit near the flood plain bed in each of the velocity traverses at $(H-h)/H = 0.1$. As a result, the measurements at z coordinates 0.95, and 1.45 mm above the flood plain bed ($z = 0$) were nullified. However this discovery of faulty measurement did not explain the dip in the calculated depth averaged velocity distribution, caused by the measurements at $z = 2.45$, 3.45 and 4.45 mm ($z = 4.45$ mm was on the horizontal plane of symmetry) which *were* within the operating range of the transducer. The measurements at $z = 2.45$ and $z = 4.45$ mm are shown in Figure 3 for $\theta = 90°$.

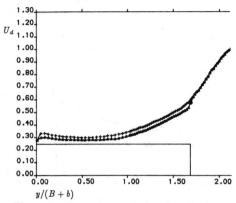

Figure 3: *Anomalous velocity distribution*

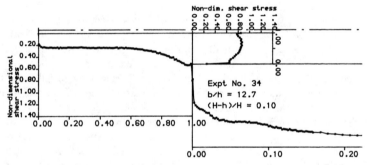

Figure 4: *Bed shear stress distribution corresponding to anomalous velocity distribution*

The distribution at $z = 3.45$ mm is not shown because at that scale it coincides with the distribution at $z = 4.45$ mm.

Figure 4 shows the corresponding boundary shear stress distribution on the flood plain bed, main channel wall and part of the main channel bed, non-dimensionalised by the cross-section mean boundary shear stress. In this case a different pressure transducer had been used, so that all of the measurements were within the range of the instrument. Although the validity of the exact shear stress values, calculated using Patel's (1965) calibration, is open to question in view of the low Reynolds number of the flow regime (discussed later), nevertheless the results clearly demonstrate a region of negligible lateral variation in dynamic pressure measured by the Preston tube, and strongly suggest a two-dimensional primary velocity field near the flood plain bed and by inference a constant bed shear stress.

Thus we had a lateral velocity distribution which remote from the flood plain

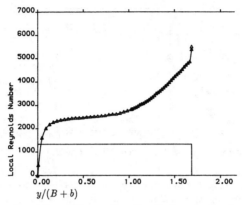

Figure 5: *Local Reynolds number distribution on the flood plain*

bed exhibited a pronounced dip, and the corresponding bed shear stress distribution which within the same lateral region was flat.

3 Discussion

The fact that the anomalous velocity distributions occurred at $(H-h)/H = 0.1$ for all three wall angles excluded the likelihood of random error as an explanation. The possibility therefore remained of systematic error, or that the results were manifestations of a real flow phenomenon.

There had been some previous difficulty, when the equipment was used for rectangular duct experiments, in achieving a sufficiently accurate cross-section geometry. As discussed by Rhodes and Knight (1994), the remedial measures implemented at that time had led to a considerable improvement in the compound duct results compared with those in the rectangular cross-section. Nevertheless, at a relative depth of $(H-h)/H = 0.1$ with the step height $h = 40$ mm, the nominal duct height on the flood plain, $2(H-h)$, was only 8.9 mm. Thus, for example, an error of 0.5 mm would have represented nearly 6 per cent of the duct height. For this reason the likelihood of the velocity distribution having been influenced by inaccuracies in the cross-sectional geometry was given serious consideration, although it was indeed difficult to reconcile this explanation with the two-dimensional bed shear stress distribution.

At the same time it was realised that the Reynolds number of the flood plain flow was quite low. Because of the absence of velocity measurements at $z = 1.45$ mm and below, only an upper bound of the Reynolds number distribution across the flood plain could be calculated. This is illustrated in Figure 5. The local Reynolds number was based on the depth mean velocity and a hydraulic

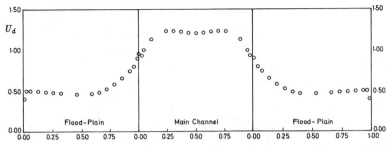

Figure 6: *Anomalous depth mean velocity distribution measured by Lai (1986)*

diameter equal to four times the ratio of the area $(H-h)y$ to the wetted perimeter $(H-h)+y$, the y datum coinciding with the flood plain outer wall. As such the hydraulic diameter gave due weight to the outer wall influence on flow in its vicinity, whilst remote from the wall the hydraulic diameter tended to the two-dimensional value of $4(H-h)$.

Figure 5 shows that the local Reynolds number was less than 3 000 for a substantial proportion of the flood plain width. This may be compared with the measurements of Patel and Head (1969) in a wide rectangular duct. They concluded that the first signs of transition from laminar to turbulent flow occurred at a Reynolds number of about 2 600 based on the criterion of skin friction, and at a Reynolds number of 2 760 according to the first signs of intermittency (Patel's and Head's Reynolds numbers have been doubled for consistency with the present definition, although they did not include the effect of the side wall in their length scale.) A tentative conclusion at this stage was that the anomalous velocity distribution was the product of inaccuracy in the duct cross-sectional geometry and the low Reynolds number flow regime (Rhodes (1991)).

However, subsequently, similar anomalous depth mean velocity distributions were observed in the results of Lai (1986). These measurements had been carried out in a smaller duct, adjustable in height and with a maximum width of 380 mm, in which various symmetric and asymmetric compound sections had been constructed. Examination of Lai's results revealed that at relative depths of approximately $(H-h)/H = 0.1$, in some cases a dip in the depth mean velocity distribution occurred on the single flood plain of the asymmetric section and on both flood plains of the symmetric cross-section. An example abstracted from Lai's results is shown in Figure 6 for a symmetric section with $(H-h)/H = 0.09$ and $B/b = 4.94$. A dip in the depth mean velocity distribution is evident on both of the flood plains.

Because the anomalous velocity distribution has now been observed in the data from two different compound ducts, it seems reasonable to seek an alternative explanation which does not attribute the effect to errors in the duct cross-sectional geometry. As already mentioned, the latter explanation is not consistent

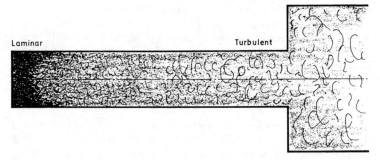

Figure 7: *Diagrammatic representation of flow regime on the flood plain*

with the observed bed shear stress distribution in any case. It is now proposed that the anomaly could be produced by the mechanism described as follows.

Proceeding from the outer wall of the flood plain, a laminar boundary layer of substantial thickness probably exists in contact with the wall. Indeed, even if the flood plain flow were fully turbulent there would be a laminar sublayer, and in the present fully developed low Reynolds number flow this might be expected to occupy a significant lateral dimension adjacent to the wall. Within the laminar boundary layer, the depth mean velocity rises steeply from the outer wall until with decreasing lateral shear it reaches a plateau. Outside this wall generated shear layer there is a region in which lateral shear is negligible, the flow is driven by the pressure gradient and largely resisted by the bed shear stress. The bed shear stress would vary little in the lateral direction, as has been observed in the results reported here.

At the other end of the flood plain, under the influence of lateral shear from the main channel, the flow is fully turbulent. Co-existence of laminar and turbulent flow in the same duct has been previously observed by Eckert and Irvine (1956), in a triangular cross-section. In the present case, the effect of the two flow regimes co-existing on the flood plain is that instabilities are transmitted laterally from the turbulent flow into the laminar flow region near the outer wall. With increasing distance from the outer wall this destabilising effect becomes more marked, so that the flow (which would now be more properly termed transitional) assumes a more nearly turbulent character. This flow condition might bear some resemblance to the buffer region in a turbulent boundary layer. A diagrammatic representation of this suggested flow regime is given in Figure 7.

For a notional flow, with a particular bulk velocity in a given cross-section, the boundary shear stress in a fully developed laminar flow regime would be less than that in a fully developed turbulent flow. The corollary of this is that, for a given pressure gradient and corresponding bed shear stress in a two-dimensional flow between parallel plates, the bulk velocity in a fully developed laminar flow would be greater than that in a fully developed turbulent flow. For transitional flow an intermediate velocity would result depending on the degree of turbulence or intermittency of the flow.

Therefore in the present case, in the absence of lateral shear, depth averaged velocity should decrease within the transitional regime as the flow develops a more turbulent character with distance from the outer wall. This is consistent with the observed results in which, with increasing distance from the outer wall, the initial steep rise in depth mean velocity is followed by a region of steadily decreasing depth mean velocity.

Proceeding still further from the outer wall, there comes a stage when the flow not only experiences the destabilising effect of turbulence but also the effect of lateral shear through the shear stress τ_{yx}, largely generated by turbulent diffusion in the mixing layer developed at the main channel–flood plain interface. This induces an opposite trend in the depth mean velocity distribution, so that at some spanwise position there is a minimum value followed by an increasing trend in depth averaged velocity with further distance from the outer wall, as observed in the present results.

This explanation may be summarised in terms of the tendency of the turbulent mixing layer at the main channel–flood plain interface to diffuse instability into a low Reynolds number flow regime further than it transmits the shear stress τ_{yx}.

So far we have not considered the additional level of complexity created by the presence of turbulence driven secondary flow on the flood plain. This is likely to occur because the salient corner at the main channel–flood plain interface is in a region of fully turbulent flow. On the flood plain where the flow depth is constant, by depth averaging the Reynolds equation (for fully turbulent flow) in the primary flow direction and invoking the continuity equation, it can be shown that

$$(H-h)\frac{\partial p}{\partial x} = (H-h)\left(\frac{d}{dy}\overline{\tau}_{yx} - \rho\frac{d}{dy}\overline{UV}\right) - \tau_b \quad (1)$$

where $\overline{\tau}_{yx}$ is dominated by the Reynolds stress component. However, provided that there is a mechanism for generating the secondary component of velocity V, equation 1 is equally applicable to transitional or laminar flows, and of course in these regimes viscous shear becomes significant in the term $\overline{\tau}_{yx}$.

In a region of the flood plain where the lateral shear stress τ_{yx} has fallen to zero or near zero, following rearrangement the equation for τ_b becomes

$$\tau_b = -(H-h)\left(\frac{\partial p}{\partial x} + \rho\frac{d}{dy}\overline{UV}\right) \quad (2)$$

where in both equations the overbar refers to depth averaged values. The term $\rho d\overline{UV}/dy$ represents the interaction of the primary flow field with the component of secondary flow in the spanwise direction, and Shiono and Knight (1991) found in compound open channel flow the term to be substantially constant on the flood plain. If this result were applicable to a compound duct, it would mean in the present case that τ_b, though modified by secondary flow, would still be constant in the spanwise direction, as actually observed.

An additional effect of the spanwise component of secondary flow in the present context might be to augment the lateral transport of both instability and damping, as the current occurs in both positive and negative y directions. Also, given the dip in depth mean velocity, secondary flow would produce some redistribution of the primary velocity field.

4 Summary and Proposed Further Work

Anomalous measurements, consisting of a pronounced dip in the depth mean velocity distribution on the flood plain of a compound duct have been reported. It has been shown that random error is highly improbable, and that systematic error due to inaccuracy in the duct geometry is also unlikely. In view of the low Reynolds number distribution on the flood plain, the tentative conclusion is that the phenomenon is the result of co-existence of laminar and turbulent flow regimes on the flood plain, with an intermediate transitional zone. A reasonable explanation has been presented based upon this premise.

Currently, work is in progress by the first author using a newly constructed compound duct. The aim is to reproduce the dip in the depth mean velocity distribution on the flood plain, and test the validity of the explanation presented here. Attention will be given to the analysis of vertical velocity profiles and local velocity time series, and flow visualisation may be undertaken.

5 Acknowledgements

The experimental work was carried out in the School of Civil Engineering, The University of Birmingham and funded by the Science and Engineering Research Council (SERC Grant No GRD/21622).

Appendix I

References

[1] **Eckert, E.R.G. and Irvine, T.F.** (1956). Flow in corners of passages with noncircular cross sections. *Trans. ASME*, 709-718.

[2] **Lai, C.J.** (1986). Flow resistance, discharge capacity and momentum transfer in smooth compound closed ducts. *Ph.D. Thesis*, The University of Birmingham.

[3] **Patel, V.C.** (1965). Calibration of the Preston tube and limitations on its use in pressure gradients. *J. Fluid Mech.*, 23, 185-208.

[4] **Patel, V.C. and Head, M.R.** (1969). Some observations on skin friction and velocity profiles in fully developed pipe and channel flows. *J. Fluid Mech.*, **38**, 181–201.

[5] **Rhodes, D.G., Lamb, E.J., Chance, R.J. and Jones, B.S.** (1991). Automatic measurement of boundary shear stress and velocity distributions in duct flow. *J. Hydraulic Res.*, IAHR, **29**(2), 179–187.

[6] **Rhodes, D.G.** (1991). An Experimental Investigation of the Mean Flow Structure in Wide Ducts of Simple Rectangular and Compound Trapezoidal Cross-Section, Examining in Particular Zones of High Lateral Shear. *Ph.D. Thesis*, The University of Birmingham.

[7] **Rhodes, D.G. and Knight, D.W.** (1994). Distribution of Shear Force on the Boundary of a Smooth Rectangular Duct. *J. Hydraulic Eng.*, ASCE. (To be published: tentative schedule July 1994).

[8] **Shiono, K. and Knight, D.W.** (1991). Turbulent open-channel flows with variable depth across the channel. *J. Fluid Mech.* **222**, 617–646. (and 1991, **231**, 693.)

Appendix II

Notation

The following symbols are used in this paper:

b	semi-width of main channel bed (Figure 1)
B	dimension from centre-line of main channel bed to flood plain outer wall (Figure 1)
h	step height at flood plain–main channel interface (Figure 1)
H	depth of flow in main channel, ie. maximum semi-depth of duct (Figure 1)
p	time averaged local pressure
U	time averaged local velocity in the x direction
U_d	depth mean velocity divided by cross-section mean velocity
V	time averaged local velocity in the y direction
x	cartesian coordinate in the primary flow direction
y	cartesian coordinate in the horizontal cross-stream direction
z	cartesian coordinate in the vertical cross-stream direction
τ_b	local bed shear stress
τ_{yx}	local shear stress acting on y plane in x direction

Hydrodynamic Forces in Hydraulic Jump Stilling Basins

António N. Pinheiro[1]
António C. Quintela[2]
Carlos M. Ramos[3]

Abstract

The knowledge of hydrodynamic forces acting on hydraulic jump stilling basin slabs has been a research topic for the past four decades.

Published methodologies for computing the hydrodynamic forces are comparatively analysed as far as their purposes and application domains are concerned. It is shown that available criteria do not always lead to similar results. Some remarks about the use of those methodologies are presented.

Guidelines for a research programme on hydrodynamic forces, presently in execution, based on simultaneous pressure measurements in several points of the stilling basin floor, are presented.

Introduction

Although a considerable amount of work concerning the characterization of the pressure field in a stilling basin beneath the hydraulic jump has been done, not many authors present design criteria to evaluate the hydrodynamic forces acting on the slabs of the basin floor. The establishment of widely acceptable criteria should be the main goal of present and future studies, so as to improve the stability design of stilling basin slabs.

---

[1] Pinheiro, Assistant, Instituto Superior Técnico (IST), Civil Engineering Department, Av. Rovisco Pais, 1096 Lisboa Codex.
[2] Quintela, Hydraulics Full Professor, IST.
[3] Ramos, Senior Investigator, Laboratório Nacional de Engenharia Civil, Av. do Brasil, 1799 Lisboa Codex.

A comparative study of the available methodologies could not be found in the literature, which may be considered an important lack. In fact, the whole of published results are based upon force or pressure measurements, with different measuring devices and approaches and so, it would be important to find explanations for the differences that naturally will turn out.

From the bibliographic review, three different methodologies for determining the hydrodynamic forces on slabs were selected to be applied to a case study. Due to short available space, it is not possible to reproduce in this paper the figures with the published results, which would make it much more comprehensive and easy to discuss.

Methodologies presentation

Following the research programme developed in the ex-USSR concerning pressure and force fluctuations beneath an hydraulic jump (Preobrazhenskii 1958), Yuditskii (1960) published a complete set of results. The slabs used in his studies were square and non movable, being their position beneath the jump varied by means of moving the weir position. The slabs were suspended and the forces were obtained through the measurement of the suspension wires deformation. Maximum amplitude of force fluctuations was obtained through a 1 min oscilograph record. The slabs dimensions did not exceed $L_s/L_j=0.4$, where L_s = slab dimension along the flow direction and L_j = length of jump. Yuditskii (1960) presented figures representing the following equation

$$\frac{A^*}{h_c} = f\left(\frac{E_0}{h_c}; \frac{x_s}{L_j}; \frac{L_s}{L_j}\right) \qquad (1)$$

where A^* = maximum amplitude of pressure head fluctuation on an area for a specific time interval, h_c = critical depth, E_0 = flow head upstream of the toe of the jump and x_s = longitudinal space coordinate of the slab gravity center relatively to the jump upstream section.

Toso (1986) developed an extensive experimental research programme whose main results were presented by Toso and Bowers (1988). For Froude numbers between 2.94 and 10.00, Toso (1986) presented C_p^- and C_p^+ coefficients (maximum positive and negative values of pressure head fluctuations, for a certain runtime, divided by the velocity head upstream of the jump) obtained for 10 min runtimes and a 50 Hz acquisition frequency, and suggested maximum values for those coefficients - $C_{p'max}$.

Toso (1986) suggested the following methdology to compute the maximum hydrodynamic forces on the stilling basin floor:

- estimate the maximum pressure fluctuation considering the suggested maximum values - $C_{p'max}$;
- estimate the macroturbulence time scale and vortex convection velocity according to available data (Toso 1986);
- compute the longitudinal scale of maximum pressure fluctuations and multiply this one for 1.6 to estimate the transverse scale;
- compute the force fluctuation considering a pyramidal pressure diagram centered on the point where the highest pressure fluctuation was registered and with base dimensions equal to the computed macroturbulence scales.

A few comments to Toso's work may be presented:

- only a limited number of experiments were made, which limits the scope of application of the methodology;
- the proposal for considering a unique $C_{p'max}$ much greater than the two extreme values $C_{p'}^-$ $C_{p'}^+$ is conservative and, as far as $C_{p'}^-$ is concerned, this will turn out to be cost increasing for assuring slabs stability;
- the runtime influence upon the maximum values was carried out only for the area of the stilling basin where the maximum pressure pulses were detected;
- a constant ratio between longitudinal and transverse macroturbulence scales is a non suitable assumption;
- the hydrodynamic force computing methodology does not seem fundamented and is difficult to apply to a rectangular slab with dimensions different from the macroturbulence scales.

Farahoudi and Narayanan (1991) presented the results of direct hydrodynamic force measurements on slabs of hydraulic jump stilling basins for Fr=4, 6, 8 and 10. The authors presented figures with the mean value, c_F, and the standard deviation of force fluctuations, c'_F, with respect to $0.5\rho V_1^2 S$, as functions of Fr, of slab dimensions and of the distance of the upstream slab edge to the toe of the jump - x_j. Several experiments were made for different slab lengths L_s, with B_s/h_1=1, and for different widths B_s, with L_s/h_1=1. However, the experiments do not cover the whole jump length.

The development of the boundary layer is not mentioned. However, regarding the experimental installation characteristics, it is probable that the boundary layer is non developed or very little developed.

Farhoudi and Narayanan (1991) proposed the following methodology to compute the standard deviation for the force actuating in a certain slab:

- obtain c'_{FL} (c'_F for $B_s/h_1=1$) for the value L_s/h_1 of the slab and c'_{FB} (c'_F for $L_s/h_1=1$) for the value B_s/h_1 from the figures presented in the paper referred to;
- compute $c'_F = (c'_{FL} \cdot c'_{FB})^{0.5}$.

These authors mentioned that the observed maximum force fluctuations are about ±3.5 the respective rms value.

Though not being explicitly mentioned, a similar procedure is considered applicable to compute c_F.

Bribiesca and Mariles (1979) and Fiorotto and Rinaldo (1992) also present design criteria for hydrodynamic forces. However, those results are considered not so interesting as the ones briefly described.

Hydrodynamic forces. Comparative analysis

The following conditions are considered for defining a case study: $Fr=6$ and 9, with $h_1=1$ m; square slabs with $L_s=7$ and 10 m; slabs centered on the stilling basin axis, positioned with the center at the middle section of the jump or with the downstream edge at the jump downstream section. This corresponds to eight different cases, whose hydrodynamic forces shall be calculated for the three methodologies previously mentioned. The main flow characteristics are in Table 1.

Table 1 - Flows characteristics.

Fr	V_1 (m/s)	E_0 (m)	h_c (m)	h_2 (m)	$L_j=5.67(h_2-h_1)$ (m)
6	18.78	18.99	8.00	3.30	39.7
9	28.17	41.49	12.24	4.33	63.7

The first difficulty arises from the different criteria that those methodologies use to characterize the hydrodynamic forces and their fluctuations. In fact, as long as Yuditskii (1960) obtained the maximum amplitude of the pressure head on a given area registered in a runtime of 1 min, A^*, Toso (1986) searched for the *maximum maximorum* value of that fluctuation and Farhoudi and Narayanan (1991) chose a statistical characterization of the forces, considering a normal distribution, and presented estimates for the force mean value and standard deviation. So, the design engineer faces a variety of criteria, which, most probably, may lead him to a conservative and non economic solution.

The parameters concerning the methodology proposed by Yuditskii (1960) are presented in Table 2.

Table 2 - Hydrodynamic force fluctuations according to Yuditskii (1960).

Fr	E_0/h_c	L_s (m)	L_s/L_j	x_s/L_j	A^*/h_c	A^* (m)	$A^*/2$ (m)
6	5.8	7	0.18	0.50	0.59	1.95	0.98
				0.91	0.32	1.06	0.53
		10	0.25	0.50	0.39	1.29	0.65
				0.87	0.24	0.79	0.40
9	9.6	7	0.11	0.50	1.31	5.67	2.84
				0.95	0.64	2.77	1.39
		10	0.16	0.50	1.06	4.59	2.30
				0.92	0.54	2.34	1.17

The fluctuation value to be used in the stability design of the slab design is $A^*/2$, which must be subtracted to mean value of the pressure head so as to obtain the minimum force acting on the slab upper face. The author does not give any indication on how to compute the mean value. The hydrostatic pressure distribution caused by the jump longitudinal profile was considered.

Toso's methodology presented a *maximum maximorum* value approach, which is considered as interesting, but also has a few faults mainly concerning the force fluctuations computation on a slab as previously mentioned. The coefficients C_p^+ and C_p^- corresponding to the eight situations obtained from Toso (1986) are presented in Table 3. As the author focused his study on the extreme fluctuations, the presented results are not applicable for slabs located far from the maximum pressure fluctuations area.

Toso's methodology is difficult to use because, as it can be concluded from the values L_x and L_y (L_y=1.6 L_x) presented in Table 3, the pressure pyramid base dimensions only by coincidence are similar to the ones of a certain slab. In case the slab dimensions significantly exceed the macroturbulence scales the estimation of F' becomes even more subjective.

Table 3 - Coeficients C_p^+ and C_p^- and L_x according to Toso (1986).

Fr	L_s (m)	x_j/h_1	C_p^+ Non dev.	C_p^+ Fully dev.	C_p^- Non dev.	C_p^- Fully dev.	$\tau V_1/h_1$	τ (s)	V_v/V_1	V_v (m/s)	$L_x=\tau V_v$ (m)
6	7	16.4	0.35	0.26	0.35	0.41	23	1.22	0.34	6.4	7.8
		32.7	-	-	-	-	-	-	-	-	-
	10	14.9	0.37	0.28	0.37	0.38	21	1.12	0.36	6.8	7.6
		34.7	-	-	-	-	-	-	-	-	-
9	7	28.4	0.27	-	0.26	-	43	1.53	0.28	7.9	12.1
		-	-	-	-	-	-	-	-	-	-
	10	26.9	0.27	-	0.26	-	39	1.38	0.28	7.9	10.9
		-	-	-	-	-	-	-	-	-	-

The hydrodynamic force fluctuations computed according to Toso (1986) are presented in Table 4. The coefficients C_p^- and $C_{p'max}$ and the pyramidal pressure distribution are applied only for the slabs with dimensions less than the respective macroturbulence scales.

Table 4 - Hydrodynamic force fluctuations computed according to Toso (1986).

Fr	x_j/h_1	L_s (m)	L_x (m)	L_y (m)	$F'/\gamma A$ (C_p^-) Non dev.	$F'/\gamma A$ (C_p^-) Fully dev.	$F'/\gamma A$ (C_{pmax}) Non dev.	$F'/\gamma A$ (C_{pmax}) Fully dev.
6	16.4	7	7.8	12.5	1.53	1.79	4.37	3.93
	14.9	10	7.6	12.2	-	-	-	-
9	28.4	7	12.1	19.4	1.65	-	5.71	-
	26.9	10	10.9	17.4	2.62	-	9.06	-

The values of c_F and c'_F (Farhoudi and Narayanan 1991) are presented in Table 5. The procedure used to obtain those coefficients was the one suggested by the authors and already described. The standard deviation was multiplied by a factor of 1.645, corresponding to a probability of the negative values being exceeded of 0.05, so as to give more meaning to the comparison of the different methodologies.

The results obtained with each of the three methodologies are presented together in Table 6, so as to make the comparison easier.

Table 5 - Hydrodynamic forces. Mean values and standard deviations computed according to Farhoudi and Narayanan (1991).

Fr	$V_1^2/2g$	L_s (m)	x_j/h_1	c_F	c'_F	$F/\gamma A$	$\sigma_{F'}/\gamma A$ (m)	1.645 $\sigma_{F'}/\gamma A$ (m)
6	18.0	7	16.4	0.33	0.034	5.94	0.61	1.00
			32.7	0.43	0.023	7.74	0.41	0.67
		10	14.9	0.31	0.025	5.58	0.45	0.74
			34.7	0.44	0.012	7.92	0.22	0.36
9	40.5	7	28.4	0.20	0.029	8.09	1.17	1.92
			-	-	-	-	-	-
		10	26.9	0.20	0.025	8.09	1.01	1.66
			-	-	-	-	-	-

Table 6 - Results of the three methodologies.

Fr	L_s (m)	x_j/h_1	$A^*/2$ (m)	$F'/\gamma A$ (C_p^-) Non dev.	$F'/\gamma A$ (C_p^-) Fully dev.	$F'/\gamma A$ $(C_{p\,max})$ Non dev.	$F'/\gamma A$ $(C_{p\,max})$ Fully dev.	1.645 $\sigma_{F'}/\gamma A$ (m)
6	7	16.4	0.98	1.53	1.79	4.37	3.93	1.00
		32.7	0.53	-	-	-	-	0.67
	10	14.9	0.65	-	-	-	-	0.74
		34.7	0.40	-	-	-	-	0.36
9	7	28.4	2.84	1.65	-	5.71	-	1.92
		56.7	1.39	-	-	-	-	-
	10	26.9	2.30	2.62	-	9.06	-	1.66
		53.7	1.17	-	-	-	-	-

The results analysis suggests the following comments:

- the three methodologies do not present the same application range, being the data of Yuditskii (1960) the one which covers a more wide range of possibilities;
- this last methodology has the disadvantage of being deterministic, without presenting a genuine maximum fluctuation value approach ;
- the work of Toso (1986) does not make reference to the hydrodynamic force mean value and the number of experimented situations is scarce;

- the force fluctuations computed with $C_{p'max}$ are extremely large comparatively to the other results;
- the values obtained with $C_{p'}^*$, although being of similar magnitude of the other methodologies ones, seem to show some inconsistency if compared with the values of the other methodologies;
- the results of Farahoudi and Narayanan (1991) have a small longitudinal coverage of the jump length;
- the statistical approach used by Farahoudi and Narayanan (1991) seems reasonable, though a comparison with a maximum fluctuation approach for the whole lenght of the jump could prove rather interesting;
- there exists a reasonable similarity between $A^*/2$ and $1.645\sigma_{F'}/\gamma A$, specially for $Fr=6.0$.

Guidelines for a research programme on hydrodynamic forces on hydraulic jump stilling basin slabs

A research programme on hydrodynamic pressures and forces beneath the hydraulic jump is presently being carried out by Instituto Superior Técnico (IST) and Laboratório Nacional de Engenharia Civil (LNEC). The experimental installation is able to produce flows for $6 \leq Fr \leq 10$, with fully developed upstream boundary layers. The flow may be artificially aerated by means of an air slot, placed upstream of the jump, connected to an air pump. An equipment with eight simultaneous channels for pressure measurements acquisition is available.

The main goal of this research programme is to establish, for a stilling basin without accessories, the hydrodynamic forces to be considered in the stability design of slabs. The hydrodynamic forces shall be characterized for the whole length of the hydraulic jump and for the different slab dimensions, square or rectangular.

The forces shall be statiscally characterized by their mean and standard deviation obtained this last one computed according to the methodology presented by Spoljaric et al. (1982). Simultaneaously, an extreme value approach for forces, similar to Toso's approach for point pressure measurements shall be carried out and compared with the statistical characterization results.

The comparison between the obtained results and the one of the methodologies presented in these paper is expected to clarify the use that may be made with each of them.

Resulting from the computation of point pressure fluctuations cross correlations, necessary to determine force fluctuations standard deviation, an improved knowledge of those cross correlations variation beneath the hydraulic jump is also expected to be obtained.

The influence of flow aeration in the pressure field is also to be investigated. In case its hypothetical beneficial influence can be shown, further investigation for stilling basins with accessories could prove rather promising.

Acknowledgements

The writers wish to thank to IST and LNEC for the financial and logistic support that have been provided for this study, with special emphasis to the latter institution as far as the experimental installation and measurement equipment development are concerned. The first author also wishes to thank Junta Nacional de Investigação Científica e Tecnológica (JNICT) for the grant he has been given for the preparation of his Ph.D. Thesis.

Appendix I. References

Bribiesca, J. S. and Mariles, O. F. (1979). "Experimental analysis of macroturbulence effects on the lining of stilling basins". *Proc. 13th Congress, ICOLD*, New Delhi, Vol.3, 85-103.
Farhoudi, J. and Narayanan, R. (1991). "Force on slab beneath hydraulic jump". *J. Hydr. Engrg.*, ASCE, 117(1), 64-82.
Fiorotto, V. and Rinaldo, A. (1992). "Fluctuating uplift and lining design in spillway stilling basins". *J. Hydr. Engrg.*, ASCE, 118(4), 578-596.
Preobrazhenskii, N. A. (1958). "Laboratory and field investigations of flow pressure pulsation and vibration of large dams". *Proc. 6th Congress, ICOLD*, New York, Question Nº 21, R.110, 1-23.
Spoljaric, A., Maksimovic, C. and Hajdin, G. (1982). "Unsteady dynamic force due to pressure fluctuations on the bottom of an energy dissipator. An example". *Proc. Int. Conf. on the Hydraulic Modelling of Civil Engineering Structures*, Coventry, C2, 97-107.
Toso, J. W. (1986). *The Magnitude and Extent of Extreme Pressure Fluctuations in the Hydraulic Jump*. Ph.D. thesis, University of Minnesota.
Toso, J. W. and Bowers, E. C. (1988). "Extreme pressures in hydraulic jump stilling basins". *J. Hydr. Engrg.*, ASCE, 114(8), 829-843.
Yuditskii, G. A. (1960). "Experimental study of force fluctuations downstream of a weir". Translation from russian to portuguese from *Izvestya VNIIG*, 65, 117-124, 1960, Laboratório Nacional de Engenharia Civil, *Translation 520*, Lisbon, 1985.

Appendix 2. Notation

c_F = ratio between the mean value of the force acting on an area S of the stilling basin floor and $0.5\rho V_1^2 S$;

c'_F = ratio between the rms value of force fluctuations on an area S of the stilling basin floor and $0.5\rho V_1^2 S$;

g = gravitational acceleration;

h_1 = flow height upstream the jump;

h_2 = sequent depth of h_1;

h_c = critical depth;

x_s = longitudinal space coordinate of the slab gravity center relatively to the jump upstream section;

x_j = longitudinal space coordinate of the slab upstream edge relatively to the jump upstream section;

A' = maximum amplitude of pressure head fluctuation on a point for a specific time interval;

A^* = maximum amplitude of pressure head fluctuation on an area for a specific time interval;

B_s = slab width;

$C_{p'}$ = rms value of pressure head fluctuations divided by the velocity head upstream of the jump;

$C_{p'max}$ = suggested *maximum maximorum* value of pressure head fluctuations divided by the velocity head upstream of the jump;

$C_{p'}^+$ = maximum positive and negative values of pressure
$C_{p'}^-$ head fluctuations, for a certain runtime, divided by the velocity head upstream of the jump (Toso 1986);

E_0 = flow specific energy upstream of the jump;

F = mean value of hydrodynamic force;

F' = force fluctuation relatively to its mean value;

Fr = Froude number;

L_s = slab dimension along the flow direction;

L_j = length of jump;

L_x, L_y = macroturbulence longitudinal and transverse integral scales;

V_1 = flow mean velocity in the upstream section of the jump;

V_v = vortex convection velocity;

S = area;

γ = unit weight of water;

ρ = density of water;

τ = macroturbulence time scale.

Measurements of the Hydrodynamic Lift
and Drag Forces Acting on Riprap Side Slope

A.F. Ahmed[*] and F.S. El-Gamal[*]

Abstract

Direct measurements have been made of lift and drag forces acting on a 20.1 mm spherical particle placed within the riprapped 1.5 H: 1 V side slope of a 10 m long channel which was protected with a rock layer of 1.5 diameter equivalent thickness. A specially fabricated electro-mechanical load beam cell comprising four identical strain gauges was applied, for various flow rates, which allowed simultaneous recordings of the hydrodynamic lift and drag forces acting on test particle. To establish the appropriate sphere diameter as well as its location at the level of maximum wall shear, preliminary experiments were conducted for various flow rates. In addition, in order to assess the significance of employing a spherical particle during these measurements, the load beam was used to measure the hydrodynamic forces acting on four different wooden blocks simulating real particle shapes. The force measuring equipment, basic techniques, calibration and experimental procedures as well as the results obtained are the subject of the present paper.

Introduction

The major forces that affect stability of a single particle, and consequently the stability of the side slope, are the gravity force and the hydrodynamic lift and drag forces. The gravity force is the resultant of the particle weight acting vertically downward, and the effective lift pressure due to its submergence in water which acts vertically upwards. This can be equated with the submerged weight of the particle.

---

[*] Senior Researchers, the Hydraulics and Sediment Research Institute, Delta Barrage (13621), Egypt.

As the flow passes the particle a drag force parallel to the flow direction is exerted on the particle. While due to the difference in static pressure between the upper and lower faces of the particle a hydrodynamic lift force is generated, which acts perpendicular to the flow direction and the side slope plane. These hydrodynamic forces were found to be randomly fluctuating around a certain mean values (Einstein and El-Samny 1949). According to the prior studies, the drag and lift forces may be expressed as :

$$F_D = 1/2 \; C_D \; \rho \; u^2 \; D^2_{50} \quad (1)$$

$$F_L = 1/2 \; C_L \; \rho \; u^2 \; D^2_{50} \quad (2)$$

In which F_D and F_L are the mean drag and lift forces respectively; C_D and C_L are the drag and lift coefficients; u is some characteristic velocity; D_{50} is the representative diameter of the particle and ρ is the mass density of the water. Due to the irregularity of the particles comprising the riprap layer, non of the local velocities at any location above the particle could be adequately used as a characteristic velocity in Eqs.(1 and 2). Therefore, in the current study, the value equal to the average flow velocity was assumed.

So far many attempts were made to measure the hydrodynamic forces, which were mostly carried out on stream beds consisting of closely packed hemispheres or a single sphere resting on top of several layers of identical spheres (Einstein and El-Samny 1949; Chepil 1958; Coleman 1967; Watters and Rao 1971). This, in other words, means that no attempt has been made to measure the forces acting on a side slope riprap particle.

Basis of Approach

Assume a sphere subjected to drag and lift forces, denoted as F_D and F_L respectively, as schematically shown in Figure (1-A). This force system can be simplified to a beam, both ends fixed, and having span L, as shown in Figure (1-B), which is subjected at the middle of its span to lift force F and moment M. The resulting bending moment, due to the combined action of the applied loads, can be evaluated as the algebraic sum of that due to force F and moment M. Consequently the resulting bending moment distribution can be separately worked out as shown in Figure (1-C). Suppose two similar strain gauges were firmly bonded on both sides of the beam so as to be at equal distance X from the mid span which are denoted as L and R in Figure (1-C). The resulting moment at these points can be worked out as follows :

$$M_L = M_{LF} + M_{LM} \quad (3)$$

$$M_R = M_{RF} + M_{RM} \quad (4)$$

In which M_L and M_R are the total resultant moments at points L and R respectively; M_{LF} and M_{RF} are the resultant moments due to the force F at points L and R respectively and M_{LM} and M_{RM} are the resultant moments due to the moment M at points L and R respectively. It is obvious from Figure (1-D) that :

$$M_{LF} = M_{RF} \tag{5}$$

$$M_{LM} = -M_{RM} \tag{6}$$

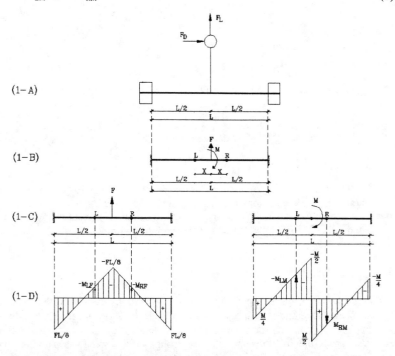

Figure 1. Bending Moment Distributions on the Load Beam

Substituting the external force F and moment M in Eqs.(3 and 4) yields :

$$F = f_1 (M_L + M_R) = f_1 (M_{LF} + M_{RF}) \tag{7}$$

$$M = f_2 (M_L - M_R) = f_2 (M_{LM} - M_{RM}) \tag{8}$$

In which f_1 and f_2 are parameters which have to be obtained theoretically or by calibration. On the other hand, for the case of a homogeneous beam with a rectangular cross section, similar to that used in the measuring devise, the resultant bending moment at any point along the beam was reported by Reeve (1975) as:

$$M_C = Z E \epsilon \tag{9}$$

In which ϵ is the applied strain which is the strain gauge reading; E is the modulus of elasticity of the material of the beam and Z is the elastic modulus of the cross section which for rectangular section of b width and h thickness is equal to $(bh^2/6)$. Applying Eq.(9) at points L and R respectively yields :

$$M_L = (bh^2/6) E \epsilon_L \tag{10}$$

$$M_R = (bh^2/6) E \epsilon_R \tag{11}$$

In which ϵ_L and ϵ_R are the strain that resulted at points L and R respectively. Substitute Eqs.(10 and 11) into Eqs.(7 and 8) respectively, one can obtain:

$$F = f_3 (\epsilon_L + \epsilon_R) \tag{12}$$

$$M = f_4 (\epsilon_L - \epsilon_R) \tag{13}$$

Eqs.(12 and 13) are the final formulae which can be applied to determine F and M values. In those equations the terms ϵ_L and ϵ_R are the strain gauge readings obtained at points L and R respectively, where the parameters f_3 and f_4 can be easily evaluated either theoretically or through the calibration process.

Force Measuring Device

Due to the fluctuating nature of the affecting hydrodynamic forces as well as their expected magnitudes, a specially fabricated electro-mechanical load beam cell was applied. The basic idea was to convert the simultaneous left and right signals, resulting from the hydrodynamic forces acting on the spherical particle, into simultaneous values of the lift and drag forces. This design was therefore specifically aimed to establish the relationships between those hydrodynamic forces and the output signals from the strain gauges.

The load cell, as schematically shown in Figure (2), consists of a simple beam made of flexible phosphor bronze. The beam was clamped at both sides by four stainless steel (s.s.) blocks which were firmly attached to a s.s. plate. A s.s. rod of 1.54 mm diameter and 100.7 mm length was then soldered at the mid point of the metallic beam and with a right angle through a base plate as shown in Figure (2). The other end of the rod was soldered into a threaded nipple which can be used to mount either the fabricated spherical or non-spherical particle. Two similar load cells were designed to enable simultaneous left and right signals to be recorded which would be in turn converted into simultaneous time series of the lift and drag forces. Each load cell, consists of a bridge of four (FLA-6-17) type strain gauges. Two of them were fixed into a dummy gauge holder, whereas the other two were firmly soldered onto the upper and lower faces of the beam at an equal distance from its mid span.

It is also worth mentioning that the two load cell bridges were designed in such way as not to alter the zero reading due to any variation in the water temperature (Norton 1969). This means, in other words, that as long as the four strain gauges contained within one bridge were kept at the same water temperature, the temperature would not have any effect on the zero reading.

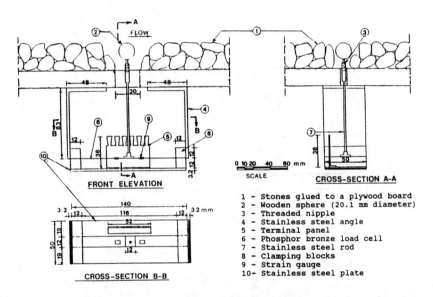

1 - Stones glued to a plywood board
2 - Wooden sphere (20.1 mm diameter)
3 - Threaded nipple
4 - Stainless steel angle
5 - Terminal panel
6 - Phosphor bronze load cell
7 - Stainless steel rod
8 - Clamping blocks
9 - Strain gauge
10- Stainless steel plate

Figure 2. Details of the Load Beam Assembly

Calibration of the Load Beam

The set-up shown in Figure (3) was employed to establish the relationships between the displayed output of the signal conditioner, which was connected to the load cells, and the applied lift and drag forces. The process was carried out in such a way as to simulate the testing condition. The load beam was submerged into a still water with almost the same temperature as that expected during the experimental work. Two stainless steel helical springs were separately calibrated to obtain a relationship between the applied static force and the corresponding elongations. The helical springs were then utilized, as shown in Figure (3), to apply various possible combinations of static forces which represent lift and drag. To measure the left and right strains (ϵ_L and ϵ_R) due to the applied static forces, the two load beam cells denoted as left and right were connected to a signal conditioner via two channels. More details of the calibration procedure and results are given by Ahmed (1988).

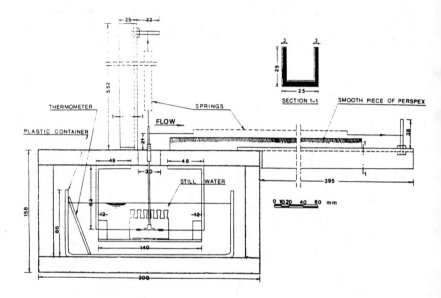

Figure 3. Calibration Set-up of the Load Beam

Particle Size Modelling

The size of the spherical particle employed in the force measurements was experimentally determined by considering the real particle configuration. The particle was designed in such a way as to reduce its bulk specific gravity to unity which consequently makes its submerged weight equal to zero. This design was accomplished by drilling a few holes in the sphere to extract some timber material then the holes were sealed. The sphere was then coated with a thin layer of fine sand to simulate the natural roughness of the rock surface.

For the purpose of placing the manufactured particle at the point of maximum side slope shear, a preliminary experiment was conducted. An artificially roughened surface was utilized so as to simulate the side slope boundary roughness similar to that technique adopted by Ghosh and Roy (1970). Using a Preston tube, which was manufactured according to the design recommended by Ippen and Drinker (1962), the location of the maximum wall shear for different flow rates was defined. This was mostly found at a vertical distance of 0.29 y to 0.35 y above the channel bed, where y is the flow depth, which is in accordance with the results obtained by Lane (1955).

Model Description and Testing Procedure

General layout of the model is shown in Figure (4). The investigation was conducted in a recirculating flume which had a working length of 21.4 m, a rectangular cross-section of 1.37 m wide and 0.61 m deep. A model of a trapezoidal cross section channel of 10 m long, 0.5 m bed width and two side slopes 1.5 H: 1 V was constructed in the flume. This channel (See Figure 4) was mainly constructed with a sand base, a sheet cloth filter and a protective layer consisting of uniform rock particles with a mean size of 20.1 mm and a thickness equivalent to 1.5 particle diameters. Twelve runs covering a wide range of flows were performed. During each test the simultaneous values of drag and lift as well as the necessary information about the flow conditions were recorded. More details are given by Ahmed (1988).

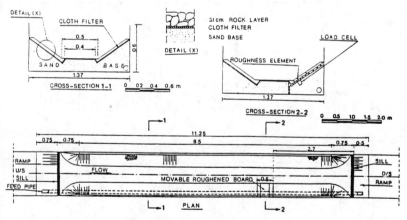

Figure 4. General Layout of the Model

Furthermore, in order to assess the effect of particle shape and orientation, four wooden particles with different shapes were made for measuring the hydrodynamic forces. The particle shapes were randomly selected similar to four real particles used in the model. Applying the same technique utilized before, the forces acting on all of these particles were acquired for only one flow condition.

The Results

The results obtained in the tests with the spherical particle revealed that the hydrodynamic lift and drag forces are randomly fluctuating around mean values, whereas the distributions of the instantaneous values were found to be approximately normal (See Figure 5).

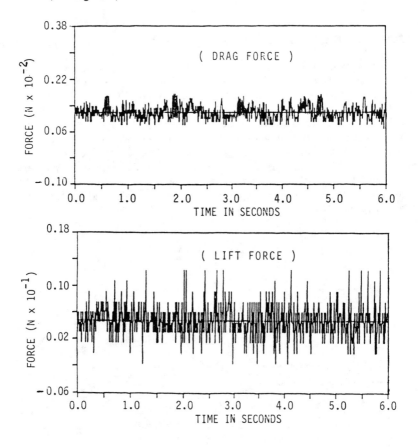

Figure 5. Recorded Fluctuations of Drag and Lift Forces
(Run No. 4 - Q = 0.065 m^3/s)

Further, relationships have been developed between both the lift and drag coefficients of Eqs.(1 and 2) and the ratio of lift to drag. Correlation was confirmed between these parameters and the values of both the relative roughness parameter (R/D_{50}) and the Reynolds number (R_e). These were found to be defined by simple power equations of the following forms :

$$C_D = 0.011 \, (R/D_{50})^{-0.27} \tag{14}$$

$$C_L = 0.063 \, (R/D_{50})^{-0.56} \tag{15}$$

$$C_D = 0.062 \, (R_e)^{-0.17} \tag{16}$$

$$C_L = 2.381 \, (R_e)^{-0.36} \tag{17}$$

$$\beta = 5.272 \, (R/D_{50})^{-0.29} \tag{18}$$

$$\beta = 38.40 \, (R_e)^{-0.19} \tag{19}$$

In which R is the hydraulic radius and β is the ratio of lift to drag forces. The results obtained for all non-spherical particles at the same flow condition revealed that the magnitude of the measured forces, (See Table 1), is dependent on the shape and the projected area of the particle with respect to drag and lift forces as well as the forme of flow fields around the particle. This obviously due to the difference between the projected area of the particle in perpendicular directions to the lift and drag forces.

Table 1. Results of Force Measurements on Non-Spherical Particles

Description	Particle No.			
	(1)	(2)	(3)	(4)
[Drag force]				
Mean F_D (N x 10^3)	2.65	2.07	1.75	1.84
standard deviation x 10^3	0.44	0.36	0.23	0.21
Drag intensity (DI) %	16.76	17.36	12.91	11.67
F_D (non-sphere)/F_D (sphere)	2.18	1.71	1.45	1.51
[Lift force]				
Mean F_L (N x 10^3)	6.79	4.07	4.42	5.27
Standard deviation x 10^3	2.91	2.39	2.57	2.60
Lift intensity (LI) %	42.80	58.71	58.19	49.35
F_L (non-sphere)/F_L (sphere)	1.42	0.85	0.93	1.10
Lift to drag ratio	2.56	1.96	2.52	2.87

Summary and Conclusions

For the purpose of formulating stability criteria for side slopes, a specially fabricated measuring device was employed to acquire simultaneous values of lift and drag forces acting on simulated spherical and non-spherical particles placed within the riprap layer. The measurements acquired for a 20.1 mm spherical particle confirmed the correlation between both lift and drag coefficients and the values of both the relative roughness parameter and the Reynolds number. Contrarily, the measurements obtained for four different non-spherical particles revealed a diversity of the results. These led to the conclusion that employing non-spherical particle for formulating stability criteria for the side slope protective layer is unjustified.

References

Ahmed, A.F. (1988)."Stability of Riprap Side Slopes in Open Channels," Thesis presented to University of Southampton, U.K., in partial fulfillment of the requirements for the degree of Doctor of Philosophy.

Chepil,W.S. (1958)."The Use of Evenly Spaced Hemispheres to Evaluate Aerodynamic Forces on a Soil Surface," Trans. American Geophysical Union, 39 (3), 397-404.

Coleman, N.L. (1967)." A Theoretical and Experimental Study of Drag and Lift Forces Acting on a Sphere Resting on a Hypothetical Stream Bed," Proc. of the 12th Congress of IAHR, Vol. 3, Fort Collins, C22-1-C22-8.

Einstein,H.A. and El-Samni, E.A. (1949)." Hydrodynamic Forces on a Rough Wall," Review of Modern Physic, 21 3) ,520-524.

Ghosh, S.N. and Roy, N. (1970)."Boundary Forces on a Rough Wall," Proc. Hyd. Div., ASCE, Vol.96, No.HY4, pp.967-994.

Ippen, A.T. and Drinker, P.A. (1962)."Boundary Shear Stresses in Curved Trapezoidal Channels", Proc. Hyd. Div., ASCE, Vol.88, No.HY5, pp.143-179.

Lane, E.W. (1955)."Design of Stable Channels," Trans. ASCE, Vol. 120, 1234-1260.

Norton, H.N. (1969)." Handbook of Transducers for Electronic Measuring Systems," Prentice - Hall, Inc., Englewood Cliffs, N.J.

Reeve, A. (1975)."Transducers", Welwyn Strain Measurement Ltd. Basingstoke, England.

Watters, G.Z. and Rao, M.V.P. (1971). " Hydrodynamic Effects of Seepage on Bed Particles", Proc. Hyd. Div., ASCE, Vol. 97, No.HY3, pp.421-439.

Recent Applications of Acoustic Doppler Current Profilers

K. A. Oberg[1] and David S. Mueller[2], M.ASCE

ABSTRACT

A Broadband acoustic Doppler current profiler (BB-ADCP) is a new instrument being used by the U.S. Geological Survey (USGS) to measure stream discharge and velocities, and bathymetry. During the 1993 Mississippi River flood, more than 160 high-flow BB-ADCP measurements were made by the USGS at eight locations between Quincy and Cairo, Ill., from July 19 to August 20, 1993. A maximum discharge of 31,400 m^3/s was measured at St. Louis, Mo., on August 2, 1993. A BB-ADCP also has been used to measure leakage through three control structures near Chicago, Ill. These measurements are unusual in that the average velocity for the measured section was as low as 0.03 m/s. BB-ADCP's are also used in support of studies of scour at bridges. During the recent Mississippi River flood, BB-ADCP's were used to measure water velocities and bathymetry upstream from, next to, and downstream from bridge piers at several bridges over the Mississippi River. Bathymetry data were collected by merging location data from Global Positioning System (GPS) receivers, laser tracking systems, and depths measured by the BB-ADCP. These techniques for collecting bathymetry data were used for documenting the channel formation downstream from the Miller City levee break and scour near two bridges on the Mississippi River.

INTRODUCTION

Acoustic Doppler current profilers (ADCP's) have been in use for more than 10 years, primarily in the study of ocean currents and estuaries. Within the last 5 years, ADCP's have been used to measure streamflow, especially in rivers or canals where conventional discharge-measurement techniques are either very expensive or impossible. Recently, a more advanced ADCP, known as a Broadband ADCP (BB-ADCP) has been developed that can measure depths and velocities in shallow waters (as shallow as 1.4 m) and with a high degree of vertical resolution (0.25 m).

The purpose of this paper is to summarize recent applications of BB-ADCP's by the USGS. The principles of operation are introduced and the application of BB-ADCP's to the measurement of

---

[1]Hydrologist, U.S. Geological Survey, 102 E. Main St., Urbana, IL 61801
[2]Hydrologist, U.S. Geological Survey, 2301 Bradley Ave., Louisville, KY 40217

high, low, and unsteady flows, water velocity near bridge piers, and channel bathymetry are discussed.

PRINCIPLES OF OPERATION

A BB-ADCP can be used to measure vertical profiles of water velocities from a moving boat. Water velocities are measured by the BB-ADCP, which transmits pairs of short, phase-encoded acoustic pulses along four narrow beams at a known, fixed frequency (from 75-1,200 kHz, depending on the transducer; see figure 1). These beams are positioned 90 degrees apart horizontally and at a known angle (typically 20 or 30 degrees) from the vertical. The BB-ADCP detects and processes the echoes from successive volumes (depth cells) along the beams. The time-lag change and difference in frequency (shift) between successive echoes is proportional to the relative velocity between the BB-ADCP and suspended material in the water that reflects the pulses back to the BB-ADCP (back scattering). This frequency shift is known as the Doppler effect. The BB-ADCP uses the Doppler effect to compute a water-velocity component along each beam. By means of simple trigonometric relations and water velocities calculated from adjacent beams, the BB-ADCP can compute water velocity and direction. BB-ADCP software can be used to set measurement parameters, such as depth-cell size and time-lag length.

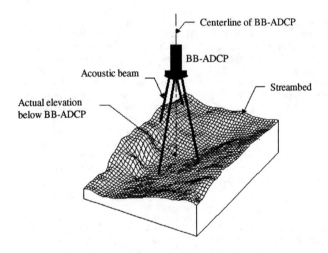

Figure 1.-- Example of Broadband acoustic Doppler current profiler and depth measurement

Because water-velocity measurements are made relative to the movement of the boat, the velocity of the boat also must be accurately measured. This can be accomplished by several methods. The most common method is called bottom-tracking and is performed by the BB-ADCP. The boat velocity relative to the river bottom is computed by the BB-ADCP using a flux-gate compass and the results of measurements of the Doppler shift of acoustic pulses reflected from the river bottom. Bottom-tracking measurements can be made with greater accuracy than water-velocity measurements because a longer pulse is used for bottom-tracking and return echoes from the bottom are much stronger than echoes from most particulates suspended in the water column. In addition to measuring boat velocity, the depth of the river is estimated from the amplitude of the bottom-track echoes (echoes returned from the bottom). Alternative methods for measuring the boat velocity include real-time differential global-positioning systems (DGPS) and long-range navigation systems (LORAN). A range-azimuth laser tracking system, or a nonreal-time DGPS, could also be used, but the software used to collect BB-ADCP data does not currently support such positioning systems.

When the BB-ADCP is being used to measure discharge, it transmits a series of acoustic pulses known as pings. Pings for measuring water velocities are known as water pings. Pings for measuring the boat velocity are known as bottom-tracking pings. Normally, water pings and bottom-tracking pings are interleaved during transmission. A group of these interleaved water and bottom-tracking pings are referred to as an ensemble. The number of water and bottom-tracking pings per ensemble is set by the user. An ensemble is analogous to one vertical in a conventional discharge measurement. For example, a typical ensemble is composed of five water pings and four bottom-tracking pings. The velocities and depths measured for each ping are averaged to yield a single velocity profile and depth for each ensemble. Increasing the number of pings per ensemble slows down the rate at which the BB-ADCP makes measurements but does not necessarily increase measurement accuracy. In a conventional discharge measurement, velocity is measured at two points in the vertical unless the depths are less than 1.5 m. The BB-ADCP can measure velocities every 0.25 m in the vertical, so that one ensemble for a vertical 10 m deep may contain as many as 34 velocity measurements.

A single discharge measurement, called a transect, is a collection of ensembles for a measuring section. A typical BB-ADCP transect will contain at least 60 ensembles, whereas a conventional discharge measurement will typically consist of 25 verticals. When measuring under relatively steady flow conditions, five or more transects are typically made at each measuring location. The mean of all the transect discharges is then used as the measured discharge. This procedure is not used when measuring under unsteady-flow conditions, where a single transect may be used as a discharge measurement. For a detailed description of the BB-ADCP and its application to streamflow measurement, the reader is referred to RD Instruments[3] (1989), Gordon (1989), RD Instruments (1993), and Simpson and Oltmann (1993).

The BB-ADCP cannot measure water velocities near the top and bottom of the water column. Water velocities near the surface cannot be measured for two reasons: the BB-ADCP must be deployed so that transducers remain under water during a measurement. In addition, the physical characteristics of the transducers are such that accurate velocity measurements cannot be made within 0.50 m of the transducer. Water velocities also cannot be measured near the bottom of the water column because of a phenomenon known as side-lobe interference. The return signals from particulates near the bottom are distorted by echoes from the riverbed directly below the BB-ADCP. (See Gordon (1989, p. 929) and Simpson and Oltmann (1993, p. 6) for more details on side-lobe interference and its effect on water-velocity measurements.) For a 1200-kHz BB-ADCP with a 20-degree beam angle, about 6% of the profiling range is lost because of side-lobe interference. The BB-ADCP software automatically rejects water-velocity data collected from beyond about 94% of the distance to the bottom.

The BB-ADCP can use two methods for estimating the water velocities near the surface or near the bottom: the constant method or the power method. With the constant method, the velocity at the surface or the bottom is assumed to equal the velocity of the first or last measured depth cell, respectively. This method is considered inappropriate for estimating the velocities near the bottom because it does not accurately represent typical vertical-velocity distributions for open-channel flow.

---

[3] Use of firm names in this paper is for identification purposes only and does not constitute endorsement by the U.S. Geological Survey.

In open-channel flow, the velocity approaches zero as the bottom is approached. The power method is based on a power-velocity distribution law. In this method, a least-squares fit of the measured water velocities is obtained using the power-velocity distribution law. The exponent of the function can be selected by the user. The exponent is typically set to 1/6, based on the 1/6 power law suggested by Chen (1989). The function is then used to estimate velocities in the unmeasured part of the water column. Conceptually, the power method is better for estimating the unmeasured part of the water column near the bottom. For the BB-ADCP measurements discussed in this paper, the constant method was used to estimate the unmeasured part of the water column near the surface and the power method with an exponent of 1/6 was used to estimate the unmeasured part of the water column near the bottom.

The BB-ADCP cannot measure water velocities near either edge of the section being measured. One reason is that the BB-ADCP's cannot accurately measure velocity in shallow water (less than 1.4 m for BB-ADCP applications discussed in this report). If the unmeasured discharge area is assumed to be triangular, the velocity for the unmeasured section V_e, is estimated by the equation of Simpson and Oltmann (1993, p. 9),

$$V_e = 0.707 \, V_m , \qquad (1)$$

where V_m is the mean velocity at the first or last BB-ADCP measured subsection. The assumption that the unmeasured flow area is triangular is reasonable for many river cross-sections where the bottom gradually slopes upwards towards the shore. However, sometimes the edge of water is a vertical wall, such as a sea wall. As the BB-ADCP approaches a vertical wall, the acoustic beams impinge the wall and cause a false bottom return. The distance at which the acoustic beam impinges the wall depends on the depth of water near the wall and the orientation of the transducers on the BB-ADCP relative to the wall. Velocities for the unmeasured edge sections could then be estimated by setting $V_e = V_m$. However, this is not entirely accurate because the velocity does decrease to zero as the wall is approached. Therefore, for the measurements described in this report, velocities near vertical walls were estimated by the following equation:

$$V_e = 0.91 \, V_m . \qquad (2)$$

The coefficient of 0.91 was estimated from data presented in Rantz and others (1982, p. 82) showing the relation of the distance from a smooth wall expressed as a ratio of the depth and mean vertical velocity. The edge discharge may then be estimated using the equation

$$Q_e = (C * V_m * L * d_m), \qquad (3)$$

where Q_e = estimated edge discharge,
         C = coefficient equal to 0.707/2 or 0.91, depending on channel shape,
         L = distance to the shore from the first or last BB-ADCP measured subsection, and
         d_m = depth at the first or last BB-ADCP measured subsection.
d_m is measured by the BB-ADCP. L is measured with a tagline or a similar measuring device.

RECENT APPLICATIONS OF BB-ADCP'S

Measurement of High Flows

During the 1993 Mississippi River flood, more than 160 high-flow BB-ADCP transects were made by the USGS at eight locations between Quincy and Cairo, Ill., from July 19 to August 20, 1993. Streamflow measurements were made using a 300-kHz and a 1200-kHz BB-ADCP. Initial attempts to measure discharge on July 19 at St. Louis, Mo., using a 1200-kHz BB-ADCP were unsuccessful. The BB-ADCP could not bottom-track continuously across the cross section being measured. It would lose bottom-track where the depths were deepest and water velocities highest. The 1200-kHz BB-ADCP apparently did not have enough acoustic energy to penetrate the water column and mobile bed layer at this location.

On the basis of this experience, the USGS contracted with the U.S. Army Corps of Engineers, New Orleans District, for the use of the 300-kHz BB-ADCP, beginning on July 26, 1993. During the next week, measurements were made with the 300-kHz and 1200-kHz BB-ADCP's. Discharges measured by the 1200-kHz BB-ADCP at St. Louis on July 26-27 were approximately 30% less than discharges measured by conventional techniques and by the 300-kHz BB-ADCP. The probable reason for this discrepancy was the inability of the 1200-kHz unit to penetrate the mobile bed. This hypothesis was tested on July 26-27. Each BB-ADCP was tied in turn to a barge anchored near the main navigation channel approximately 7 km downstream from St. Louis, and each was used to measure water and bottom-track velocities. Bottom-track velocities from the 1200-kHz BB-ADCP indicated that the boat was moving upstream at a rate of approximately 0.60 m/s at this location. These measured bottom-track velocities confirmed the hypothesis and indicated the speed of the mobile bottom layer at this location. For this reason, water velocities measured by the 1200-kHz unit at this location were biased low because the BB-ADCP bottom-tracking algorithm assumes a stable bed. The bottom-tracking measurements from the 300-kHz BB-ADCP correctly indicated that the boat was not moving while attached to the barge. Preliminary results of comparisons between measurements made with the 300-kHz and 1200-kHz units indicate that the 1200-kHz unit can be used on the Mississippi River upstream from the confluence of the Mississippi and Missouri Rivers. These results may be attributed to lower sediment concentrations and a more stable bed in the Mississippi River above the confluence with the Missouri River.

A maximum discharge of 31,400 m^3/s was measured at St. Louis, Mo., on August 2, 1993, with the 300-kHz BB-ADCP. Preliminary results show that 300-kHz BB-ADCP measurements at St. Louis were within 1% of concurrent current-meter measurements made during July 26-August 20, 1993. During August 2-10, 1993, the 300-kHz BB-ADCP was the only method available for measuring discharge at St. Louis, because the monorail used to make conventional streamflow measurements on the upstream side of the bridge was damaged. At about 10 p.m. on August 1, 1993, one of several boats moored near The Arch, in downtown St. Louis, broke loose and hit the deck of the Poplar Street Bridge, damaging the monorail and making conventional measurements impossible for a time.

Both BB-ADCP's were used to measure streamflow in the breach in the Len Small Levee. This breach is known as the Miller City levee break. The Len Small Levee failed near Miller City about 55 km upstream from Cairo, Ill., on July 15, 1993, eventually allowing water from the Mississippi River to cross a high-amplitude meander bend and to reenter about 24 km upstream from Cairo. BB-ADCP measurements were made in response to concerns expressed by several State agencies that this flow might cause the formation of a meander cutoff. The 300-kHz BB-ADCP was used to measure the difference in streamflow in the Mississippi River upstream and downstream from the levee break. The 1200-kHz BB-ADCP was used to measure stream discharge and velocities, and channel bathymetry within the break. BB-ADCP measurements of flow in the levee break itself were difficult to make because of the very turbulent flow through the breach. Measuring conditions were especially difficult during the 10 days before and after the peak Mississippi River stage on August 10, 1993. Despite these poor streamflow-measurement conditions, the 1200 kHz BB-ADCP was used to measure velocities and channel bathymetry from July 31 to October 25, 1993. Kinematic differential GPS was used to obtain the horizontal position of the boat while making measurements with the 1200-kHz BB-ADCP in and downstream from the levee break. A subsequent section discusses the techniques for making bathymetric measurements with a BB-ADCP.

Measurement of Low Flows

Low-flow BB-ADCP measurements were made to determine the amount of leakage through three controlling structures near Chicago, Ill.: the Chicago River Controlling Works (CRCW), the Wilmette pumping station (Wilmette), and the Thomas J. O'Brien Lock and Dam (O'Brien). The

CRCW includes the Chicago Lock and two sets of sluice gates located in harbor walls near the mouth of the Chicago River (fig. 2). These measurements are unusual in that the average velocity for the measured section was often less than 0.10 m/s. Few instruments can accurately measure velocities this low. Price AA meters have been fitted with an optic head to increase the accuracy of low-flow-velocity measurements. However, optic-head meters are not rated for velocities less than about 0.05 m/s and cannot resolve the flow direction, except at the surface. Although acoustic point-velocity meters can be used to measure velocities less than 0.05 m/s, measurement time is not less than that required for a conventional discharge measurement. The BB-ADCP was considered advantageous because of its ability to make measurements of discharge quickly from a moving boat.

The accuracy of water-velocity measurements with a BB-ADCP depend on the speed of the boat, among other factors. The rule-of-thumb is that the average boat velocity should be approximately equal to or less than the average water velocity. It was necessary, therefore, to control the velocity of the boat more accurately than could be done with a motor. A two-speed sailing winch was mounted on the boat used for BB-ADCP measurements, and a rope was stretched across the measuring section. The winch was used to pull the boat across the measuring section at a fairly constant speed.

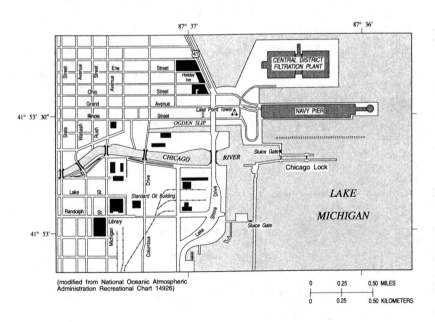

Figure 2. Location of the Chicago River Controlling Works, the Chicago Lock, and the Chicago River, Chicago, Illinois.

Using these techniques, measurements of leakage as low as 1.4 m^3/s were obtained at Wilmette and 0.7 m^3/s at O'Brien. During leakage measurements at the Chicago Lock in July 1993, the mean leakage through the river gate of the lock was 5.8 m^3/s; the standard deviation was 0.3 m^3/s. Average velocities for these leakage measurements were about 0.03 m/s. Averaged measured velocities in the Chicago River were even lower, primarily because the cross-sectional flow area in the Chicago River is approximately double that in the Chicago Lock. Although the Chicago River measurements were more variable than measurements made in the Chicago Lock, the measured velocities and discharges were not unrealistic. The greater variability in the Chicago River measurements may be attributed to the very low velocities and the consequent measurement errors. However, the variability in flow may be real. For example, during a day when the wind was blowing upstream, it was visually apparent that the wind was actually "pushing" water near the surface upstream. This visual observation was confirmed by the BB-ADCP, which indicated a shear in the vertical-velocity profile, with the top water-layer discharge being consistently in the upstream direction. It would not be unusual to have some minor oscillation of flow as a result of wind and wave action at this location.

Measurement of Unsteady Flows

A BB-ADCP also has been used to measure unsteady flows upstream from a hydropower facility. The USGS operates an acoustic velocity meter (AVM) 11 km upstream from the Lockport Powerhouse on the Chicago Sanitary and Ship Canal. The AVM has been operated since 1988 at this location in partial fulfillment of requirements made by a U.S. Supreme Court decree regulating the amount of water diverted from Lake Michigan by the State of Illinois. Conventional discharge measurements are difficult at this site because of the unsteady flows. A conventional discharge measurement requires at least 90 minutes, yet the discharge in the canal can triple in 15 to 20 minutes. The BB-ADCP is, therefore, being used to refine the calibration of the AVM. Figure 3 shows the results of a set of BB-ADCP transects made in July 1993 at the AVM. Each BB-ADCP discharge plotted in figure 3 is for an individual transect. The AVM discharges shown are the average of 1-minute AVM discharges for the time period of each corresponding BB-ADCP transect. The mean difference between BB-ADCP and AVM discharges is less than 3%.

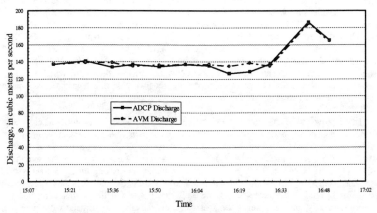

Figure 3.--Discharge measured in the Chicago Sanitary and Ship Canal with Broadband acoustic Doppler current profiler and acoustic velocity

Streambed Scour at Bridges

Detailed field measurements are needed to extend the knowledge of processes responsible for scour of the streambed at bridges. Prior to the development of the BB-ADCP, velocity measurements were normally limited to single-point measurements of magnitude only. Two-dimensional velocities could be measured by a vertical or horizontal axis meter with a flux-gate compass mounted in the weight or by an electromagnetic velocity meter. The capability of the BB-ADCP to measure a three-dimensional velocity profile from a moving deployment platform significantly increases the detail at which the hydraulics associated with scour at bridges can be measured.

The primary difference between collecting of velocity and bathymetric data and making discharge measurements with a BB-ADCP is the horizontal-positioning requirement. The exact position of each ensemble of data for a discharge measurement is not critical; the software for the BB-ADCP automatically computes the discharge of the section traversed. However, for analysis and modeling of hydraulic and sediment-transport processes, the position of each ensemble is very important. GPS and a range-azimuth laser tracking system were used to measure the horizontal positions of each ensemble.

Initially for bridge-approach- and exit-reach data, a nondifferential GPS was used to measure the location of the start and end of each transect. With the start and end of each transect recorded in the field notes, the location of each ensemble can be calculated from the bottom-track data collected by the BB-ADCP by a user-developed postprocessing routine. The accuracy of the nondifferential GPS was estimated to be about 30 m in the horizontal. This accuracy is marginally acceptable for bridge-approach- and exit-reach surveys and is unacceptable for detailed hydraulic and bathymetric data near the bridge.

Real-time kinematic differential GPS (DGPS) was used for streambed-scour measurements in cooperation with the U.S. Army Corps of Engineers, St. Louis District. Because the base station and transmission towers had been set up quickly to facilitate work on the recent Mississippi River flood, the accuracy of the differential correction had not been determined. However, the accuracy of the position data collected is believed to be less than 1 m. The software used with the BB-ADCP is compatible with a DGPS and allowed the position data to be recorded concurrently with the ensemble of data from the BB-ADCP. The software also allowed either bottom track or DGPS to be used as the vessel-velocity reference. Visual comparisons of the vessel speeds measured by DGPS and a 600-kHz BB-ADCP during data collection showed very good correlation. Although the DGPS positions and ensemble data are recorded concurrently, they are recorded in different files. To determine the exact position of each ensemble requires the user to develop a postprocessing program to combine the information recorded in the two files.

DGPS proved to be an accurate and efficient method for positioning detailed velocity and bathymetric data during bridge-approach- and exit-reach surveys. Near bridges, however, the bridge structure often blocked the sky sufficiently to cause the roaming DGPS unit to lose 'lock' on a sufficient number of satellites. A range-azimuth laser tracking system was used to measure the position of the BB-ADCP near bridges. The range-azimuth laser tracking system was set up on shore, and a target on the vessel was manually tracked. Because the range-azimuth laser tracking system is not directly compatible with the software for the BB-ADCP, the position of the boat was recorded separately, and notes were taken to associate the velocity measurements with the recorded position data. Again, a user-written postprocessing routine was required to correlate the recorded positions with the ensembles collected by the BB-ADCP. The range-azimuth laser tracking station worked well at the bridges, but the need to survey multiple setups to collect bridge-approach- and exit-reach surveys would have been time consuming. Therefore, the combination of DGPS and the range-azimuth laser tracking systems provided an accurate and efficient method of measuring the horizontal position of the boat while collecting data with the BB-ADCP.

Detailed velocity data were collected in the bridge-approach and exit reaches and near the bridge, however, velocity measurements of the turbulence and vortices adjacent to the bridge piers could not be made. The processing algorithm used in the BB-ADCP assumes that the water velocity along a horizontal plane passing through the four beams is uniform. The turbulence and vortices adjacent to the piers were often smaller than the area bounded by the four beams; thus, the flow measured by one beam was not continuous with flow measured by another beam. Additional research and signal processing will be required before three-dimensional velocity-profile measurements in the vortices adjacent to the piers can be made in the field.

Bathymetric surveys can be made with the BB-ADCP if accurate position data are collected and correlated with each ensemble. Each ensemble collected by the BB-ADCP includes velocity, instrument attitude, and water depth. Although the BB-ADCP corrects the measured velocity and bin depth for the instrument attitude, the depth of water measured by the individual beams is not corrected. When the BB-ADCP is measuring discharge, the BB-ADCP software simply uses the average of the depths measured along each of the four beams as the depth of flow for each ensemble. This averaging technique may yield an average depth that is not representative of the depth below the BB-ADCP (figure 1). Because the beam depths and instrument attitude data are stored, a postprocessing routine can be developed that could correct the depth and position of each beam depth for the instrument attitude. Use of such a postprocessing routine would yield four depths and positions for each ensemble of data, thereby providing very detailed bathymetric data.

SUMMARY

A Broadband acoustic Doppler current profiler (BB-ADCP) is a relatively new instrument that has a number of potential uses, including the measurement of stream discharge and velocities, and bathymetry. During the 1993 Mississippi River flood, more than 160 high-flow BB-ADCP measurements were made by the USGS at eight locations between Quincy and Cairo, Ill., from July 19 to August 20, 1993. A maximum discharge of 31,400 m^3/s was measured at St. Louis, Mo., on August 2, 1993. Low-flow BB-ADCP measurements also have been made to determine the amount of leakage through three control structures near Chicago, Ill. Discharges as low as 1.4 m^3/s and velocities as low as 0.03 m/s have been measured. BB-ADCP's are also being used in support of studies of scour at bridges. During the recent Mississippi River flood, BB-ADCP's were used to measure water velocities and bathymetry upstream from, next to, and downstream from bridge piers of several bridges over the Mississippi River. Bathymetry data also were collected by merging location data from GPS receivers, a range-azimuth laser tracking system, and depths from the BB-ADCP. Preliminary results indicate that a combination of DGPS and the laser tracking system are needed to make detailed bathymetric and velocity measurements for use in assessments of scour at bridges. Nonreal-time DGPS was used in conjunction with a BB-ADCP to collect bathymetry data for documenting channel formation downstream from the Miller City levee break.

REFERENCES

Chen, Cheng-lung. 1989. "Power law of flow resistance in open channels--Manning's formula revisited." <u>International Conference on Channel Flow and Catchment Runoff: Centennial of Manning's Formula and Kuichling's Rational Formula</u>, Charlottesville, Virginia, May 22-26, 1989, p. 817-848.

Gordon, R.L. 1989. "Acoustic measurement of river discharge." <u>J. Hydr. Engrg.</u>, ASCE, 115 (7) p. 925-936.

Rantz, S.E., and others. 1982. Measurement and computation of streamflow, Volume 1, Measurement of discharge: U.S. Geological Survey Water-Supply Paper 2175, 631 p.

RD Instruments. 1989. Acoustic Doppler current profilers, Principles of operation: A practical primer: RD Instruments. San Diego. 36 p.

RD Instruments. 1993. Direct-reading broadband acoustic Doppler current profiler technical manual: RD Instruments. San Diego. 52 p.

Simpson, M.R., and Oltmann, R.N. 1993. Discharge-measurement system using an acoustic Doppler current profiler with applications to large rivers and estuaries: U.S. Geological Survey Water-Supply Paper 2395, 32 p.

ACOUSTIC-DOPPLER VELOCIMETER (ADV) FOR LABORATORY USE

Atle Lohrmann[1], Ramon Cabrera[2], and Nicholas C. Kraus[3], M. ASCE

Abstract

A remote-sensing, three-dimensional (3D) velocity sensor has been developed and tested for use in physical model facilities. The sensor is based on the Acoustic Doppler principle and measures the 3D velocity at a rate of 25 Hz in a sampling volume of less than 1 cm^3. The sensor is mechanically rugged, can be used in open-air physical model facilities, and can be readily moved from location to location with minimum setup time. A convenient interface operated on a PC allows instrument setup, file naming, and visual monitoring of the three flow components from simple menus. The instrument can replace several types of flow measurement instruments such as propeller gauges, electromagnetic gauges, Pitot tubes, LDVs, and hot-film anemometers, thereby simplifying procedures for technicians while providing continuous digital records at user-specified sampling rates.

Introduction

The development of the Acoustic-Doppler Velocimeter (ADV) was initiated under contract by the U.S. Army Engineer Waterways Experiment Station (WES) in 1992 to satisfy the need for an accurate current meter that can measure 3D dynamic flow in physical models. Such facilities include uniform-flow flumes, spillway models, 2D and 3D shallow-water wave basins, and outdoor near-prototype-scale riprap facilities. A requirements analysis conducted by two

---

[1] Senior Oceanographer, SonTek, 7940 Silverton Ave., #105, San Diego, CA 92126.
Tel (619) 695-8327, Fax: (619) 695-8131.

[2] Senior Engineer, SonTek.

[3] Director, Conrad Blucher Institute for Surveying and Science, Texas A&M University - Corpus Christi, TX 78412, Tel: (512) 994-2646, Fax: (512) 994-2715

laboratories at WES resulted in a development project to identify and implement a technology that would allow sub-centimeter resolution and minimum 25 Hz sampling rate while achieving a commercial unit cost of less than $10,000. In addition, primary importance was attached to achieving measurements close to solid boundaries, easy integration with existing data-acquisition systems, and minimum need for recalibration.

To meet the requirements, a "bistatic" (focal-point) ADV was designed and built. The Doppler measurement technique, in contrast to the acoustic travel-time and electromagnetic techniques, has the advantage of being inherently drift-free and does not require routine recalibration. Also, acoustic pulses do not suffer the range limitations of optical (light) pulses in turbid water.

In the following sections, we describe the measurement principle and the instrument calibration, and then examine the size and shape of the sampling volume in some detail. We also report preliminary results from specialized applications such as ultra-low flows (<1 cm/s), characterization of turbulent parameters, and the possibility of using ADVs to make concurrent measurements of velocity and sediment concentration.

Principle of operation

Fig. 1 shows an overview of the ADV as implemented by SonTek (ADV-1). The system has three main modules: measurement probe, signal conditioning module, and signal processing module. In the present implementation [Kraus et al. 1994], three 10-MHz receive elements are positioned in 120° increments on a circle around a 10-MHz transmitter. The probe is submerged in the flow and the receivers are slanted at 30° from the axis of the transmit transducer and focus on a common sampling volume. The volume is located either 5 or 10 cm from the probe to reduce flow interference.

The transducer array is mounted at the end of a thin stem to

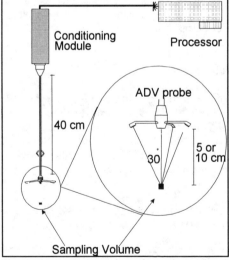

Fig. 1 The ADV-1 consists of a probe, a signal conditioning module, and a processing module.

facilitate installation of the instrument in typical laboratory setups, and to minimize flow interference by keeping the main bulk of the instrument away from the measured flow. The other end of the stem is attached to a 50-mm diameter, 300-mm long waterproof housing containing the signal conditioning module. This module contains all of the sensitive analog electronics that permit detection of the weak backscattered signals.

Digital signal processing and system control is performed by a single circuit card that fits into a 16-bit IO-slot of an IBM AT-compatible computer. This card contains a powerful signal processor that can perform the required computations for real-time estimation of 3-axis velocities at output rates up to 25 Hz. Electrical power for the signal conditioning module and for driving the transmit transducers is derived from the AT power bus so that no external power supply or other external components are needed.

Fig. 2 shows the basic measurement technique employed by an ADV. The system operates by transmitting short acoustic pulses along the transmit beam. As the pulses propagate through the water, a fraction of the acoustic energy is scattered

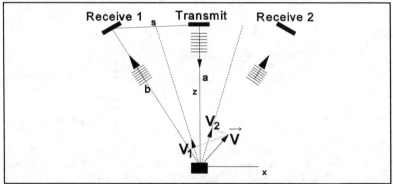

Fig. 2 Doppler Measurement Technique

back by small particles suspended in the water (e.g., suspended, sediments, small organisms, etc.). The three receivers detect the "echoes" originating at the sampling volume, which are Doppler shifted due to the relative velocity of the flow with respect to the probe. The Doppler shift observed at each receiver is proportional to the component of the flow velocity (V_1 and V_2) along the bisector of the receive and transmit beams. Doppler shifts measured at the three receivers thus provide estimates of flow velocity along three different directions, which are then combined geometrically to obtain the three orthogonal components of the water velocity vector **V**.

The ADV-1 measures the three orthogonal components of the velocity vector within a common sampling volume defined by the interception of the transmit and receive beams. The system uses pulse-to-pulse coherent Doppler techniques ([Miller and Rochwarger 1972], [Zrnic 1977], and [Lhermitte and Serafin 1984]) to achieve short term velocity errors of 1-10 mm/s at rates up to 25 Hz, even in highly variable flow conditions with peak velocities as high as 2.5 m/s. The processing algorithms are self-adaptive to avoid pulse-to-pulse interference when operating close to the surface or to a solid boundary.

Calibration

The calibration of the ADV is done in two steps. First, the exact probe geometry is determined. Although designed to a specification of 30°, the three angles between the transmit transducer, sampling volume, and receive transducers will vary slightly from probe to probe. The small variations in angle must be included in the calibration, and each probe has its own unique calibration table. The data are generated during factory testing and can be expressed as a transformation matrix T between the measured velocity V_i with components V_1, V_2, and V_3 and the earth-referenced velocity vector V with orthogonal components V_x, V_y, and V_z:

$$V = \begin{bmatrix} \sin\phi_1\cos\theta & \sin\phi_1\sin\theta & \cos\phi_1 \\ \sin\phi_2\cos(\theta+2\pi/3) & \sin\phi_2\sin(\theta+2\pi/3) & \cos\phi_2 \\ \sin\phi_3\cos(\theta+4\pi/3) & \sin\phi_3\sin(\theta+4\pi/3) & \cos\phi_3 \end{bmatrix} * V_i \qquad (1)$$

where ϕ_i represent the three calibration angles, and θ is the rotation around the z-axis (heading). Rotation around the x- and y-axis (pitch and roll) has been disregarded. The system is calibrated by running the probe three times in a tow tank; each time with a different heading. Detailed analysis of the full set of equations shows this technique to provide estimates of the calibration angles that are accurate to within 0.1°. The method does not require a high-precision mounting fixture and is robust with respect to small pitch and roll angles (<5°).

The geometry of the probe does not change unless it is physically damaged. Normally, any deformation can be detected by simple inspection but it is also possible to detect changes by using software utilities. If, for example, a receiver arm is twisted or bent, the receiver elements are no longer focused on the sampling volume. Consequently, the signal strength from the deformed receiver will display significantly reduced signal strength compared to the undamaged receiver arms. Because the signal strength is both displayed during data collection and recorded to file, probes that are bent will be detected. For deformations that are too small to be detected by a reduction in signal strength, there may be a small effect on the calibration. The geometry of the probe arms and the implementation of the signal processing, however, prevent significant impact on the calibration. With reference to Fig.2, the total distance from the transmit to the receive transducer (s) is unlikely to change even if the probe arm is bent. Also, the total travel time for the signal (a+b) will not change since the sampling time is determined by the hardware. Any error in the calibration angle is thus subject to the following constraints:

$$s = constant\ (= 25mm)$$
$$a+b = constant\ (= 108mm)$$
(2)

When solving the resulting systems of equations, it becomes apparent that small deformations in the receiver arm only result in very small errors in the horizontal velocity. For example, the calibration error is less than 1% even if the change in angle as a result of the deformation is as large as 5%.

The second part of the calibration is contained in the normal data collection procedure. Before each run, the speed of sound is calculated based on site-specific temperature and salinity values. The speed of sound can vary from 1440 m/s in cold, fresh water to 1540 m/s in warm, salt water. Without calibration, a nominal speed of sound of 1490 m/s could lead to errors as large as 3.3%. With routine calibration prior to each use, the error due to the speed of sound is limited to about 0.2% if the operator knows the temperature to within 1°C.

Sampling volume

The size of the sampling volume is determined by the length of the transmit pulse, the width of the receive window, and by the beam pattern of the transmit and receive beams. The first set of parameters defines the vertical extent of the sampling volume whereas the second set determines the width or horizontal extent. The latter can be approximated by the size of the transmit ceramic because a 7-mm ceramic vibrating at 10 MHz operates in the acoustic near-field out to 100 mm. In the near-field, the beam pattern is approximately cylindrical and the standard, 3-dB beam width is approximately equal to the size of the ceramic. In reality, this picture is simplified and the correct horizontal shape can only be derived by numerically integrating the joint nearfield beam patterns. The results show the horizontal weight function to be slightly asymmetric toward the receive transducer. The horizontal extent is thus slightly different for each of the three beams. From numerous velocity measurements, however, we have not been able to detect any results that would indicate that these differences are significant

The vertical extent of the sampling volume is usually more important than the horizontal extent because of the precise positioning required in boundary layer experiments. For acoustic backscattering systems, the sampling volume is defined as the convolution between the transmit pulse and the receive window. In a typical configuration of the ADV, the transmit pulse is shorter than the receive window and the resulting weight function is flat, has a baseline of 9 mm, and the midpoint of the sampling volume can be positioned as close as 4.5 mm from a solid boundary.

To get closer to a boundary, we can reduce the length of the transmit pulse or reduce the size of the receive window. The first is done at the penalty of a reduction in signal strength. The effect of this reduction is the same as the penalty for a decreasing the receive window: an increase in the Doppler noise. To estimate the mean current profile, however, we can reduce the Doppler noise by averaging over time. Both strategies are thus available and the smallest sampling volume has a vertical extent of about 0.6 mm. This means that the ADV, at least theoretically, can make measurement where the midpoint of the sampling volume is less than 1 mm from a solid boundary. In the case of a movable bed, the distance should be increased by 1 or 2 mm to allow for lateral variation in the bed profile and to permit bed movement within the averaging period.

Test results

Since the construction of the first prototype in January of 1993, several ADVs have been subjected to an extensive series of tests to verify their performance in a wide variety of operational conditions. Besides accuracy and temporal resolution, these tests addressed specialized topics such as flows less than 1 cm/s, response to the probe moving in and out of the water (such as might occur during wave motion), measurement of turbulent kinetic energy and Reynolds stress, response to high sediment concentrations, and directional response. The results from selected tests are briefly discussed here.

The first evaluation of the system was performed at WES [Kraus et al. 1994], where tests were conducted in a spillway model, a 2D wave basin, a 3D wave basin, a ship navigation model, and a near-prototype rip-rap circular channel. In these tests the measurements collected with the ADV were compared to those gathered with other velocity sensors routinely used at WES such as Laser Doppler Velocimeters (LDV), acoustic travel time sensors, pitot tubes, and surface height gauges. The ADV was successful in all tests, providing data that agreed well with those gathered by other velocity sensors. These tests also demonstrated the versatility and simplicity of use of the system. The ADV was easily installed, typically within a few minutes, in a variety of physical models, from small wave flumes to stream-size outdoor channels. It was also capable of operating in a wide range of flow regimes that included the breaker zone in a 3D wave basin and a hydraulic jump in a spillway model. Most comparisons had to be made based on the statistical characteristics of the velocity data such as the mean, standard deviation, and power spectra. This was due to the short spatial and time scales of the flow in the models, which resulted in slight disagreement in the instantaneous velocity measured by sensors placed at different locations. Other factors contributing to discrepancies in the instantaneous velocity were differences in the sampling volume and response time.

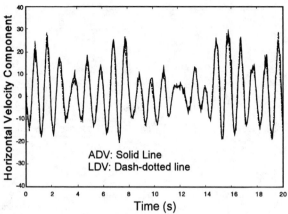

Fig. 3 - Horizontal velocity from ADV and LDV in wave flume.

Fortunately, the 2D wave basin provided an opportunity for a direct comparison of instantaneous velocities. In this test, the sampling volumes of the ADV and that of a 2D Dantec LDV were aligned visually so that both instruments would measure flow velocities from approximately the same volume of water. The still-water depth in the wave channel was 0.3 m, and the sampling volume of both instruments was located approximately 0.15 m below the mean water surface. The LDV was set to sample at 50 Hz and the ADV at 25 Hz. Fig. 4 compares measurements of the horizontal component of wave orbital velocity measured with the ADV (solid line) and the LDV (dash-dotted line). The broadband waves had a mean period of 1.0 s and an RMS wave height of 0.15 m. The plot shows good visual agreement in shape and peak of the oscillatory signals from the two instruments. Linear regression was performed on the data sets, generating a slope of 1.03 and an offset of 0.11 cm/s. Good agreement was also obtained for the weaker vertical velocity component.

Tow-tank tests

Tow-tank tests of the ADV were conducted at NorthWest Research Associates, Inc (NWRA) in Bellevue, Washington. The NWRA channel is 9.75 m long, 0.91 m wide, and 0.91 m deep. The ADV probe was attached to a tow carriage mounted on top of the channel and driven by an electrical motor. The speed of the carriage is determined independently by a tachometer, and by the time required for a 25.4-cm long bar, mounted horizontally on the carriage, to pass three sets of optical sensors located on the channel wall (giving three additional speed measurements). The standard deviation of these four speed estimates was calculated to be 0.2% over the range 1 to 250 cm/s.

The purpose of these tests was to verify the linearity of the ADV velocity measurements over its full dynamic range, ± 2.5 m/s. As with any instrument, the

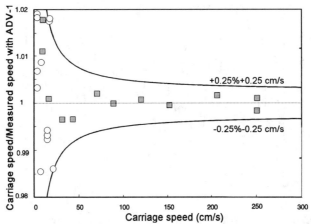

Fig. 4 - ADV Linearity over full range. Data collected at NWRA (squares) and at USGS/Stennis Space Center (circles).

importance of linearity lies in the calibration. As describe previously, the velocity vector is computed from the three measured along-beam velocities by using a transformation matrix that incorporates the relative position and orientation of the four acoustic transducers. If the system is linear, the transformation matrix can be determined empirically by calibrating the system at one speed, and will remain constant as long as the physical dimensions and geometry of the probe remain unchanged.

Fig. 4 shows the ratio between carriage speed and along-channel velocity component over carriage speeds from 10 to 250 cm/s. The output is linear to better than 0.25% over a range of 15 to 250 cm/s. Strong linearity assures a highly accurate 3D velocity measurement system over the full velocity range and makes it possible to calibrate the instrument in a single-speed tow channel.

Below 15 cm/s, the test at NWRA showed a small bias of the ADV toward lower velocities. The bias was caused by a hardware problem that was subsequently corrected. Later low-speed tests at the tow tank operated by the U.S. Geological Survey at the Stennis Space Center showed (Fig. 4 - circles) that the bias was no longer present and that the ADV can be accurate to within $\pm 0.25\%$ ± 0.25 cm/s over its full velocity range if properly calibrated.

Performance at low flows

An example of the ADV capability for measuring low-velocity flows is given by the time series of velocities shown in Fig 5. This data set was collected with the ADV measurement probe immersed in a 1-gallon bucket of water, collecting data at a rate of 25 Hz. The oscillations were initiated by tilting the bucket slightly then letting it rest. The plot shows how well the ADV can resolve a 2-Hz sinusoidal signal with an amplitude of less than 1 cm/s.

A special mode of operation has recently been developed that allows the ADV to measure very low flows with accuracies better than 0.1 mm/s over velocity ranges of up to a few cm/s. This mode of operation was tested at the borehole calibration facility at the U.S. Geological Survey in Denver. During the tests, the probe was submerged in a

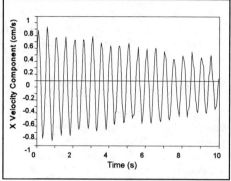

Fig. 5 - Time series of one horizontal velocity component collected with an ADV in a small bucket of water.

cylinder of diameter 0.1 m. The flow is controlled with a pump and the setup is designed to simulate the conditions in a well or in a borehole. The simulation is complicated by thermally driven circulation that persists until the air and water temperature have stabilized. The convective currents are of the same order as the net transport rate, and we subtracted 1.8 mm/s from the measured vertical velocity to account for the circulation at the position of the probe. After this correction, the 3-minute mean vertical velocity was plotted on a loglog scale (Fig. 6). The

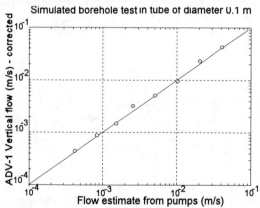

Fig. 6 - Calibration at low flows. The solid line is the 1:1 curve and the circles depict the data points.

agreement between the calibrated pumps and the ADV is quite good all the way down to the minimum test velocity of 0.4 mm/s. Toward the end of the day, when the pumps were turned off and the water was in approximate thermal equilibrium, the measured vertical velocity was 0.04 mm/s and the mean horizontal velocity

0.015 mm/s (= 0.003 ft/min = 4.5 ft/day). This gives credence to the hypothesis of a 1.8 mm/s thermally driven circulation. Also, these minimum velocities are close to what we previously have seen in laboratory experiments and they probably represent a good estimate of the sensitivity of the ADV.

Turbulence measurements

The relatively high temporal resolution and small sampling volume of the ADV implies that it is possible to measure field and prototype scale turbulence on a routine basis. Practical tests have also shown that the resolution is sufficient to capture a significant fraction of the turbulent kinetic energy and Reynolds stress within the bottom boundary layer of most laboratory flumes.

The short-term error, or the Doppler noise, is an inherent part of all Doppler-based backscatter systems. It represents a fundamental aspect of the measurement process and is related to the random distribution of the particles that contribute to the acoustic echo. The magnitude of Doppler noise is a function of the signal strength but also depends strongly on the flow condition itself. Unfortunately, it is not always easy to predict its magnitude in advance of an experiment because there always is a degree of uncertainty with respect to the exact flow conditions.

There are three aspects of the Doppler noise that should be noted: a) the noise can be reduced by averaging over independent realizations, b) the noise from two independent channels is uncorrelated, and c), the noise is white in the spectral domain. The possibility of reducing the noise by averaging is important because it means that the magnitude and the importance of Doppler noise will diminish as the averaging period increases.

For the purpose of describing the effect of Doppler noise on turbulence measurements, we separate the measured velocity components V_i into a "mean" velocity and a fluctuating part that contains turbulent energy V_i' and Doppler noise V_{di}:

$$V_i = \overline{V}_i + V_i' + V_{di}, \qquad \overline{V_i' + V_{di}} = 0, \qquad \overline{(V_i' + V_{di})^2} = \overline{V_i'^2} + \overline{V_{di}^2} \qquad (3)$$

The first two expressions are definitions, whereas the third expression incorporates the fact that the time-averaged product between natural fluctuations and the Doppler error is zero. This is an important aspect of the noise and allows the calculation of quantities that are smaller than the Doppler noise.

The relationship between the beam velocities with subscript i and the orthogonal velocity components can be described by the matrix $\mathbf{A} = inv(\mathbf{T})$ with elements a_{ij}. The turbulent kinetic energy can then be expressed as:

$$|V'^2| = \sum (\overline{(AV_i')^2} + \overline{(AV_{di})^2}) \qquad (4)$$

The Doppler noise enters directly into the equation and the measured energy will be biased high. The bias scales with $(V_{di}/V_i')^2$ and will be small in high-energy situations such as boundary layer flow. In oscillatory flow, the turbulence level can be estimated in the same way as the dissipation rate by analyzing the energy level of the saturation range in the velocity spectrum. Because the Doppler noise is white it is easily identified as a "noise floor" (See fig. 7) and its signature is a flattening of the spectrum (S_u, S_v, or S_w) as we approach the Nyquist frequency (f_N) at 12.5 Hz in the case of a standard ADV-1. In the worst case this may take place around 4-5 Hz for the horizontal velocity but more typically in the range 5-10 Hz. For the vertical velocity, the noise floor rarely plays any role below the Nyquist frequency, and this is the component of choice in case of isotropic turbulence. For fully developed turbulence with a well-defined spectral equilibrium range we can extract the turbulent kinetic energy (or the dissipation rate) by transforming the frequency spectrum to a wave number spectrum and then using the appropriate scaling.

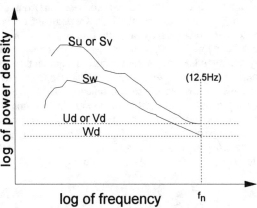

Fig. 7 - Typical power spectrum for the ADV. The Doppler noise is higher for the horizontal components (Ud, Vd) than for the vertical component (Wd).

The Reynolds stresses are typically smaller than the turbulent kinetic energy and often smaller than the Doppler noise. If the receive beams were orthogonal, i.e. $V_i = V$, this would not be a problem because the Doppler noise in two independent channels is uncorrelated and the estimate of the Reynolds stress would be independent of the magnitude of the Doppler noise. In the case of an ADV, however, the receive beams are not orthogonal and Reynolds stress must be derived using the matrix **A**. If we let the x, y, and z-coordinates be defined by the probe geometry, we can write:

$$\overline{u'w'} = a_{11}a_{31}(\overline{V_1'^2} + \overline{V_{d1}^2}) + a_{12}a_{32}(\overline{V_2'^2} + \overline{V_{d2}^2}) + a_{13}a_{33}(\overline{V_3'^2} + \overline{V_{d3}^2}) + \\ (a_{11}a_{32} + a_{31}a_{12})\overline{V_1'V_2'} + (a_{11}a_{33} + a_{31}a_{13})\overline{V_1'V_3'} + (a_{12}a_{33} + a_{32}a_{13})\overline{V_2'V_3'} \qquad (5)$$

The first line contains three terms that all involve the total variance of the along-beam velocity. The second line only contains terms with cross-products between beams and the uncorrelated Doppler noise has no contribution. If we sum up the constants in the first line ($a_{11}a_{32} + a_{12}...$) we find that the sum is zero, and it would be simple to conclude that the contribution from the first line is zero. In an experiment conducted by the U.S. Geological Survey, St. Petersburg, in a flume located at the University of South Florida, this proved not to be true. Instead, the contribution was evenly split between the first and the second line, implying that the asymmetry in the along-beam variance is important. The contribution of the Doppler noise, however, is zero as long as the Doppler noise is the same in all beams. Because the Doppler noise is a function of the signal strength this means that the signal strength should be the same in all three beams. This turned out not to be the case of the ADV used in the St. Petersburg experiment, where a constant, positive offset can be observed at low flows in Fig 8. At higher flows, the slight asymmetry in the Doppler noise becomes negligible and the match between the data collected with a Dantec LDV and the ADV is quite good.

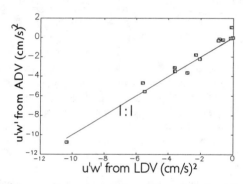

Fig. 8 - Turbulent stress measurements from flume. The straight line depicts a 1:1 relationship between the LDV and the ADV.

In sum, an ADV can be used to measure turbulent kinetic energy in excess of the Doppler noise and/or indirectly estimate the energy as long as the saturation range in the spectrum can be resolved. The sensitivity to Reynolds stress is not limited by the magnitude of the Doppler noise but by how well the Doppler noise is balanced in the three receive channels.

Scattering measurements

There are two aspects of scattering measurements that are of interest here. The first affects the velocity measurements directly through the functional relationship between the signal strength and Doppler noise. If the signal strength is too weak, as may be the case in quiescent basins or in stationary flumes, the Doppler noise becomes high and the introduction of small particles (seeding) may be required. To ensure that this information is available both during the data collection and during postprocessing, the software displays and stores the signal-to-noise ratio (SNR) of the scattered signal. The noise level N is measured at the beginning of each measurement sequence and is close to the electronic noise level for the ADV. The signal strength I is measured each time the receivers are turned on to measure the velocity and the signal-to-noise ratio is defined as:

$$SNR = 10 \log_{10}\left[\frac{I}{N}\right] \qquad (6)$$

As mentioned previously, the Doppler noise varies with the SNR. In our experience, an SNR of 15 dB is sufficient to obtain reasonably low-noise data at 25 Hz, whereas a ratio of about 5 dB is sufficient to obtain good quality data at 1 Hz.

The second reason for measuring the scattered intensity I is the potential for extracting information about the concentration of particles in the sampling volume [Hay, 1991]. Although the details can be algebraically complex, a simplified version of the problem can be expressed as:

$$I = I_0 C S_f S_a \frac{e^{-2(\alpha_w + \alpha_s) r}}{r^2} \qquad (7)$$

where I_0 is the transmitted intensity, C is the particle concentration, α_w is the water absorption, α_s is the attenuation due to particle scattering, r is the acoustic propagation path, S_a holds all the particle specific parameters (size, elasticity, and density), and S_f contains all the system specific parameters such as transducer size, efficiency, probe geometry, receive sensitivity, as well as the full nearfield expression for the pairs of transmit and receive beams.

From the equation, it can be seen that the scattering strength increases in proportion to the concentration. Unfortunately, the echo is also modified by increased particle attenuation, and a plot of concentration versus intensity [Schaafsma and der Kindern, 1985] will show a near-linear relationship for low and intermediate concentrations (<1 g/l) and then exhibit strong non-linear features before the attenuation finally becomes strong enough to reduce the intensity I below an acceptable level of SNR. The span of concentrations over which it is possible to measure velocity is thus much larger than the span over which the measured intensity is proportional to concentration.

It should be added that the above equation is only valid if the particle size distribution remains constant. If ka<1, where k is the acoustic wave number and a is the particle radius, the scattering is proportional to the square of the particle volume and thereby sensitive to the exact size distribution. For ka>1 (a>24 μm in case of a 10 Mhz system) the scattering is proportional to the surface area. This size-dependency is the same as that of an optical backscatter sensor (OBS) and the scattering characteristics for the two instruments are similar for "large" particles.

Conclusions

The ADV has proven to be a versatile and accurate 3-D velocity sensor suitable for most applications in physical models. The overall assessment from previous studies [Kraus et al 1994] shows the ADV to have: (1) somewhat less temporal and spatial resolution but comparable accuracy to a laser-Doppler velocimeter (LDV); (2) added benefit of providing full three-axis velocity measurement; (3) good penetration in turbid water; (4) significant improvements in operational simplicity and versatility as compared to LDVs; (5) one-tenth the cost of a conventional LDV, and (6) ruggedness and convenience of measurement in difficult-to-reach areas in the flow compared to time-of-flight acoustic flow meters or LDVs. The accuracy of the system has been reaffirmed in additional tow tank tests, and the linearity has proven to be within $\pm$ 0.25%.

In addition, the ADV has been shown to be suitable for low-flow applications. Calibration at the borehole facility at U.S. Geological Survey in Denver suggests that reliable data can be obtained down to the minimum test velocity of 0.4 mm/s and possibly an order of magnitude lower. Estimates of turbulent kinetic energy are limited by the Doppler noise but alternative estimation techniques can be employed to take advantage of the fact that Doppler noise is white and eliminate bias in the energy or dissipation estimate. Comparison with an LDV shows that the Reynolds stresses can be accurately determined even at levels below the Doppler noise and that care must be taken to construct systems with a balanced SNR. Finally, the linearity between SNR and the logarithmic concentration of suspended sediments points to the possibility of measuring the velocity and concentration in a single volume. Although limited by size-sensitivity and particle attenuation at high concentrations, this methodology holds promise and will be further explored in upcoming experiments. These experiments will take advantage of the recent development of a PC-independent electronics module that allows integration with a self-contained data acquisition system and permits increased use in short-term field applications.

Acknowledgements

We sincerely appreciate the assistance of numerous WES staff members during the requirements analysis and tests of the ADV. We also appreciate the assistance of Dr. Yogesh Agrawal and Mr. Lee Piper at NWRA in the tow tank test. Dr. John Haines and Dr. Guy Gelfenbaum of USGS in St. Petersburg collected and analyzed the turbulence data.

References

Hay, A., 1991: "Sound Scattering from a particle-laden, turbulent jet", *J. Acoust. Soc. Am*, Vol 90, No 4, 2055-2074.

Kraus, N. C., Lohrmann, A., and Cabrera, R., 1994: "New Acoustic Meter for Measuring 3D Laboratory Flows", *J. Hydraulic Engineering*, Vol 120, No.3 , 406-412.

Lhermitte, R. and Serafin, R., 1984: "Pulse-to-pulse coherent Doppler signal processing techniques", *J. Atmos. and Oceanic Technol.*, 1, 293-308.

Miller, K.S. and Rochwarger, M.M., 1972: "A covariance approach to spectral moment estimation", *IEEE Trans. Inform. Theory*, IT-18, 588-596.

Schaafsma, A.S. and der Kindern, W.J.G.J, 1985: "Ultrasonic instruments for the continuous measurement of suspended sand transport", *Symp. on Measuring Techniques in Hydraulic Res.*, Delft, 22-24 April, 1985

Zrnic, D.S., 1977: "Spectral moment estimates from correlated pulse pairs", *IEEE Trans. Aerosp. and Electron. Syst.*, AES-13, 344-354.

FIELD PERFORMANCE OF AN ACOUSTIC SCOUR-DEPTH MONITORING SYSTEM

By Robert R. Mason, Jr.[1], Member ASCE, and D. Max Sheppard[2]

Abstract

The Herbert C. Bonner Bridge over Oregon Inlet serves as the only land link between Bodie and Hatteras Islands, part of the Outer Banks of North Carolina. Periodic soundings over the past 30 years have documented channel migration, local scour, and deposition at several pilings that support the bridge. In September 1992, a data-collection system was installed to permit the off-site monitoring of scour at 16 bridge pilings. The system records channel-bed elevations at 15-minute intervals and transmits the data to a satellite receiver. A cellular phone connection also permits downloading and reviewing of the data as they are being collected. A digitally recording, acoustic fathometer is the main component of the system. In November 1993, current velocity, water-surface elevation, wave characteristics, and water temperature measuring instruments were also deployed at the site. Several performance problems relating to the equipment and to the harsh marine environment have not been resolved, but the system has collected and transmitted reliable scour-depth and water-level data.

Introduction

Scour processes at bridges are very difficult to predict as they are dependent on the physical properties of the bridge and many environmental factors, such as sediment transport rates. Some bridges are known to be scour-critical and may need to be continuously monitored pending remedial repair or replacement (Johnson and Jones, 1993). A variety of scour-depth monitoring systems have been proposed for this task ranging from simple mechanical devices to elaborate bottom-penetrating sonar and radar systems (Jarrett and Boyle, 1986; Skinner, 1986; Gorin and Haeni, 1989; Butch, 1991). The most efficient and cost-effective way to monitor scour depends on the type and location of the structure, hydraulic and environmental conditions, and sediment characteristics and concentrations (Fenner, 1993).

This paper describes the acoustic scour-depth monitoring system used at the Herbert C. Bonner Bridge, Oregon Inlet, North Carolina, including its design, its performance, and the conditions under which the system has operated. Instruments used to collect ancillary hydrographic data also are described, and some preliminary

---

[1] Hydrologist, U.S. Geological Survey, 3916 Sunset Ridge Road, Raleigh, NC 27607.
[2] Professor, Coastal and Oceanographic Engineering Department, University of Florida, Gainesville, FL 32611.

results from the monitoring system are presented. Although the scour-depth monitoring system was installed in September 1992, most of the discussion is limited to data collected in November 1993, the period for which there are concurrent scour-depth and other hydrographic data.

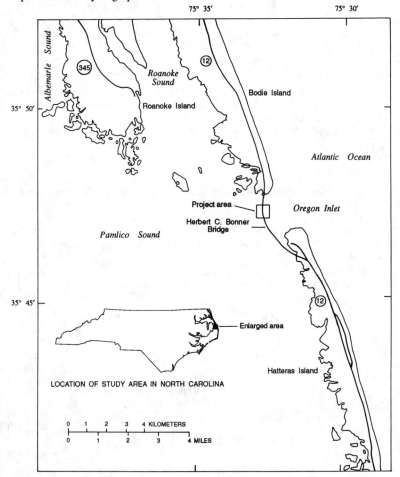

Figure 1. Location of Herbert C. Bonner Bridge and Oregon Inlet, North Carolina.

Background

Oregon Inlet is a narrow, shallow, and dynamic waterway connecting Albemarle and Pamlico Sounds to the Atlantic Ocean and separating Bodie and Hatteras Islands, two in a chain of barrier islands commonly referred to as North Carolina's Outer Banks (fig. 1). Since its formation by a severe Atlantic storm in 1846, the inlet has frequently shifted in depth, width, and location as a result of persistent southward migration of Bodie and Hatteras Islands; local scour produced by tide and wind-driven currents and waves; deposition due to long-shore sediment transport; and transport of sand from shoals in the sound during periods of strong easterly currents (Dolan and Lins, 1986). The rate of southward migration of the inlet varies greatly but has averaged 21 meters (m) per year from 1846 to 1993 (U.S. Department of Transportation, 1993). Currently (January 1994), the inlet is approximately 1.6 kilometers (km) wide and ranges from 2 to 3 m deep, except for the navigation channel, which is maintained at about 12 m deep. The channel bed is composed of thick deposits of coarse sand with little to no vegetative cover.

Oregon Inlet is spanned by the Herbert C. Bonner Bridge. Built in 1962 and located 1,200 m inland of the Atlantic Ocean and partly within the surf zone, the bridge is 4 km long, 11 m wide, and crests 20 m over the mean low-water level of Oregon Inlet. The Bonner Bridge is supported by 240 pile bents each capping clusters of 8-12 driven piles, which penetrate 15-27 m into the channel sediments. The bridge is the main component of the only land link from the mainland to Hatteras Island and carries an average daily traffic load of 6,900 vehicles during the summer months.

Since completion of Bonner Bridge in 1962, the North Carolina Department of Transportation (NC-DOT) has made periodic soundings of Oregon Inlet to monitor channel migration, deposition, and local scour. These data indicate that some sections of Oregon Inlet have scoured and filled through an 11-m range (fig. 2). In 1978, NC-DOT discovered that, as a result of scour, several pilings were penetrating only 2.1 m

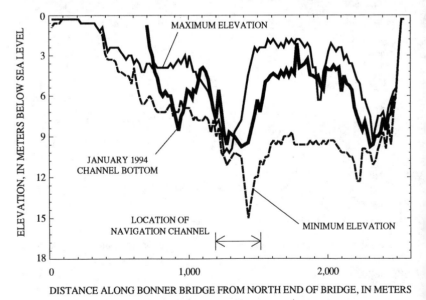

Figure 2. Cross section of Oregon Inlet at Bonner Bridge showing maximum and minimum channel-bottom elevations and January 1994 channel-bottom elevation.

into the channel bottom. From that year to 1992, scour-preventive and remedial actions to protect the bridge have cost over $9 million. In response to this history of channel instability, the U.S. Geological Survey (USGS) and the NC-DOT joined in a cooperative effort to develop and install a data-collection system to permit continuous remote monitoring of scour depth at 16 bridge pilings. The system was installed in September 1992.

A second data-collection program began in November 1993 and ended in January 1994. This effort consisted of deployment and operation of three instruments that measure and record current magnitude and direction, wave frequency and direction, water-surface elevation, and water temperature near the Bonner Bridge. This program was a cooperative effort between the Florida Department of Transportation (F-DOT), the University of Florida, Coastal and Oceanographic Engineering Department (UFCOE), the NC-DOT, and the USGS. In conjunction with the scour-depth data, the resulting hydrographic data can enable evaluation of theoretical scour-depth predictive equations.

Scour-Depth Monitoring System

The main component of the Bonner Bridge scour-depth monitoring system is a Datasonics PSA 902[a] digitally-recording, acoustic fathometer operating 16 transducers, each generating a 200-kHz acoustic beam with a conical 10-degree spread. At the time of deployment in 1992, each transducer was mounted at least 1.5 m above the channel bottom and between 1.8 and 4 m below the water surface. These depths were chosen to minimize interference from entrained air, surface turbulence (waves), suspended sediment, and debris. Transducers are secured to the inside of 7-degree battered pilings on both ends of eight consecutive pile bents (fig. 3).

The fathometer is configured as two separate channels, each controlling eight transducers. A time-varying gain circuit, one for each channel, can be adjusted to calibrate the transducers on that channel to an overall (group) optimal setting. However, the transducers cannot be calibrated individually.

The distances between the transducers and the channel bottom (scour depths) are recorded at 15-minute intervals and are output through a standard RS-232 port to a Vitel VX1004 programmable electronic datalogger and then transmitted to a Geostationary Earth Orbiting Satellite (GOES) receiver. The data logger is also equipped with a Mitsubishi CDL 100 cellular phone and computer modem for downloading of data and for changing software attributes remotely.

On November 4, 1993, three additional hydrographic instruments were installed near the Bonner Bridge in the project area (fig. 1). Two of these instruments were Endeco current meters (type 174SSM) that measure and record water temperature as well as current magnitude and direction. The third instrument was a Seadata wave-tide recorder that measures and records current magnitude and direction, wave frequency and direction, and water-surface elevation. The current-meter instruments were located in line with three consecutive piers and approximately 30 m east (ocean side) of the bridge.

---

[a]Any use of trade, product, or firm names is for descriptive purposes only and does not imply endorsement by the U.S. Government.

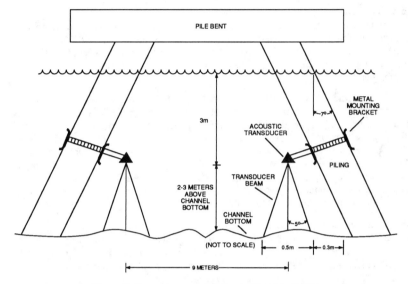

Figure 3. Transducer mounting on Bonner Bridge.

Monitoring System Performance

The performance of an acoustic scour-depth monitoring system is influenced by its design, its mounting, and by the environment in which it operates. Mounting considerations include the rigidity of the mounting and the location of the transducer relative to the piling, the channel bottom, and the water surface. Environmental factors, which can influence system performance, include suspended debris, suspended sediment, entrained air, biological growth, varying bed-slopes and bed-forms, and possibly, electronic interference from radios, radars, power lines, and other sources.

Four calibration checks were made during the period from November 17, 1992, to May 25, 1993. In each case the distances from the transducers to the channel bottom (scour depths) were physically measured by NC-DOT scuba divers. Some transducers were mounted closer to the channel bottom than were others. This difference and likely differences in bed-form and local bed-slope beneath each transducer probably resulted in slightly different acoustic-signal properties and rates of attenuation. Thus, the resulting operational settings were not necessarily optimum for each individual transducer. Some transducers, therefore, could be expected to perform more reliably than others. Overall, however, the correlation between the physically measured and acoustically sensed data is very good. The results of the calibration checks for transducer 10 are typical of the overall correlations (fig. 4).

Fathometer records for transducer 13 typify the variability of scour and fill during the data-collection period from August to November 1993 (fig. 5). On August 30, the channel bed beneath transducer 13 was 2 m below the transducer. This depth increased to 2.6 m due to erosion from currents associated with Hurricane Emily (August 31 to September 1, 1993). Following the passage of the hurricane through September 12, there was a net deposition of approximately 0.5 m of sediment as indicated by a sustained reduction in scour depth. From October 12 to November 14, there was a net erosion of approximately 1 m of sediment. Periods of significant fill occurred on September 18 (0.4 m), November 15 (0.9 m), and November 27 (0.2 m), 1993. In addition to scour associated with Hurricane Emily, scour also occurred on October 12 and November 28. On November 30, the channel bed was 2.0 m below the transducer, having traversed a minimum to maximum change in depth of 1.2 m.

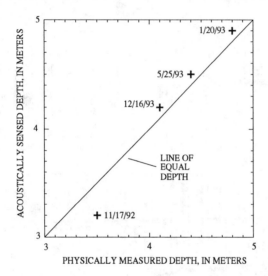

Figure 4. A typical result of transducer-calibration checks.

For the purposes of this paper, the reliability of the Bonner Bridge scour-depth monitoring system during November 1993 was evaluated using two performance characteristics: (1) the percentage of transmitted acoustic signals that were not recorded by the datalogger (failed signal return) and (2) the consistency of computed distances from one reading to the next (table 1). For the 16 transducers, the percentage of failed signals during November ranged from about 6 percent for transducers 13 and 14 to nearly 100 percent for transducer 7 (table 1). Failed signals are shown in the record for transducer 13 as lines extending to the top of the graphs (fig. 5).

The cause of the failed signals is unknown. Attempts to correlate signal dropouts or no returns with hydraulic phenomena such as tidal current magnitude and direction have been unsuccessful. However, strong daily cycles are evident in the frequency of failure of transducers 1 and 8. In both instances the highest frequency of failure occurs from 700 hours to 1700 hours, when the solar panel battery-chargers are in operation. Intermittent current surges resulting from battery recharging operations during daylight could be the source of the problem with these transducers. Modifications to the system to remedy this problem are planned. Failures of the

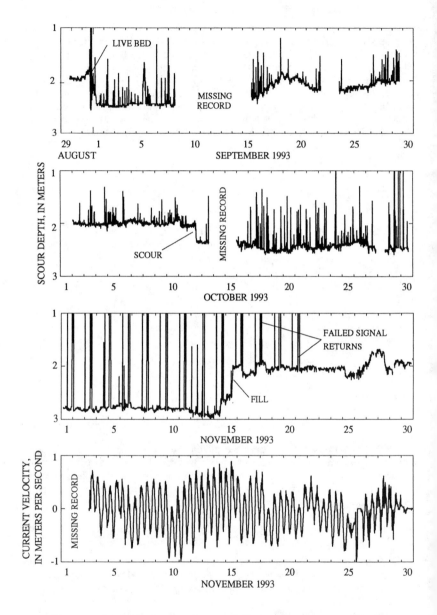

Figure 5. Scour depth at transducer 13 for September-November 1993 and tidal current velocity for November 1993.

remaining transducers appear to be random in a temporal sense, but are highly correlated to one another. When a failure occurs on a normally reliable transducer, transducer 2 for instance, the remaining transducers usually also fail. This correlation is probably the result of communication anomalies between system components.

Table 1.--*Percentage of failed signal returns and consecutive readings differing by selected ranges for each transducer, November 1993*
[m, meter]

Transducer no.	Failed signal returns (percent)	Percent of consecutive readings differing by indicated ranges			Mean distance to channel bottom (m)
		0.25-0.49 (m)	0.50-0.99 (m)	1.0-1.5 (m)	
1	81.9	14.5	37.3	28.6	0.80
2	8.8	4.2	17.6	10.0	3.67
3	8.8	4.1	8.8	4.4	.45
4	8.8	7.0	25.1	13.6	2.13
5	8.8	2.9	12.9	8.4	2.36
6	8.8	8.8	3.9	1.5	1.38
7	99.6	0	81.8	81.8	5.10
8	32.1	19.2	27.6	27.6	2.26
9	55.7	1.8	1.0	.1	2.15
10	14.4	2.4	4.6	6.3	2.90
11	12.8	2.1	6.3	2.0	1.68
12	9.1	1.3	7.7	4.8	3.12
13	6.4	.10	.30	.20	2.24
14	6.4	.20	.30	.50	2.07
15	68.6	0	.1	3.7	2.10
16	66.0	.2	.6	3.9	2.72

Movement of sediment after episodes of scour or fill is detected by the fathometer as a change in scour depth. Although changes are expected, they are usually small (<0.25 m) for consecutive 15-minute readings. To evaluate measurement consistency, the differences between consecutive scour-depth measurements were computed, and the percentage of these differences exceeding selected thresholds (0.25 m, 0.5 m, 1.0 m, and 1.5 m) were summarized for each transducer (table 1). The magnitude and frequency of differences between consecutive readings were examined for correlations with current magnitude and direction and temporal patterns, but none were detected.

Large differences between consecutive readings also can be caused by multiple echoes of the same transducer signal. Multiple echoes occur when a transducer return signal is emitted from the transducer, bounces off the channel bottom, then off the transducer, and back off the channel bottom for a second time before it is detected by the fathometer. Multiple echoes result in computed depths that are twice the actual depth to the channel bottom (fig. 6).

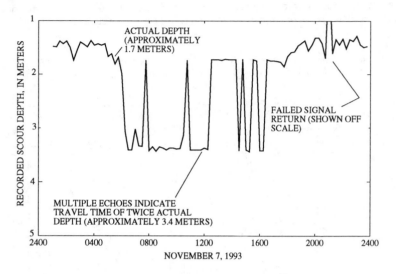

6. Depth to channel bottom for transducer 3 showing effects of multiple echoes.

Large differences in depth readings also can be the result of suspended debris. An object could enter the signal path and cause abnormal readings. A similar condition could also result from high concentrations of suspended sediment as during live-bed conditions. Such conditions were not noted during the period when the current meters were deployed (November 1993); however, a live-bed condition could have developed during the passage of Hurricane Emily (fig. 5). As the hurricane currents abated, some of the suspended sediment was transported, leaving the channel bed approximately 0.6 m below its previous elevation.

The performance of the other hydrographic instruments was likely hampered by the movement of suspended and bed-load sediment. Prior to deployment of the current meters and wave-tide recorder on November 3, 1993, a bathymetric survey of an area approximately 300 m by 300 m on either side of the bridge was conducted. On January 25, 1994, 83 days after the instruments were deployed, an attempt was made to retrieve the instruments and download the data. While making a second bathymetric survey of the same area, it became evident that a significant quantity (2 to 3 m) of sand had been deposited in the study area since November. The dive to recover the instruments verified this fact. The Seadata meter was completely buried, and the two Endeco meters were just above the channel bottom. Tethers on the Endeco meters allowed them to move up and remain above the deposited sediment. The meters did, however, suffer some damage. The impellers on both instruments were missing. The top of the Seadata meter that was originally 2 m above the channel bottom is believed to be buried under 0.5 m of sediment and is yet to be recovered. Current velocity data for November, downloaded from one of the Endeco meters, are shown in figure 5. There is little, if any, apparent correlation between these velocities and changes in scour depths for the same period.

Conclusions

Remote collection of acoustically sensed scour-depth data in a dynamic tidal environment such as Oregon Inlet is possible. Even though there is considerable unexplained variation within the Bonner Bridge scour-depth data, they do demonstrate the overall trends of scour and deposition. However, the Bonner Bridge data-collection system deployed is susceptible to errors arising from failed signal returns, multiple echoes, live-bed conditions, and suspended objects. Modifications of the data-collection system hardware and software are underway to improve the reliability and accuracy of the data. In addition, deployment of current meters and other hydrographic sensors are expected to yield more information about the effect of the Oregon Inlet environment on system performance. Assuming the Seadata velocity meter can be recovered and that it functioned properly during November 1993, the complete data set will include scour depth, wave frequency and direction, water-surface elevation, and water temperature.

References

Butch, G.K., 1991, Measurement of bridge scour at selected sites in New York, excluding Long Island: U.S. Geological Survey Water-Resources Investigations Report 91-4083, 17 p.

Dolan, Robert, and Lins, Harry, 1986, The Outer Banks of North Carolina: U.S. Geological Survey Professional Paper 1177-B, 46 p.

Fenner, T.J, 1993, Scoping out scour: Civil Engineering, v. 63, no. 3, p. 75-77.

Gorin, S.R., and Haeni, F.P., 1989, Use of surface-geophysical methods to assess riverbed scour at bridge piers: U.S. Geological Survey Water-Resources Investigations Report 88-4212, 33 p.

Jarrett, R.D., and Boyle, J.M., 1986, Pilot study for collection of bridge-scour data: U.S. Geological Survey Water-Resources Investigations Report 86-4030, 89 p.

Johnson, P.A., and Jones, J.S., 1993, Merging laboratory and field data in bridge scour: Journal of Hydraulic Engineering, v. 119, no. 10, p. 1176-1181.

Skinner, J.V., 1986, Measurement of scour-depth near bridge piers: U.S. Geological Survey Water Resources Investigations Report 85-4106, 33 p.

U.S. Department of Transportation, Federal Highway Administration, and North Carolina Department of Transportation, 1993, Administrative action, draft environmental impact statement and draft section 4(F) evaluation: Federal Highway Administration FHWA-NC-EIS-93-01-D, p. 3.1, 3.30-3.38.

COMPARISON OF CURRENT METERS USED FOR STREAM GAGING

Janice M. Fulford[1], Kirk G. Thibodeaux[1], and William R. Kaehrle[1]

Abstract

The U.S. Geological Survey (USGS) is field and laboratory testing the performance of several current meters used throughout the world for stream gaging. Meters tested include horizontal-axis current meters from Germany, the United Kingdom, and the People's Republic of China, and vertical-axis and electromagnetic current meters from the United States. Summarized are laboratory test results for meter repeatability, linearity, and response to oblique flow angles and preliminary field testing results. All current meters tested were found to under- and over-register velocities; errors usually increased as the velocity and angle of the flow increased. Repeatability and linearity of all meters tested were good. In the field tests, horizontal-axis meters, except for the two meters from the People's Republic of China, registered higher velocity than did the vertical-axis meters.

Introduction

An ideal current meter, whether mechanical or electromagnetic, should respond instantly and consistently to any changes in water velocity, and should accurately register the desired velocity component. Additionally, the meter should be durable, easily maintained, and simple to use under a variety of environmental conditions. Mechanical-current meters measure velocity by translating linear motion into angular motion. The two types of mechanical current meters, vertical-axis and horizontal-axis, differ in their maintenance requirements and performance because of the difference in their axial alignment. Mechanical meter performance depends on the inertia of the rotor, friction in the bearings, and the ease with which water turns the rotor. Electromagnetic current meters measure velocity using Faraday's Law, which states that a conductor (water) moving in a magnetic field (generated by the probe) produces a voltage that varies linearly with the flow velocity. Electrodes in the probe detect the voltages generated by the flowing water. Performance for electromagnetic current meters depends on the probe shape, location of the electrodes on the probe, and the construction of the meter electronics.

Many studies of current-meter performance have been conducted by researchers (Thibodeaux, 1992). Most of these studies were published before 1960, prior to the development of electromagnetic current meters for stream gaging. Yarnell and Nagler's (1931) study on mechanical current meters is one often referenced. The previous studies used mechanical meters that have since been modified and rarely investigated the performance of electromagnetic current meters. Recent laboratory (Fulford and others,

---

[1]Hydrologist, U.S. Geological Survey, Bldg. 2101, Stennis Space Center, MS 39529.

a) Upper meter, Price type-AA; lower meter optic Price type-AA

d) Ott C-31, standard impeller; impellers clockwise from top, plastic, A, and R.

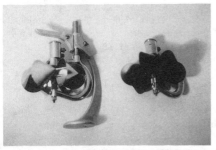

b) Left to right, winter Price type-AA, metal; winter Price type-AA, polymer

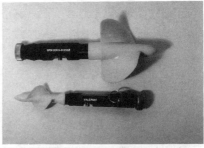

e) Top, Valeport BFM001; bottom, Valeport BFM002.

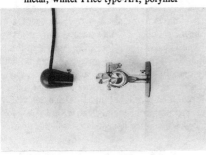

c) Left to right, Marsh McBirney 2000; Price pygmy

f) Top, PRC LS25-3A, metal; bottom, PRC LS25-3A, plastic

Figure 1. Photographs of tested meters. Vertical-axis meters a,b, horizontal-axis meters d-f and vertical-axis and electromagnetic meter c.

378   HYDRAULIC MEASUREMENTS AND EXPERIMENTATION

1993) and field testing has been completed by the U.S. Geological Survey (USGS) on several current meters. A comparison of the laboratory and field tests for fourteen of these meters is presented in this paper.

Meters Tested

All meters tested measure one component of flow velocity for a small volume of the total flow measuring section. Total flow or discharge is determined with these meters by making multiple velocity measurements in the section and multiplying each measured velocity by its contributing flow area. Comparisons are presented for five mechanical vertical-axis meters, Price type-AA, optic Price type-AA, Price pygmy, winter Price type-AA, and winter Price type-AA with polymer rotor; eight mechanical horizontal-axis meters, Ott C-31[2] with metal, plastic, A, and R impellers, Valeport BFM001 and BFM002, and People's Republic of China (PRC) LS25-3A with metal and plastic impellers; and one electromagnetic meter, the Marsh McBirney 2000. These meters are shown in figure 1. All meters tested use battery powered electronic devices to either count the meter revolutions or to measure the voltage. Maintenance of the electronic devices consists of checking batteries for proper voltages and replacing or recharging when needed.

The vertical-axis meters tested have six conical cups fixed to a hub that rotates a vertical shaft. Vertical-axis meters do not present a symmetrical profile to flow velocities. Velocities angled in the vertical plane impinge on a meter profile that is very different from the meter's horizontal profile. These meters have few parts and are relatively easy to maintain and clean. The bearings are located in an air pocket to prevent contamination from silt and sediment. Disassembly for cleaning requires the removal of the shaft and rotor assembly from the yoke. Daily cleaning and oiling is recommended for vertical-axis meters.

The horizontal-axis meters tested all have screw type impellers that rotate about a horizontal axis. Unlike the vertical-axis meters, the horizontal-axis meters present a symmetrical profile to velocities in the measurement section. Maintenance requirements vary widely among the horizontal-axis meters. The Ott C-31 and the PRC meters require disassembly of numerous parts, cleaning, and oiling between discharge measurements. Both of these meters have a complex ball bearing assembly that is sealed in oil to provide lubrication and exclude sediment. The PRC meter is similar in construction to the Ott, but has three times the number of internal parts. In contrast to the Ott and PRC meters, the Valeport is simpler and has fewer parts. Cleaning is recommended with clean water between discharge measurements. The Valeport meters' bearing surface is inside the impeller nose and uses water as the lubricant.

The electromagnetic meter tested has no moving parts and presents a symmetrical tear drop shape to the velocities in the measurement section. Cleaning is recommended with clean water and mild soap to remove dirt and nonconductive grease and oil from the probe's electrodes and surface. The zero reading should be checked periodically in still water. In contrast to the mechanical meters, rinsing the probe with clean water after a measurement is usually the only maintenance needed.

---

[2]Use of brand names in this report is for identification purposes only and does not constitute endorsement by the U.S. Geological Survey.

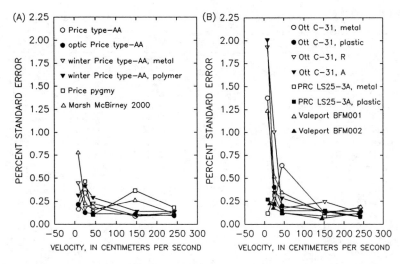

Figure 2. Percent standard error computed by velocity for (A) vertical-axis and electromagnetic meters and for (B) horizontal axis meters.

Laboratory Tests

Laboratory testing of the meters included repeatability testing and oblique flow response testing and was conducted in the jet tank at the USGS Hydraulic Laboratory Facility, at Stennis Space Center, Mississippi. The laboratory testing for the mechanical meters has been previously described by Fulford and others (1993) in a paper comparing vertical- and horizontal-axis current meters. Parts of this previous work are presented here with the addition of the tests for the electromagnetic meter. Test results for the electromagnetic meter are included in the vertical-axis meter figures. The repeatability test measures how repeatable or consistently a meter measures velocity. For each meter, ten measurements were made at each of five velocities, 7.62, 24.38, 45.72, 152.46 and 243.84 cm/s. Standard errors for each test velocity were computed for the mechanical meters from the meter revolutions per second (r/s) and for the electromagnetic meter from the readings displayed on its electronic readout device. Percent standard errors were computed at a test velocity by dividing the standard errors by the mean and multiplying by 100. Plots of percent standard error versus the test velocity are shown in fig. 2. For all meters except the Price pygmy and Marsh McBirney, percent standard errors decrease with increasing velocity. The vertical-axis meters have the most consistent response of the meters tested. For the five velocities tested the percent standard errors for the vertical-axis meters are less than 0.5% and for the horizontal-axis current meters are less than 0.75% for velocities greater than 24.38 cm/s. The Marsh McBirney meter has percent standard errors less than 0.5% except for the lowest velocity tested (7.62 cm/s) where the percent standard error is 0.78%. At the lowest test velocity (7.62 cm/s) the metal Ott, Ott A, Ott R, and Valeport BFM001 had percent standard errors from 1.2 to 2.0% and the PRC meters, the plastic impeller Ott, and the Valeport BFM002 had percent standard errors less than 0.5%.

Table 1.-Linear response of meters; root mean squared (RMS) errors and regression coefficients determined from repeatability data.
[cm/s, centimeters per second; cm/rev, centimeters per revolution]

Meter	RMS error (cm/s)	slope (cm/rev)	intercept (cm/s)	meter type
Optic Price type-AA	0.524	67.391	-0.427	vertical
Price type-AA	1.067	68.976	-0.579	vertical
Winter Price type-AA/metal	0.622	69.921	-0.122	vertical
Winter Price type-AA/polymer	1.716	79.004	1.707	vertical
Price pygmy	1.634	32.034	-1.676	vertical
PRC LS25-3A /metal	0.567	19.934	0.274	horizontal
PRC LS25-3A /plastic	0.527	19.903	-0.061	horizontal
Valeport BFM001	1.234	26.518	-0.061	horizontal
Valeport BFM002	0.735	11.003	1.737	horizontal
Ott C-31 /metal	1.545	25.603	2.652	horizontal
Ott C-31 /plastic	1.372	25.451	0.213	horizontal
Ott C-31 /R impeller	1.402	25.085	2.316	horizontal
Ott C-31 /A impeller	1.999	12.834	3.871	horizontal
Marsh McBirney 2000	0.875	30.450	1.341	electromagnetic

Repeatability data were also fitted using linear regression to determine a meter's linear response. For each meter, the 50 repeatability measurements (velocities of 7.62, 24.38, 45.72, 152.46 and 243.84 cm/s) were regressed against the reference velocity. The root mean squared errors (RMS) and computed regression coefficients are listed by meter in table 1. RMS ranged from 1.999 cm/s for the Ott with A impeller to 0.524 cm/s for the optic Price type-AA. The Marsh McBirney meter RMS (0.875 cm/s) was larger than the two PRC meters, the optic Price type-AA, metal winter Price type-AA and the Valeport BFM002. All meters had RMS less than 2.000 cm/s and for velocities less than 7.62 cm/s percent standard errors smaller than 0.75%. However, the vertical-axis and electromagnetic meters had better repeatability and smaller standard errors at velocities less than or equal to 45.72 cm/s.

The oblique flow response test is a measure of how accurately a meter measures the appropriate vector component of the flow. This test is also known as the cosine response because an ideal meter would register the cosine component of an angled flow. Each meter was tested at speeds of 7.62, 24.38, 45.72, 152.46 and 243.84 cm/s and at flow angles ranging from 90° to -90° in increments of 10°. Positive angles were flows directed downward onto the vertical-axis meters or from center to right side for horizontal-axis meters. At each combination of velocity and angle, two velocity measurements were made with each meter. Because only the meter and not the actual flow could be angled, the vertical-axis meters were positioned with the axis perpendicular to the force of gravity when testing for response to vertical angles of flow. This insured a consistent loading of the meter bearings throughout the oblique flow tests. Tilting the vertical-axis meters in the vertical plane would load the meter bearings differently for each angle tested and produce a varying error in the test.

Percent error was computed as $100 \times [\text{revs/sec}_\alpha \div (\cos\alpha \times \text{revs/sec}_0) - 1]$ for the mechanical meters. Subscripts α denotes the angle of flow and 0 straight flow. The electromagnetic meter percent error was computed using the display device reading instead

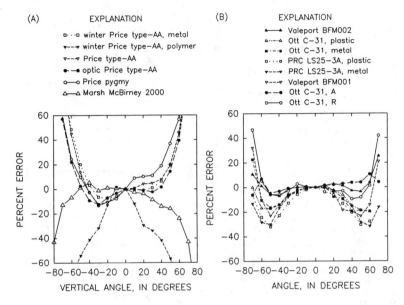

Figure 3. Average response for oblique flows for (A) vertical-axis and electromagnetic meters and for (B) horizontal-axis meters.

of r/s. Vertical-axis and electromagnetic meters were tested for response to vertical and horizontal angles of flow. Due to limited length of this paper, only the results of vertical-angle testing are shown. Stream gagers are unable to correct for errors due to vertically angled flow during field use of meters.

All tested meters under-registered and over-registered velocity depending on the angle and flow speed. In figure 3 are plots of average percent error for the five test velocities versus angle. Only the angles between ±80° are shown in figure 3 because any registration of velocity at ±90° results in a large error. The winter Price type-AA meter with polymer rotor under-registered for all angles tested. The other vertical-axis meters over-registered for positive angles and under-registered for angles between -40° and 0° the flow velocity. The electromagnetic meter over-registered for angles of -50°, -40°, -20°, and -10°, and under-registered for all other angles. All horizontal-axis meters stalled for flow angles greater than 70° and except for the Ott with the A or R impeller and the Valeport BFM002, tended to under-register the velocity for most angles. At angles between ±10° the vertical-axis meter errors range from -3.30% to -0.17% for the optic Price type-AA and from -7.87% to 8.92% for the Price pygmy. At angles between ±10° errors for horizontal-axis meters range from 0.58% to 0.91% for the Ott with plastic impeller and from -2.02% to -3.77% for the PRC meter with plastic impeller. The electromagnetic meter errors range from -2.565% to 0.699% at angles between ±10°. At larger angles of ±30° the vertical-axis meter errors range from -6.71% to 1.01% for the winter Price type-AA with metal rotor and from −31.83% to -33.97% for the winter Price type-AA with polymer rotor. For the horizontal-axis meters the errors range from -0.68% to 2.95% for the Ott with A

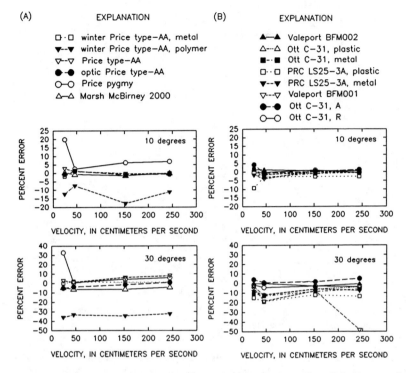

Figure 4. Response for 10° and 30° oblique flows for (A) vertical-axis and electromagnetic meters and for (B) horizontal-axis meters.

impeller and from -12.87% to -13.19% for the PRC meter with plastic impeller. The electromagnetic meter errors are -1.24% at -30° and -8.68% at 30°.

Error due to oblique flows increases slightly with velocity, except at the lowest velocity tested. At the lowest velocity tested, errors are larger than or nearly equal to the errors found for the highest velocity tested. In fig. 4 are plots of percent error versus jet velocity for 10° and 30° flows. Because the horizontal-axis meters stalled at low velocities in oblique flows, test results for 7.62 cm/s are omitted from the plot. All meters tested have larger errors at larger angles of flow. The Ott meters, equipped with component impellers (A and R) designed to register the cosine component of angled flow, have the smallest errors and the Price pygmy and the winter Price type-AA with polymer cup have the largest errors in oblique flow. The electromagnetic meter has smaller errors than the vertical-axis meters for most angles tested. Unlike the horizontal-axis meters, the vertical-axis and electromagnetic meters have an obvious asymmetrical response to vertical angles of flow. The asymmetrical responses are probably caused by flow disturbances that result from the contact chamber at the top of the vertical-axis meters and from the signal cable exiting the top of the electromagnetic meter probe.

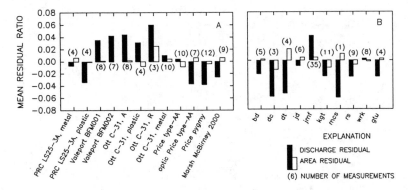

Figure 5. Mean percent residual grouped by (A) meter and by (B) stream gager.

Field Tests

Laboratory testing approximates and does not duplicate the field conditions in which current meters are used. In the field, meters are subjected to changing velocities and to an unknown range of flow angles. Meters in the field may not be subjected to the entire range of flow angles tested in the laboratory. Field testing is necessary to help interpret the importance of the laboratory findings. Field testing was done in 4 sections at the floodplain facility at the Stennis Space Center in Mississippi, and at five USGS gaging locations in Colorado and Wyoming. Three floodplain sections were located along a grassed half-trapezoid channel section. The fourth section was in the riprap bottomed exit channel of the facility. The locations in Colorado and Wyoming were mountainous streams with sand, gravel, and cobble bottoms. Discharge measurements were made using USGS stream-gaging procedures (Rantz, 1982). Winter Price meters were not used during field testing. Due to time constraints every meter was not used at every location. The Ott C-31 with metal impeller, Price type-AA, Price pygmy, Marsh McBirney, and Valeport BFM001 were used at every location except for the fourth floodplain section where the Valeport BFM002 was used instead of the BFM001. The remaining meters were used whenever possible. Although multiple stream gagers used various meters to measure the discharge, only three stream gagers (the authors) made discharge measurements at each of the sites. Usually a current meter was used once to measure discharge at a site. Meters were not rotated among gagers intentionally and some meters were used by only one stream gager throughout the field tests. All measurements were made by wading and meters were positioned in the water using the USGS top-setting wading rod. For the horizontal-axis meters, adapters were fabricated to allow their use with the top setting rod.

Discharge was not determined at any of the sites by means other than current meter measurements. As a result, meter performance can only be compared relative to the other meters. Of the total 86 discharge measurements, 41 were made in the floodplain and 45 were made in Colorado and Wyoming. Mean flow depths ranged from approximately 30.48 to 60.96 centimeters. Discharges ranged from 0.765 to 6.003 cubic meters per second (m^3/s) for the sites in Colorado and Wyoming and from 1.841 to 2.124 m^3/s for the floodplain sites in Mississippi. All fourteen meters operated satisfactorily during the

384 HYDRAULIC MEASUREMENTS AND EXPERIMENTATION

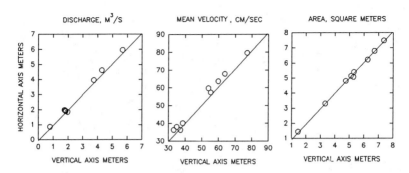

Figure 6. Test location averages of discharge, mean velocity, and area for horizontal-axis meters plotted versus averages for vertical-axis meters.

field tests and functioned in the 110°F water of the Hot River in Wyoming. However, the Valeport BFM001 nose cone unscrewed during measurements and had to be tightened in midstream. It was observed that the horizontal-axis meter impellers shed grass and other vegetation somewhat better than the vertical-axis meters and that grass did not prevent the electromagnetic meter from registering velocity.

Residual ratios were computed for discharge and area for each measurement location as residual ratio = (measured value$_i$ - mean value$_i$) ÷ mean value$_i$, where i is a location and the mean value$_i$ is computed from all the measurements at location i. Mean residual ratios of discharge and area were computed for each of the meters tested and for each stream gager. Bar charts of the mean residual ratio of discharge and area computed for the meters tested and for the stream gagers are in figure 5. Because discharge is the product of velocity and area, area is a possible source of error and was included in the analysis. Also included in figure 5 is the number of measurements made with each meter and by each stream gager.

Because all depths were measured using a top setting rod, small mean area residual ratios were computed for the meters (fig. 5A) and for the stream gagers (fig. 5B). Mean area residual ratios ranged from -0.008 to 0.025 for the meters and from -0.015 to 0.019 for the gagers. The mean discharge residual ratios for most meters and stream gagers are at least twice as large as the mean area residual ratios. Mean discharge residual ratios ranged from -0.038 to 0.060 for the meters and from -0.061 to 0.041 for the gagers. Except for the PRC meters(-0.007 metal, -0.034 plastic), horizontal-axis meters have positive mean discharge residual ratios. The vertical-axis meters, except for the Price type-AA(0.004), have negative residual ratios. The Marsh McBirney electromagnetic meter has a negative residual ratio of -0.026. The Ott C-31 and Price type-AA had the smallest discharge residual ratios. Because each gager did not use every meter at each measurement location, the mean residual ratios represent not only meter bias in discharge measurements but stream gager bias as well. Conversely, because some meters were used by one gager, gager bias is represented in the chart for the meters.

Groat(1918) found that vertical-axis meters over-register velocity and horizontal-axis meters under-register. However, for the field test data, most of the horizontal-axis meters over-registered the flows and most of the vertical-axis meters under-registered the flows.

The two PRC meters under-registered and the Price type-AA over-registered. Averages of discharge, mean velocity, and area for the horizontal-axis meter measurements are plotted versus the averages for the vertical-axis meters measurements in figure 6. The average areas are distributed about the line of perfect agreement between the meters in figure 6. The discharge and mean velocity are distributed above the line of agreement in figure 6 because horizontal-axis meters usually measured more velocity than the vertical-axis meters.

Summary

All meters tested had good repeatability (small percent standard errors) and similar linearity of response (RMS < 2.000 cm/s). For all meters tested, repeatability and response to oblique flow is poorest at the lowest test speed of 7.62 cm/s. Two horizontal-axis meters with component impellers, the Ott C-31 meter with A and R impellers had the smallest error in oblique flows. Except for the winter Price type-AA with the polymer rotor, the vertical-axis meters over- and under-register oblique flows with angles between $\pm 40°$. The Marsh McBirney electromagnetic meter also over and under-registers oblique flows with angles between $\pm 40°$. Horizontal-axis meters tended to under register oblique flows with angles between $\pm 40°$. The magnitude of error for horizontal-axis meters and the electromagnetic meter is usually smaller than those for vertical-axis meters in oblique flows. All meters tested, except for the Price type-AA, Price pygmy, and the winter Price type-AA polymer, had absolute meter errors less than 5% for flow angles between $\pm 10°$. Of the remaining meters, only the pygmy (-7.9% to 8.9%) did not have absolute errors less than 6% for flow angles between $\pm 10°$.

In previous literature it had been concluded that vertical-axis meters over register in "turbulent" flows in comparison to horizontal-axis meters. However, for the field data collected the vertical-axis meters and electromagnetic meters registered less velocity when compared with most of the horizontal-axis meters. The exceptions were the PRC meters, which registered lower velocity in comparison to the Price type-AA and greater velocity in comparison to the optic Price type-AA and the Price pygmy. The Marsh McBirney registered lower velocity in comparison to the Price type-AA and the PRC meter with plastic impeller.

References

Fulford, Janice M., Thibodeaux, Kirk G., and Kaehrle, William R. (1993). "Repeatability and oblique flow response characteristics of current meters," Proc. of 1993 Hydraulic Engineering Conference 1993, ASCE, New York, 1452-1457p.
Groat, B.F.(1918). "Characteristics of cup and screw current meters; performance of these meters in tail-races and large mountain streams",Transactions of American Society of Civil Engineers, New York, 819-870p.
Rantz, S.E. and others (1982). "Measurement and computation of streamflow: volume1. measurement of stage and discharge, USGS Water Supply Paper 2175, 79-183p.
Thibodeaux, Kirk G. (1992). "A brief literature review of open-channel current meter testing," Proc. of Hydraulic Engineering Conference 1992, ASCE, New York, NY, 458-463p.
Yarnell, David L. and Nagler, Floyd A. (1931). "Effect of turbulence on the registration of current meters." Trans. of the ASCE, 766-860p.

LDV SYSTEM FOR TOWING TANK APPLICATIONS

R. N. Parthasarathy[1] and F. Stern[2]

ABSTRACT
The design, construction, testing, and operation of a two-dimensional LDV system for use in a towing tank is described. The primary design considerations were underwater application, minimal disturbance at the measuring location, reduced vibration effects when towed, and capability of near-wall measurements. The operation of the system is demonstrated in making velocity measurements in the turbulent boundary layer of a surface-piercing flat plate.

INTRODUCTION
Surface-piercing ship boundary layers and wakes are greatly influenced by the presence of free-surface and gravity waves. The wave pattern, wave-induced separation and wave-breaking along with the turbulence/free-surface interaction are key issues in performance prediction, propeller-hull interaction and signature reduction. Over the past ten years, extensive experimental and theoretical research concerning free-surface effects on ship boundary layers and wakes has been carried out (Stern et al., 1989, Toda et al., 1992, Stern et al., 1993, Choi and Stern, 1993) at the Iowa Institute of Hydraulic Research (IIHR). As a result, certain features of the physics of the flow such as the significance of wave-induced pressure gradients, wave/boundary layer and wake interactions and the influence of free-surface boundary conditions for laminar flow have been reasonably understood. However, issues relating to the effects of the wave-induced pressure gradients and the free surface on the structure of turbulence in the boundary layer and wake and the effects of turbulence on wave-induced separation remain unresolved.

Accurate velocity measurements close to the free surface and details of the turbulence structure in the flow field are sorely needed to resolve these issues and to make progress in the computational modeling of such flows. Instrumentation with adequate frequency response and spatial resolution is required to make these measurements. While hot-films are a possible choice, they suffer from calibration problems. Moreover, the water in the towing tank cannot be maintained clean, resulting in frequent breakage of the hot-films due to impact with dirt particles. In recent years, laser-Doppler velocimetry (LDV) has emerged as an instrument with

---

[1] Assistant Professor, School of Aerospace and Mechanical Engineering, The University of Oklahoma, 235, Felgar Hall, 865 Asp Avenue, Norman, OK 73019

[2] Associate Professor, Department of Mechanical Engineering and Research Engineer, Iowa Institute of Hydraulic Research, The University of Iowa, Iowa City, IA 52242

adequate frequency response and spatial resolution to study the structure of turbulence in various flow configurations. In addition, LDV does not suffer from any calibration constraints and is non-intrusive. It was, therefore, decided to design and use a LDV system for velocity measurements in the towing tank.

While numerous studies of various flow configurations using LDV have been reported in the past, relatively few investigations have been conducted that involved the towing of an LDV system. This is primarily due to the problems encountered when towing in a water tank: space and weight restrictions, power requirements, vibration and humidity problems, and seeding problems, to name a few. The recent development of fiber-optic probes has facilitated the use of LDV in previously hard-to-reach regions. Kakugawa et al. (1991) discuss one such system that was used to make measurements in the wakes of ship models in the towing tank at the Ship Research Institute, Mitaka, Japan. They used a two-channel DANTEC fiber-optic probe mounted in a housing with a side window. Water invasion into the probe due to inadequate sealing and high humidity, vibration, and seeding deficiency were some of the problems encountered. It was concluded that due to the low efficiency in the measurement, LDV cannot be used as a tool for routine measurements in towing tanks.

This paper reports the design, installation and testing of a two-dimensional LDV system in the towing tank at IIHR. The paper begins with an overview of the design and construction of the system, followed by a description of the testing of the system. Finally, sample measurements in the towing tank are reported.

DESIGN AND CONSTRUCTION

Design Considerations
As mentioned above, the use of LDV in a towing tank has more restrictions than other standard uses due to the fact that the probe is immersed in the water and is towed during use. The system should be designed such that the disturbance created at the measurement region by the probe should be as small as possible. The system should facilitate easy cleaning of the exposed optical components. Vibration of the probe should me minimized as it leads to significant beam wandering and erroneous data. Since only limited time is available for measurements during each run (due to the length of the towing tank and acceleration and deceleration times), data rates need to be maximized. This requires consideration of external seeding methods.

Towing Tank Arrangement
The experimental work was carried out in the towing tank at IIHR, which is 91.4 meters long, 3 meters wide and 3 meters deep. The towing tank was originally equipped with two instrumented carriages, one driven and the other trailing. To facilitate the use of LDV in the towing tank, the carriage and trailer were modified to handle the weight of the LDV equipment and the cables. The drive carriage housed the computer, signal processor, oscilloscope and the traverse controller, and was used to tow two trailers. One trailer carried the laser and the sending optics while the model, the fiber-optic probe and its traverse were mounted on the other. Walkways were provided between the trailers to facilitate viewing when the assembly was towed. The towing carriage was cable-driven by a 15 hp motor fitted with a tachometer-feedback speed regulator and was capable of speeds of 2.5 m/s with an accuracy of ± 10 mm/s.

Foil-Plate Model

The flow configuration of interest was the turbulent boundary layer and wake of a surface-piercing flat plate with waves present on the free surface. The foil-plate model was similar to that used by Stern et al. (1989, 1993). By varying the depth of immersion of the airfoil, waves of different steepnesses could be generated. The flat plate was constructed from clear plexiglass to provide optical access for the LDV system; it was 1.2 m long, 0.5 m wide and 12.7 mm thick. It had a rounded leading edge and a sharp trailing edge. In order to induce turbulent flow, a row of cylindrical studs of 0.8 mm height and 3.2 mm diameter was fitted with 9.5 mm spacing on the plate at 60 mm from the leading edge. The water line was placed 0.2 m from the top of the plate and several station lines were scribed on the sides of the plate. The overall arrangement of the carriage, trailers, flat plate and the airfoil is illustrated in Figure 1.

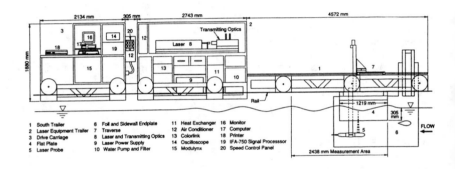

Figure 1. Schematic of the towing tank experimental apparatus

LDV System

The LDV system consisted of a 5 Watt Argon ion laser (INNOVA 70 series manufactured by Coherent Inc.), a fiber-optic probe with 10 m length cable, the colorburst and colorlink systems, all manufactured by TSI Inc. In order to prevent unnecessary deposits on the plasma tube that could occur when the water in the towing tank was used to cool the laser, distilled water circulated by a heat exchanger was used. The water from the towing tank was pumped into the heat exchanger to cool the distilled water. The water intake and discharge sites were located such that minimal disturbance was caused in the flow.

The fiber-optic probe was cylindrically shaped and located in a housing with a prism assembly at the end that deflected the beams at $90°$ to the probe. Different designs of the housing were discussed with the manufacturer of the LDV system, TSI Inc., to eliminate flow separation at the location in the housing at which the beams exited. A plexiglass model of the probe and the housing design recommended by TSI were glued together and tested in a flume. A schematic of the probe and its housing is shown in Figure 2. Dye was injected at various locations and videos were taken of the flow around the housing. Of particular interest was flow separation and the formation of bubbles at the side window and the flow disturbance caused by the probe. The flow visualization results indicated that this design performed satisfactorily. This configuration also provided easy access to clean the prism assembly periodically by spraying distilled water.

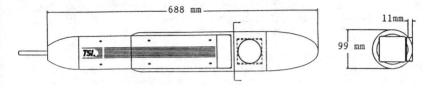

Figure 2. Sketch of the fiber-optic probe housing

The fiber-optic probe was 83 mm in diameter and 465 mm long. It had four fibers transmitting two green and two blue beams and a receiving fiber to collect the scattered light. The green and blue beams were spaced 50 mm apart at the exit of the probe and were focused with a 350 mm-focal distance lens with an aperture of 61.5 mm. A 20 mm collimating lens was used to control the initial beam diameter. The resulting fringe spacings were 3.611 and 3.425 microns for the green and blue channels respectively. A 250 mm-focal distance lens was used to collimate the light on to the receiving fiber. The scattered light was collected direct on-axis backscatter. The dimensions of the measuring volume were estimated to be 0.07 x 0.07 x 1 mm. Since the near-wall and near-free-surface signals were expected to be noisy, two receiving fibers were provided: 125 microns and 50 microns in diameter, the smaller fiber to be used for near-wall and near-free-surface measurements. In addition, masks were provided to cut off any light reflected back from the wall or free-surface. The mounts of all four transmitting fibers and the receiving fiber could be adjusted for proper beam crossing.

The fiber-optic probe could be used to measure only two velocity components at a time. The system was designed to use the probe in two orientations, illustrated in Figure 3, to measure all three velocity components.

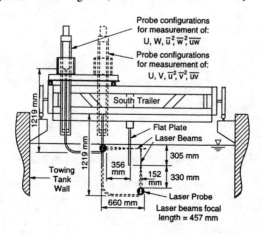

Figure 3. The two probe orientations

In the first configuration, the probe was parallel to the plate and the beams passed through the plate so that the streamwise (along the plate) and vertical velocities could be measured. In the second orientation, the probe was located underneath the plate with the beams shooting vertically up, to facilitate the measurement of streamwise and transverse velocities. Thus, streamwise velocities were measured in both configurations and served to verify the location of measurement when going from one orientation to another.

Measurements were desired in the boundary-layer and the wake regions of the plate. The desired region of interest was 30 mm in the direction transverse to the plate, 150 mm in the vertical direction, and 2.4 m along the plate. A computer-controlled modulynx traverse was chosen for the transverse and vertical directions; these traverses were fitted on a plate that could be slid on stainless steel guide rods to provide the streamwise movement of the probe. In order to minimize effects of probe vibration, there were two options: the fibers could be individually wrapped instead of having all the fibers wrapped in one cable housing or the fibers could be enclosed in an airfoil-shaped housing that was immune to vibration. In this design, the latter option was chosen. A streamlined strut, as illustrated in Figure 4, was used to support the fiber-optic cable and the probe to minimize vibration effects when towed.

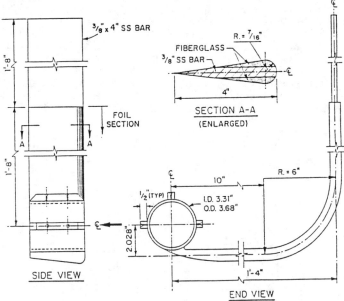

Figure 4. Design of the strut that held the fiber-optic cable

Due to the need for accurate processing of noisy signals, it was decided to use processors that would detect the Doppler shift in the frequency domain rather than conventional processors based on zero crossings. The TSI IFA 750 signal processor was chosen for this purpose. The signal processor was digitally interfaced to an IBM-compatible microcomputer using a TSI MI 750 interface and FIND software. In order to ensure that velocity measurements from both channels were made on the same particle, an electronic time window of 100 microseconds was used to determine the coincidence of the measurements.

TESTING OF THE LDV SYSTEM

Stationary calibration involved the measurement of the edge velocity of a rotating plexiglass disk of known diameter. The rate of rotation was measured independently with a photoelectric sensor and an accurate frequency counter. The differences between the velocities measured with LDV and the velocities calculated using the measured frequency and disk radius were less than 1 percent. Measurements in still water resulted in mean values of 1 to 2 mm/s with standard deviations of 5 to 7 mm/s, primarily due to electronic noise.

Next, the system was towed and the carriage speed was measured with the LDV system. It was found that the data rates were very small (on the order of 2 Hz). A seeding mechanism was constructed to disperse manually a solution of water mixed with titanium dioxide particles (nominal size of 3 to 5 microns) during one run. The particles remained suspended for two to three days. Calculations indicated that the terminal velocity of these particles in water was 0.05 mm/s and that they could adequately follow sinusoidal motions with frequencies up to 2 kHz. The addition of these seed particles resulted in coincident data rates between 5 and 10 Hz near the plate and free surface and 20 - 30 Hz away from the plate. The difference between the carriage speed measured by the LDV system and that measured by the transducer on the wheel was less than 1 percent.

To appreciate the effects of vibration, measurements of both streamwise and vertical velocities were made when the carriage was towed at 1.37 m/s (which was the maximum test speed). The measured standard deviations were on the order of 30 mm/s indicating that the standard deviations were amplified from their values in still water by the overall vibration of the probe. Various methods were attempted to reduce probe vibration. The holes for the bolts that held the probe-holder to the traverse were enlarged and fitted with rubber bushings. The outside nut on each wheel was loosened slightly so that the wheels would be able to follow the track better. A brace was added to the probe-holder to increase its stiffness. Sandbags were mounted against the traverse to absorb the vibrations. The most effective solution was the vibration absorption by strategically-placed weights on the plate-instrumentation trailer. With the added weights, the standard deviations in both the streamwise and vertical components of velocities were 20 mm/s even at the lowest position of the probe-holder (for a carriage speed of 1.37 m/s). This was only 1.5 percent of the freestream velocity and could be viewed as the equivalent of freestream turbulence encountered in wind tunnels.

A plate with a circular hole was attached to the trailer and positioned such that the measuring volume of the LDV system (the crossing point of the beams) was at the center of the hole. If the measuring volume moved a distance greater than the radius of the hole as the LDV system was towed, one or more of the beams would get blocked. Testing with this arrangement indicated that the relative movement of the measuring volume and the carriage was less than 2 mm with the added weights on the plate-instrumentation trailer. Thus, the overall spatial resolution of the system was 2 mm and freestream standard deviations were on the order of 1.5 percent for a towing speed of 1.37 m/s.

RESULTS

The conditions for the experiments were based on Stern et al. (1989 and 1993). Measurements were made for two carriage speeds: 0.46 m/s and 1.37 m/s. resulting in Reynolds numbers (based on plate length) of 0.56×10^6 and 1.64×10^6 respectively. Three wave steepnesses were studied: zero, medium, and large. The results of the zero steepness condition tests will be discussed here. The measurements for the various wave conditions are described in detail by Stern et al. (1994). Previous studies (Stern et al., 1989) had indicated that to simulate conditions corresponding to zero steepness, wave effects were actually smaller for the case of the deeply-submerged airfoil, than with the total absence of the airfoil. Apparently, the plate-and airfoil-induced waves canceled, resulting in a flat free surface, when the airfoil was submerged deep. The measurements for the zero-steepness condition serve as benchmark results for comparison with previous measurements in two-dimensional flat-plate turbulent boundary layers due to the present non-ideal conditions involving free-surface effects and low Reynolds numbers.

All measurements were averaged using 1500-4000 samples and refined by eliminating measurements outside ± 3 standard deviations. Limited number of samples contributed to the experimental uncertainties primarily. Experimental uncertainties were less than 4 percent for the mean streamwise velocities, 6 percent for the mean vertical and transverse velocities, 10 percent for velocity standard deviations, and 18 percent for the maximum Reynolds shear stresses. Positioning accuracies were better than 0.25 mm in the vertical and transverse directions and 1.5 mm in the axial direction.

A cartesian coordinate system is used with the origin at the intersection of the mean free surface and the leading edge of the plate. The x-axis is coincident with the longitudinal axis of the plate and is directed downstream, and the y- and z-axes are in the starboard and vertically downward directions respectively. The corresponding velocity components are denoted by u, v, and w, with capitals representing mean velocities. All dimensions are normalized with the plate length, and all velocities are normalized with the carriage speed, unless stated otherwise.

The wall-shear stress, shape parameter, velocity variations in terms of inner and outer variables are illustrated in Figure 5, including previous data obtained with three-hole and five-hole probes by Stern et al. (1989, 1993) and those given by White (1991). The wall-shear stress was obtained using a Clauser chart, while the boundary-layer, displacement, and momentum thicknesses were calculated using the trapezoidal rule of integration. The present measurements using LDV compare well with the previous data obtained by Stern et al. (1989, 1993). The primary discrepancy is in the shape parameter; the measured shape parameters are larger than those given by White (1991), but agree with the previous measurements. Experimental errors and free-surface effects could be the reason for this discrepancy.

The measured Reynolds-stress profiles, normalized using the shear velocity are shown in Figure 6. Also shown are the benchmark data of Klebanoff (1955). Satisfactory agreement of the present data with those of Klebanoff (1955) is displayed. However, the peak Reynolds shear stress values were not resolved due to unacceptable levels of signal-to-noise ratio for $y^+ < 70$. Also, relatively larger freestream turbulence levels are exhibited; u'w' is not zero, which provides a measure of the free-surface effects. Nevertheless, it can be concluded that the boundary layer development for z>0.02 shows the typical characteristics of a two-dimensional flat-plate turbulent boundary layer, with free-surface effects decreasing with increasing depth.

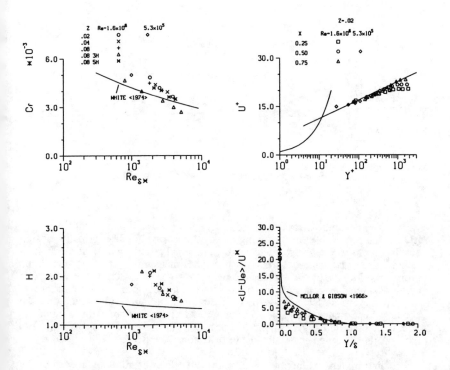

Figure 5. Measured shear-stress, shape factor, and velocity profiles
3H, 5H are previous measurements made with 3-hole and 5-hole probes

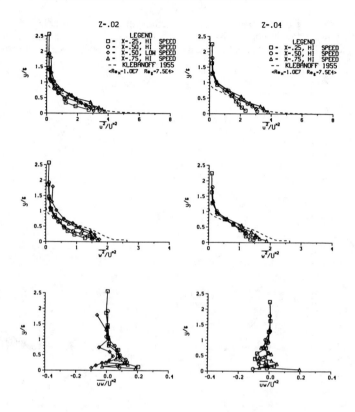

Figure 6. Measured Reynolds stress profiles

CONCLUSIONS

The design, construction, and operation of a two-dimensional fiber-optic LDV system for use in a towing tank is described. The primary design considerations were underwater application, minimal disturbance, reduced vibration effects, and reasonable accuracy for near-wall measurements. The fiber-optic probe, based on the green and blue beams of an Argon-ion laser, was enclosed in a cylindrical housing equipped with a nose cone. The beams exiting from the end of the probe were deflected by a prism assembly through a side window in the housing. The fiber-optic cable was supported in a streamlined strut to minimize effects of vibration. The system was designed for use in two configurations to facilitate the measurement of all three velocity components.

Preliminary testing indicated significant effects of vibration as seen by the large standard deviations in the measured velocities and the large movement of the measuring volume. Various methods were adopted to minimize these effects. The resulting arrangement had a spatial resolution of 2 mm, and a velocity standard deviation of 1.5 percent of the mean velocity for a towing speed of 1.37 m/s. Measurements were made in the turbulent boundary layer of a surface-piercing flat plate at Reynolds numbers of 0.56×10^6 and 1.54×10^6. The mean velocities agreed with the previous measurements made using Pitot probes; the Reynolds stresses agreed with previous measurements made in flat-plate boundary layers, thus, providing confidence in the operation of the system.

ACKNOWLEDGMENTS

The financial assistance provided by the Iowa Institute of Hydraulic Research, The University of Iowa, and the Office of Naval Research is gratefully acknowledged. The authors are also thankful to Jim Goss, Doug Houser, and Mark Wilson for their assistance in the design and construction of the system.

REFERENCES

Choi, J.E. and Stern, F. (1993), "Solid-Fluid Juncture Boundary Layer and Wake with Waves," *Proceedings of the Sixth International Conference on Numerical Ship Hydrodynamics*, Iowa City, Iowa.

Kakugawa, A., Takeshi, H. and Makino, M. (1991), "Flow Field Measurements around Marine Propellers at Towing Tank Using Fiber Optics LDV, " *Laser Anemometry*, Vol. 2, pp. 807-817, ASME, New York.

Klebanoff, P.S. (1955), "Characteristics of Turbulence in a Boundary Layer with Zero Pressure Gradient," NACA Report 1247.

Mellor, G.L. and Gibson, D.M. (1966), "Equilibrium Turbulent Boundary Layers," *Journal of Fluid Mechanics*, Vol. 51, pp. 225-253.

Stern, F., Hwang, W.S., and Jaw, S.Y. (1989), "Effects of Waves on the Boundary layer of a Surface-Piercing Flat Plate: Experiment and Theory," *Journal of Ship Research*, Vol. 33, pp. 63-80.

Stern, F., Choi, J.E. and Hwang, W.S. (1993), "Effects of Waves on the Wake of a Surface-Piercing Flat Plate: Experiment and Theory," *Journal of Ship Research*, Vol. 37, pp. 102-118.

Stern, F., Parthasarathy, R.N., Huang, H.P. and Longo, J. (1994), "Effects of Waves and Free Surface on Turbulence in the Boundary Layer of a Surface-Piercing Flat Plate," ASME Symposium on Free-Surface Turbulence, Lake Tahoe, Nevada.

Toda, Y., Stern, F. and Longo, J. (1992), "Mean-Flow Measurements in the Boundary-Layer and Wake and Wave Field of a Series 60 C_B = 0.6 Ship Model - Part 1: Froude Numbers 0.16 and 0.316," *Journal of Ship Research*, Vol. 36, pp. 360-377.

White, F. (1991) *Viscous Fluid Flow*, Second edition, McGraw Hill, New York.

EXPERIMENTAL STUDY ON UNSTEADY FLOW IN OPEN CHANNELS WITH FLOOD PLAINS

A. TOMINAGA[1], M. NAGAO[2] and I. NEZU[3]

Abstract

Hydraulic characteristics of unsteady flows were investigated experimentally in open channels with flood plains using a computer-controlled water supply system. Some noticeable features of unsteady flow structure in compound channels were revealed which were significantly different from those in single cross-sectional rectangular channels. The unsteadiness of the velocity becomes very large in a main chanel.

Introduction

In recent years, flood plains become very useful as ecological and recreational spaces in urban rivers. When one consider such utilities of flood plains for multiple purposes, it is very important to understand hydrodynamic characteristics in compound open channel flows in flood. A number of researches have been conducted on resistance law and flow structures in compound channel flows. Three-dimensional turbulent structures of compound open channel flows associated with secondary currents have been recently revealed by the use of velocity measurements and flow visualizations (e.g. Tominaga & Nezu (1991) and Imamoto & Ishigaki (1991)). However, most of them treated the flood flow as a steady uniform flow with maximum discharge. It is known that the flow in a rising stage of flood is rather different from the flow in a falling stage as shown by Nezu et al. (1993a,b) and Tu & Graf (1992). They found that the bed shear

---

[1]Associate Professor, Dept. of Civil Engineering, Nagoya Institute of Technology, Showa, Nagoya 466, Japan.
[2]Professor, the same as the above.
[3]Associate Professor, Dept. of Civil & Global Engineering, Kyoto University, Sakyo, Kyoto 606, Japan.

stress becomes considerably larger in the rising stage than in the falling stage. It is very interesting to investigate the unsteady flood flow over compound open channels, since it encounters a sudden change of cross section. In this study, time-dependent three-dimensional flow structures are measured in compound open channels when a discharge is varied according to some flood hydrographs.

Experimental Apparatus and Method

The experiments were conducted in a 13m long and 0.6m wide 0.4m deep tilting flume. Trapezoidal flood plain was set on one side of the flume. The upper and lower surface width were 0.262m and 0.304m, respectively and the height of the flood plain was 0.042m. So, the side slope of the main channel is equal to 1/2. The channel slope S was set as S=0.001.

Fig.1 shows a open-channel experiment system. The discharge is controlled by changing the rotation cycle of a water-pump motor by means of a transistor inverter. The rotating cycle is controlled by a personal computer using feedback from the signal of an electromagnetic flow meter through an A/D and D/A converter board. Arbitrary discharge hydrographs can be obtained from the computer. The water is supplied into an settling tank with mesh screens at the channel entrance. The test section was set at 9.0m downstream from the entrance.

Table 1 Hydraulic conditions

CASE	Q_b (cm³/s)	Q_p (cm³/s)	h_b(cm)	h_p(cm)	T_{hp}(s)	λ	U_{max}(cm/s)
D60	5000	20000	3.94	8.22	72	0.72	68.3
D120	5000	20000	3.93	8.47	132	0.38	60.6
D240	5000	20000	4.10	8.72	256	0.19	54.1
T60	5000	20000	2.24	5.53	68	0.75	76.5
T120	5000	20000	2.26	5.75	132	0.38	75.1
T240	5000	20000	2.29	5.87	256	0.20	75.2

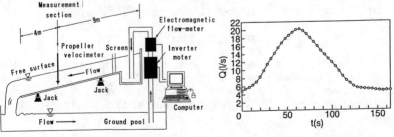

Fig. 1 Schematic description of experiment system

Fig. 2 Discharge hydrograph (T_p=60s)

The base discharge Q_b was set to $0.005 m^3/s$. In this case, the flow depth h was 3.9cm and the flow was limited in the main channel. The peak discharge Q_p was set to $0.02 m^3/s$. The discharge is increased from Q_b to Q_p linearly in a rising time T_p. After keeping the peak discharge during $(1/6)T_p$, the discharge is decreased from Q_p to Q_b in a falling time same as T_p. In the present experiments, T_p was taken as 60s, 120s and 240s. An example of given discharge hydrograph when T_p=60s is shown in Fig.2. The identity of these repeated discharge hydrograph was reasonably good. Table 1 shows the present hydraulic conditions (case D) with the preliminary experiments in a rectangular single cross-sectional channel (case T).

Velocity Measurements

Velocity was measured by a micro propeller velocimeter 3mm in diameter. Simultaneously, water depth was measured by a condenser-type water-wave gage 0.1m down stream from the velocimeter. Since the sensing point of velocimeter submerges or unsubmerges with time, the propeller velocimeter is adequate for the measurements of unsteady flood flow. The output signal of the propeller velocimeter tends to be reduced and linear relation between the velocity and the output voltage is no longer sustained near the free surface. A calibration equation was made from the preliminary experiments as a function of the submerged depth of the propeller. Then, the velocity was corrected by this equation using the flow depth obtained by the water-wave gage. Both the output signals of the velocimeter and the wave gage were recorded on floppy disks and processed by computers. The sampling frequency was 50Hz when T_p=60s and T_p=120s, and 20Hz when T_p=240s. The sampling time was 180s, 320s and 600s, respectively for T_p=60s, 120s and 240s.

In order to obtain the time-dependent three-dimensional flow structures in the whole cross section, the velocity was measured at each point repetitively during the flood event with the identical hydrograph. The velocity signals were arranged in time so as to be t=0 at the rising point of the flow depth. Then, the time was divided into consecutive 4 seconds periods and the velocity was averaged in each period. These 4s interval-averaged values of velocity are used in the following discussions.

Characteristics of Time Variation of Flow Depth

In Table 1, h_b and h_p are the base and peak flow depth. T_{hp} is the arrival time of the peak flow depth since the flow-depth is started rising at the test section. U_{max} is the maximum interval-averaged velocity during the flood. λ is an unsteadiness parameter defined by Tkahashi(1969) given as follows:

$$\lambda = \frac{(h_p - h_b)/T_p}{S\sqrt{gh_p}} \tag{1}$$

The numerator of eq.(1) means the rising speed of actual water surface and the denominator indicates the vertical component of propagation velocity of long wave with peak flow depth. When $\lambda<<1$, the flow becomes quasi steady state and when $\lambda>>1$, hydraulic serge appears. In actual floods, λ is quite small like a kinematic wave, but the effects of unsteadiness are often observed in many Japanese rivers (e.g. Fukuoka et al. (1990)). In the present experiments, λ is rather large and the flood wave includes of dynamic-wave characteristics.

Though the flow depth in the compound channel becomes much larger than that in the rectangular channel, the arrival times of the peak stage T_{hp} indicate little difference between the compound and rectangular channels. The peak flow depth hp becomes larger as increasing the arrival time T_{hp}.

Fig.3 shows the normalized time variations of flow depth in all cases. The abscissa is normalized time divided by T_{hp}. The ordinate is the dimensionless variation of flow depth Δh $(=(h-h_b)/(h_p-h_b))$. Rising processes show linear increase and are almost similar in all cases, whereas falling processes of the compound channel become to be relatively later than those in the rectangular channel.

Time Variation of Depth-Averaged Velocity

Fig.4 shows the normalized time variation of depth-averaged primary velocity U_m at the representative sections in the main channel and the flood plain. U_m is divided by the maximum velocity U_{max}. As shown in Table 1, the maximum velocity U_{max} increases with a decrease of T_{hp} in the compound channel, whereas it is not so changed in the rectangular channel. When $T_p=60s$, the velocity in the main channel attains a peak much earlier than the peaktime of the flow depth. This peak time was also much smaller than that in the rectangular channel. As increasing T_p, the peak time of the main-channel velocity becomes relatively faster, but it does not monotonically decrease after the peak time. The high velocity is maintained till the peak time of the flood plain velocity. When $T_p=240s$, the second peak appears at almost the same time as the peak in the flood plain. The flood-plain velocity attains a peak faster than Thp when $T_p=60s$. However, with an increase of T_p, the peak time of the flood-plain velocity approaches to the peak time of flow depth. These features of velocity variation implies that the turbulent mixing of the main-channel and flood-plain flow becomes to be sufficiently developed as increasing the rising time of the flood.

Fig.5 shows the depth-velocity curves at the main channel and the flood plain for each T_p. In the main channel, the velocity varies drawing a clear loop and the difference of the velocity is extremely large between the rising and falling stages. Total process of depth-velocity variation can be divided into four stages: first and second rising stage and first and second falling stage, as shown in Fig.5. With an increase of T_p, the area surrounded by the loop curve becomes smaller, but the velocity difference between the rising and falling stage is still large compared with that in the single cross-sectional rectangular channel. In the flood plain, loop property is not significant except for the case of $T_p=60s$.

Time Variation of Friction Velocity

It is indicated by Nezu et al(1993a) that the log-law is applicable even in such an unsteady flow. The log-law distribution for smooth open-channel flows is given as follows:

$$\frac{U}{U_*} = \frac{1}{\kappa}\ln\left(\frac{U_* y}{\nu}\right) + A_s \tag{2}$$

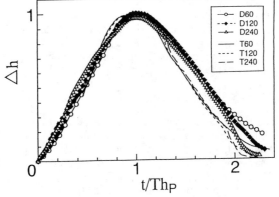

Fig. 3 Time variations of flow depth

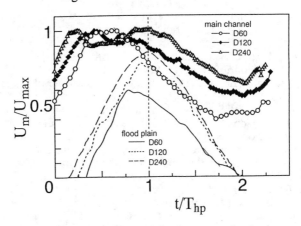

Fig. 4 Time variations of depth-averaged velocity

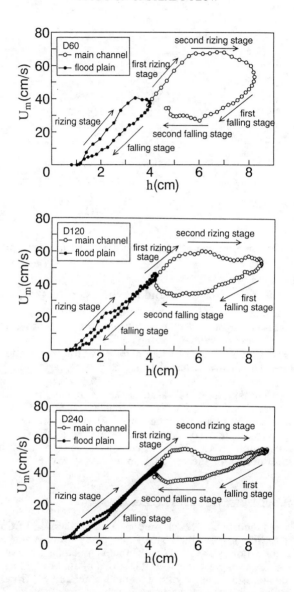

Fig. 5 Depth-velocity curves in compound channel

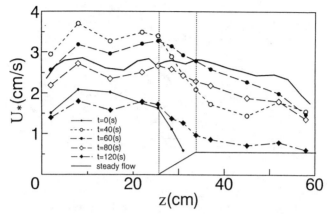

Fig. 6 Lateral distribution of friction velocity at each time (T_p=60s)

in which von Karman constant κ and integral constant A_s are taken as κ=0.41 and A_s=5.29. Though an applicability of the eq.(2) on the flood plain is suspicious about the integral constant in the present study, the friction velocity U* was roughly estimated using eq.(2).

Fig.6 shows time variation of friction velocity along the wetted perimeter when T_p=60s. The distribution of steady flow with maximum discharge of Q=0.02m³/s is also shown in this figure. At t=40s when the main-channel velocity becomes maximum, the value in the main channel considerably exceed the steady-flow value with peak discharge. On the flood plain, it becomes maximum at t=60s, but it does not exceed the steady value. It is known that the bed shear stress becomes larger in the rising stage in rectangular channels. This feature becomes more significant in the main channel with flood plain and the peak appears very earlier than the stage peak.

Time-Dependent Three-Dimensional Flow Structures

Fig.7 shows the isovels at each time period of the flood from the rising stage to the falling stage with those of the steady flow. When t=40s in the rising stage, the velocity in the main channel is much larger than the steady value and decrease rapidly towards the flood plain. So, high shear layer is formed at the interface region of the main channel and the flood plain. At the stage peak of t=80s, the structure in the main channel is very similar to the steady one, but the velocity on the flood plain is smaller than the steady one. At t=120s in the falling stage, the velocity is much smaller than that at t=40s which has almost the same flow depth. It is understood that the effects of unsteadiness appear locally when the flood passes such a complicated cross section.

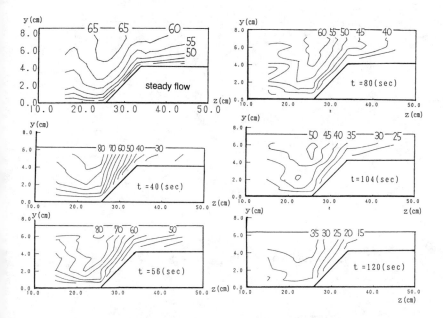

Fig. 7 Time variations of isovels in compound channel ($T_p=60s$)

Time Variation of Turbulence Intensity distribution

The evaluation methods of turbulent fluctuation in unsteady flows were discussed by Hayashi & Ooshima (1988) and Nezu et al (1993a). There are ensemble average method, Fourier transform method and moving average method to separate the mean velocity component and turbulence component from the instantaneous velocity. In the present study, since propeller velocimeter has not sufficiently high-frequency response, the moving average method was adopted for simplicity. So, the turbulence intensity obtained here is not accurate, but the characteristics of its spatial distribution can be reasonably discussed.

Fig.8 shows time variation of lateral distributions of turbulence intensity u' at y=4.8cm. The value of u' is evaluated for consecutive 10s periods because 4s is too short to evaluate the turbulence intensity. Turbulence becomes very large in the interface region of the main channel and the flood plain in the same manner as steady compound channel flows shown by Tominaga & Nezu (1991). Especially, when t=40s in the rising stage, turbulence becomes maximum. This is considered to be the result of the formation of strong lateral shear layer in the rising stage.

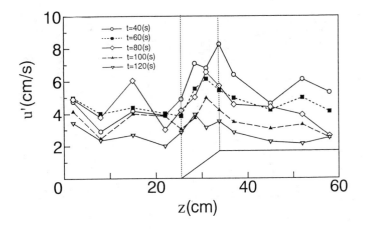

Fig. 8 Lateral distributions of turbulence intensity at each time (T_p=60s)

Conclusions

Unsteady flow structures in open channels with flood plain were investigated from the repetitive traversing measurements on the same discharge hydrograph using a computer-controlled water supply system. Unsteady flow structure in compound channels were significantly different from those in rectangular channels. In the main channel, the unsteadiness of the velocity becomes very large and the peak value of the velocity appears very earlier than in rectangular channels. In the flood plain, unsteadiness is smaller than that in the main channel. Consequently, the difference between main-channel velocity and flood-plain velocity becomes significant in the rising stage. So, the turbulence intensity becomes maximum in the interface region in the rising stage. The bed shear stress in the main channel becomes much larger than that expected from the steady flow.

Acknowledgment

The financial support of the ASAHI GRASS Foundation was awarded for this research, and it is hereby acknowledged.

References

Fukuoka et al. (1990): Study on flood flow and river-bed variation in the Hinuma River, Report of PWRI, vol.180-2, pp.1-94.

Hayashi,T., and Ooshima, M. (1988): Effects of the unsteadiness of flood waves on their turbulence structure, Proc. of 32nd Japanese Conf. on Hydraulics, pp.607-612.

Imamoto,H. and Ishigaki,T. (1991): Experimental study on the turbulent mixing in a compound open channel, 24th IAHR Congress, Madrid, vol.C, pp. 609-616.

Nezu,I. and Nakagawa, H. (1993a): Basic structure of turbulence in unsteady open-channel flows, 9th Symp. on Turbulent Shear Flows, Kyoto, vol.1, 7.1.1-7.1.6.

Nezu,I., Nakagawa,H., Ishida,Y. and Fujimoto,H. (1993b): Effects of unsteadiness on velocity profiles over rough beds in flood surface flows, 25th IAHR Congress, Tokyo, vol.A1, pp.153-160.

Takahashi,H. (1969): Theory of one dimensional unsteady flows in an prismatic open channel, DPRI, Kyoto University, vol.12B, pp.515-527 (in Japanese).

Tominaga,A. and Nezu,I. (1991): Turbulent structure in compound open-channel flows, J. Hydraulic Engrg., ASCE, vol.117(1), pp.21-41.

Tu,H. and Graf,W.H. (1992): Velocity distribution in unsteady open-channel over gravel beds, J. Hydroscience and Hydraulic Engineering, vol.10, No.1, pp.11-25.

COHERENT STRUCTURES IN COMPOUND OPEN-CHANNEL FLOWS BY MAKING USE OF PARTICLE-TRACKING VISUALIZATION TECHNIQUE

Iehisa NEZU[1]; Member, ASCE, Hiroji NAKAGAWA[1] and Ken-ichi SAEKI[2]

ABSTRACT

In this study, images of very small 100 μm diameter particles uniformly suspended in compound open-channel flows were recorded on an optic disc using a high-sensitive CCD camera and 2W Argon-ion laser slit illumination techniques. Instantaneous velocities at each point in the laser slit plane were measured from these continuous four images of one particle. This new technique called the *"Particle-Image Velocimetry* (PIV)" seems to be very powerful to investigate space-time correlation structures of coherent vortices of turbulence using the conditional sampling method of quadrant motions of velocity components because simultaneous velocity components can be obtained at all grid points of arbitrary section plane of water flows.

INTRODUCTION

Strong anisotropy of turbulence near the boundary between a main channel and flood plain in compound open-channel flow causes intermittently upward secondary currents and therefore generates three-dimensional (3-D) coherent vortices in these regions, e.g., Tamai et al. (1986), Fukuoka and Fujita (1989), Shiono and Knight (1989), Tominaga and Nezu (1991) and others. Such many researchers have pointed out significant importance about the lateral transports of momentum and suspended sediment between the main channel and flood plains by these secondary currents. Same phenomena always occur in flood rivers, and thus very important topics in river engineering. These time-averaged structure of secondary currents in compound open-channel flows has been revealed experimentally using accurate measurements with fiber-optic laser Doppler anemometers (FLDA) by Tominaga and Nezu (1991), and also these could be successfully computer-simulated using a refined algebraic stress model (ASM) by Naot et al. (1993 a and b), and Naot and Nezu (1993). Tominaga and Nezu (1993) have comprehensively reviewed such researches of compound open-channel flows which were conducted inside and outside of Japan.

---

1. Dept. of Civil and Global Environment Eng., Kyoto University, Kyoto 606, Japan.
2. HAZAMA Corporation, 2-5-8 Kita-Aoyama, Minato-ku, Tokyo 107, Japan.

It is further necessary to investigate the hydrodynamic behaviors of coherent vortices in both space and time even quantitatively in order to reveal the evolution mechanism of such vortices and the associated sediment transport in compound open-channel flows. Fortunately, a new quantitative visualization technique has recently be developed using a high-power laser-slit illumination and a CCD camera. This technique uses a *"Particle-Image Velocimetry* (PIV)" which can measure even instantaneous velocities in the laser light sheet (LLS) using image analyses by reasonable computer algorithms. Tsuda et al. (1991, 93) have tried to develop such a reasonable computer algorithm for image analyses from a CCD camera through a laser disc recorder. This visualization technique, i.e., PIV technique is very powerful to investigate large-scale coherent structures in space and time, because flow images in the LLS space are taken in the measuring time span. In contrast, the conventional laser Doppler anemometry (LDA) and fiber-optic LDA system can measure very accurately velocity fluctuations at one point, but cannot measure them simultaneously at *many* points. For example, Nezu and Nakagawa (1989) have successfully analyzed space-time correlations of coherent vortices behind step flow using two sets of LDA systems; one is the detecting LDA, and the other is the sampling LDA. Nezu and Nakagawa (1991, 93) have revealed three-dimensional (3-D) structures of coherent vortices behind dunes by the simultaneous measurements of velocity-velocity combination using both LDA and hot-film anemometers. These techniques, however, are limited as only *"two-point"* simultaneous measurements in water flows; a rake of 10 hot-wires is easily used in air.

In this paper, horizontal large-scale coherent vortices in compound open-channel flows could be measured using the *particle-image velocimetry* (PIV) system, as schematized in Fig. 1. Conditional sampling space-time correlation structures of such coherent vortices are analyzed reasonably and the evolution mechanism of turbulence vortices can be discussed because simultaneous velocity components can be obtained at all grid points of arbitrary section plane of water flows in compound open channels.

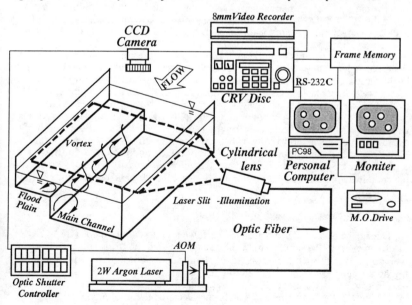

Fig. 1. Arrangements for Experimental Flume and PIV System.

PARICLE-TRACKING VISUALIZATION TECHNIQUE

Experimental Flume

Experiments were conducted in 8m long, 30cm wide and 30cm deep tilting flume. A flood-plain model was made of 2mm sand rough plate, as shown in Fig.2. Four different hydraulic conditions were chosen in two kinds of flood plains; one is rough bed and the other is vegetation model planted. $B_1/B=1/3$, $H/D=$ 1.1 to 1.4, the bulk mean velocity $U_m=27$ cm/s, Re= 1.5×10^4 and Fr= 0.3 -0.4, as indicated in Fig. 2. More detailed information is available in Nezu et al. (1994).

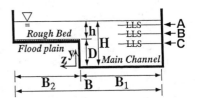

Fig. 2. Compound Open-Channel and Nomenclatures.

Visualization of Very Small Tracer-Particles in Water and Image Recorder

Schematic arrangements for the present flow visualization and image analyses are shown in Fig. 1. Very fine 100 μm powder of specific gravity 1.02 was scattered in alcohol and then mixed into the circulating water of flume. The concentration of neutrally suspended particles in water was only about 0.5ppm, and thus particles in water flow cannot, of course, be seen by human eye. However, high-power laser light illumination enables one to see these fine particles. The PIV is a great technological innovation of conventional flow visualization, e.g., see the conventional but very refined visualization technique by Utami and Ueno (1987).

In the present visualization technique, 2W high-power argon-ion laser beam was guided through an optic fiber cable and illuminated a horizontal spatial surface in the water using a pair of cylindrical lenses, as shown in Fig. 1. This illuminated section is called the *"laser light sheet"* (LLS). In the present study, three kinds of LLS (A, B and C, as shown in Fig. 2) of the different elevation were used for particle illumination. These elevations were chosen to $y_A=D+3/4h$, $y_B=D+1/4h$ and $y_C=D-1/4h$. The hydrodynamic behaviors of horizontal coherent structures just below and above the flood-plain elevation y=D were investigated intensively to reveal interaction mechanism between main channel and flood plain. This is because strong upward-tilting secondary currents have been measured with a fiber-optic LDA by Tominaga and Nezu (1991) and computer-simulated with a refined ASM by Naot et al. (1993 a and b).

Images of particles were taken on an optic disc through a high-sensitive CCD camera that was placed 70cm above the water surface. The shutter-time length of the camera can be adjusted by rotary switches in order to obtain the best images of particles. These images were taken at every 1/125s and indicated dot-like shapes clearly in the case of $U_m=$ 10- 50 cm/sec. Therefore, it is easily possible to determine the coordinate of particle. The shutter was timing-controlled and opened once in 1/60s.

IMAGE ANALYSES

General Explanation for Images of Optic Disc

One image is taken on an optic disc *"Component Recording Video"* (CRV, SONY-made) in every 1/30s step. This 1/30s image is called the *"frame image"* in a frame memory. One frame image consists of about 500 x 500 pixels. Each pixel has brightness degree which is divided into 256 grades, i.e., from 0 to 255. Images of the pixels on the odd-number lines are accumulated in the first 1/60s, and those of the other pixels in the next 1/60s. The former is called the *"odd-field* (the 1st) image" and the latter the *"even-field* (the 2nd) image".

Processing of Image Database

A frame memory board which can store four frame images was installed into a personal computer (NEC PC-98). Frame image that was read from the CRV disc was divided into two field images and rewritten as two images of 1/60s difference on the frame memory board. These divided field images have no information on the *even* (or *odd*) lines, as shown in Fig. 3.

So, the interpolated brightness degree between brightness degrees of the upper and lower pixels was placed on odd (or even) line pixels in order to make these two divided field images nearly equal in quality. After these processing, the coordinates of particles in every 1/60s field were determined and these database was saved into *Hard Disk* and *M.O. Drives* as the *"particle-location data files"*.

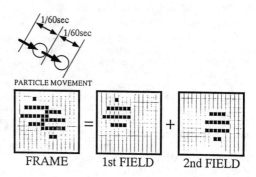

Fig. 3. Division into Two Field Images.

Tracking Particles and Determination of Velocity Vectors

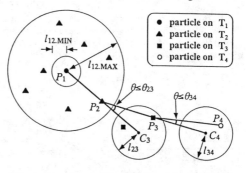

Fig. 4. Algorithm of Particle-Tracking Method.

The same particle in continuous four field-images was automatically recognized by the following algorithm, as schematized in Fig. 4. Arbitrary particle P_1 on 1st-time field-image is listed up. Secondly, the particle P_2 on the 2nd-time field-image between $l_{12,MIN}$ and $l_{12,MAX}$ far from the 1st-time field particle P_1 on the pixel screen is picked up. Thirdly, the next-time valid particle P_3 on the 3rd-time image is scanned within angle θ_{23} and distance l_{23} from the expected position C_3 at the the 3rd-time. Finally, the particle P_4 on the 4th-time image is sought within θ_{34} and l_{34} from the expected location C_4. These expected procedures of same particle recognition are conducted using a Kalman filter algorithm. The above parameters were reasonably selected so that the velocity-vector pattern-recognition might be as accurate as possible. If even one point among all four points P_1, P_2, P_3 and P_4 may not be recognized in the LLS plane, such a questionable particle is dropped out in the algorithm. In the present study, the slit thickness of the LLS was set to be 2mm.

Consequently, the particle points on the continuous four field-images P_1, P_2, P_3 and P_4 were recognized as the positions of the same particle at four different times. The velocity vector of the particle was calculated from the coordinates P_1 and P_4. That is to say, the starting point is P_1 and the corresponding terminal point is P_4. Its length means the traveling distance during 3/60s. In this manner, velocity vectors of all particles were obtained. However, these particles were located at the random points and it is difficult to measure velocity vectors at arbitrary fixed point. Therefore, the time-series of velocities on the fixed mesh points were calculated by the interpolation.

Accuracy of the Present PIV Measurements

(a) *Mean Velocity Distribution (U, W) in Compound Open-Channel Flows*

Fig. 5 (a) shows an example of time-averaged velocity distribution of streamwise component $\bar{U}(z)$ at three elevations A, B and C for cases of H/D = 1.1, 1.2, *see* Fig. 1. These flows were also measured very accurately with a new-type two-component fiber-optic laser Doppler anemometer (FLDA, 2W argon-ion laser), in which the Doppler burst signals can be processed by the "*Flow Velocity Analyzer*" (DANTEC 58N20) and IBM/PC. The FLDA data were obtained from a setting of fiber probe above the water surface, in the same manner as CCD camera shown in Fig. 1. The computer-simulation data obtained from an algebraic stress model (ASM) by Naot et al. (1993a) are also included in Fig. 5(a). These data obtained from three quite different methods are in a reasonably good agreement with each other. Fig. 5 (b) shows an example of time-averaged velocity distribution of spanwise component $\bar{W}(z)$. The PIV data coincide fairly well with the FLDA data, if one considers that it has ever been very difficult to measure the spanwise velocity component $\tilde{w}(y,z,t)$ in water flows. It can be concluded then that the advent and innovation of PIV as well as FLDA *first enabled* one to investigate turbulence phenomena like horizontal interactions and mixings in shallow waters. A very good example is the compound channel hydraulics, which is very important topics in basic and applied hydraulic engineering, as mentioned in the Introduction, *also see* the IAHR monograph of Nezu and Nakagawa (1993a).

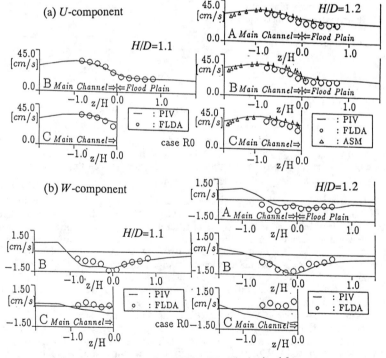

Fig. 5. Comparisons of Mean Velocities (U, W) obtained from PIV(Visualization), FLDA(Fiber-optic Laser Doppler) and ASM(Simulation) Methods.

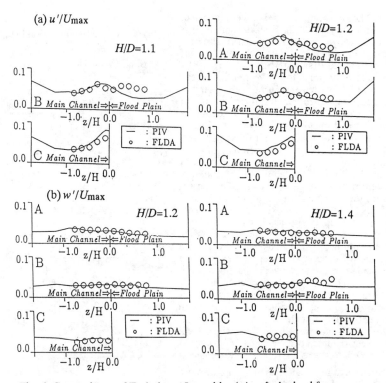

Fig. 6. Comparisons of Turbulence Intensities (u', w') obtained from PIV(Visualization) and FLDA(Fiber-opticLaser Doppler Anemometer) Methods.

Figs. 6 (a) and (b) show the spanwise distributions of turbulence intensities u' and w' normalized by the maximum main velocity U_{max}, respectively. The PIV data also coincide well with the much more accurate FLDA data. Such very good accuracy of the present flow visualization technique stems from the use of fine particles as tracers of flow. The present experiments used 100μm powder of Nylon 12 (MITSUBISHI-KASEI made). This diameter is by only one order larger than that of scattering particles used in both LDA and FLDA.

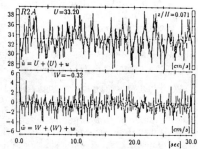

Fig. 7. Time-series of Velocity Fluctuations. $\tilde{u}_i(t) = U_i + <U_i> + u_i$

No effect of solid-water two-phase flow appear in the present innovating visualization technique, i.e., *particle image velocimetry* (PIV). For example, Fig. 7 shows the instantaneous velocity fluctuations $(\tilde{u}, \tilde{w})$. Figs. 8 (a) and (b) show some examples of velocity spectrum measured with the present PIV; (a) and (b) are the results of cases without and with vegetations planted on the flood plain, respectively.

The Kolmogoroff -5/3 power law can be indicated by a triangle in each figure. It is obviously understood that there is comparatively large scatter or noisy among the spectral distributions from the PIV database, because the PIV contains several kinds of interpolation effects in the processing of image database and in the determination of velocity vectors as mentioned previously. Nevertheless, it should be recognized that these PIV data indicate evidence of the Kolmogoroff -5/3 power law. These findings suggest strongly that the present PIV can measure turbulence even in high-shear flows with complicated geometry like compound open-channel flows. Of particular significance is that a large-scale spectral peak appears at about 1 Hz in the case of (b) vegetated flow on the flood plain. This means that a large-scale coherent vortex with about 1 Hz may appear behind vegetations planted on the flood plain. It can be concluded that the present PIV may be very powerful and advanced instruments that can measure coherent structures in planetary large-scale eddies like horizontal vortices generated in the high shear layer between the main channel and flood plains.

In contrast, the conventional flow visualization often uses particles of $10^3 \mu m$ or more as tracers of flow. It should be noted that the conventional solid-tracer visualization cannot measure turbulence quantitatively, except for the hydrogen-bubble technique, e.g., a comprehensive review of refined visualization techniques is available in the monograph of Nezu and Nakagawa (1993a).

(a) **N o** Vegetation on Flood Plain

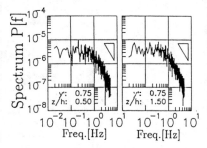

(b) **Some** Vegetation on Flood Plain

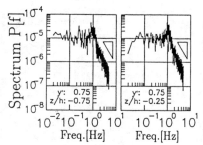

Fig. 8. Frequency Spectral Distributions. The -5/3 slope line stems from the well-known Kolmogoroff theory of *local isotopy*.

APPLICATION OF PIV TO SPACE-TIME CORRELATIONS

Conventional Space-Time Correlation Analysis

It may be the most splendid and powerful merits that the PIV can analyze easily space-time correlation structures which are, in particular, inevitable in researches of coherent turbulence structures, like bursting phenomena, vortex-pairing phenomena, evolution mechanism of some kinds of large-scale eddies in turbulent flows.

Conventional space-time correlation coefficients $C_{u_i u_j}(x, \Delta x; z, \Delta z; \tau)$ of velocity components u_i at a fixed point P and u_j at arbitrary point Q are defined, as the follows:

$$C_{u_i u_j}(x, \Delta x; z, \Delta z; \tau) = \frac{\overline{u_i(x,z,t) \cdot u_j(x+\Delta x, z+\Delta z, t+\tau)}}{u'_i \cdot u'_j} \qquad (1)$$

where, (x, z) is the coordinate of P, $(\Delta x, \Delta z)$ is the lag distance from P to Q, and τ is the lag time. u'_i is the turbulence intensity of u_i component. In contrast, conditionally sampling ensemble-averaged space-time correlation $<C_{u_i u_j}>$ is defined as follows:

$$\left\langle C_{u_i u_j}(x, \Delta x; z, \Delta z; \tau) \right\rangle = \frac{\int_T u_i(x, z, t) \times u_j(x+\Delta x, z+\Delta z, t+\tau) \times I(x, z, t) dt}{u'_i u'_j \times \int_T I(x, z, t) dt} \quad (2)$$

where, $I(x,z,t)$ is a detection function that distinguish vortex of interest at the fixed point P from background fine turbulence. The value of $<C_{u_i u_j}>$ can describe space-time correlation structures of coherent vortices in flows, as pointed out by Nakagawa and Nezu (1981). If $I(x,z,t)$ is equal to unity at every time, (2) is reduced to (1).

In order to analyze coherent structures with background fine turbulence in steady a)and unsteady flows, it is a very useful technique to separate the instantaneous velocity $\tilde{u}_i$ into the following three components.

$$\tilde{u}_i(t) \equiv U_i + <U_i>(t) + u_i(t) \quad (3)$$

where, U_i is the mean velocity, $<U_i>(t)$ is the conditionally ensemble-averaged velocity component which implies contributions of coherent vortices, and $u_i(t)$ is the background fine turbulence. An example of time series of these three separated velocity components U, $<U>$, u, and W, $<W>$, w is shown in Fig. 7. The coherent component $<U_i>(t)$ can be reasonably calculated using the Fourier transform technique of raw velocity signals $\tilde{u}_i(t)$, e.g., see Nezu and Nakagawa (1994).

It is very useful in coherent phenomena to use conditional quadrant techniques as a detection function $I(x,z,t)$, because the conditional quadrant theory can reveal such phenomena; more detailed discussion and review are available in the IAHR-monograph of Nezu and Nakagawa (1993a). The conditional quadrant sampling technique used in the present study is available in Nezu et. al. (1994).

Fig. 9 shows an example of **conventional** (*the sampling condition*=0) space-time correlation $C_{uu}(x, z)$ as a function of time lag τ. Each panel indicates the area of S_x=12cm x S_z=9cm ; the upper half area is in main channel , whereas the lower half is on the flood plain. The fixed point P is selected at near the boundary between them.

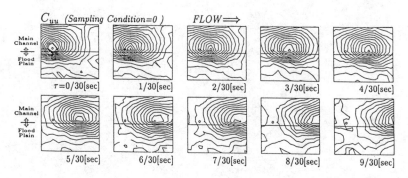

Fig. 9. **Conventional** Space-Time Correlation Contours $C_{uu}(x, z; \tau)$.

The contours of conventional correlation C_{uu} indicate the evolution of mean-scale vortex patterns very clearly; $C_{uu}=1$ at $\tau=0$ and point P, because of the autocorrelation. The convection velocity U_c of such vortex and the associate mean-scale turbulence characteristics could be obtained successfully. Finally, Figs. 10 (a), (b) and (c) show examples of **conditional** quadrant sampling correlation structures $<C_{uu}>$, $<C_{uw}>$ and $<C_{wu}>$, respectively. These database of conditional quadrant space-time correlations,

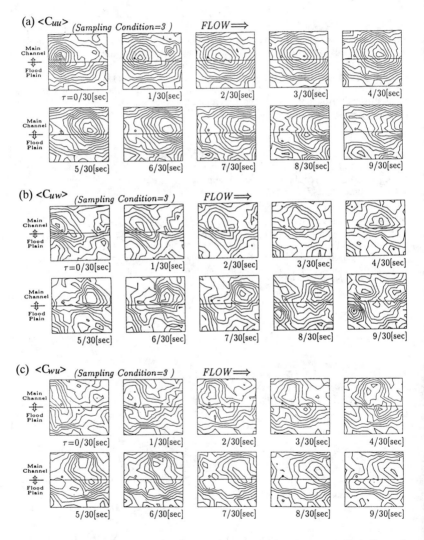

Fig. 10. **Conditional Quadrant** Space-Time Correlation Contours $<C_{u_i u_j}>(x, z; \tau)$.

in regards to all combinations of u and w including the associated scalars such as suspended sediment, are very valuable and essentially important to reveal coherent structures and the associated turbulence phenomena in water flows. More detailed results of the present study and these discussion will be published anywhere because of the limitation of pages in this Symposium Proceedings, *e.g. see* Nezu et al. (1994).

CONCLUSIONS

In the present study, images of very small 100 μm diameter particles uniformly suspended in compound open-channel flows were recorded on an optic disc using a high-sensitive CCD camera and 2W Argon-ion laser slit illumination techniques. Instantaneous velocities at each point in the laser slit plane were measured from these continuous four images of one particle. This new technique called the *"Particle-Image Velocimetry* (PIV)" seems to be very powerful to investigate space-time correlation structures of coherent vortices of turbulence using the conditional sampling method of quadrant motions of velocity components because simultaneous velocity components can be obtained at all grid points of arbitrary section plane of water flows. These PIV database first enable one to reveal the evolutionary patterns of horizontal large-scale vortices and interaction mechanisms between the main channel and flood plains in compound open-channel flows.

ACKNOWLEDGEMENTS

The financial support for the present research from the **Asahi Glass Foundation** and the valuable communication with Messrs. T. Isa and N. Tsuda at the **Nippon Steel Corporation** are hereby acknowledged.

REFERENCES

1) Fukuoka, S. and Fujita, K. (1989); J. Hydraulic, Coastal and Environmental Engineering, JSCE, No.411, II-12, pp.63-72 (*in Japanese*).
2) Nakagawa, H. and Nezu, I. (1981); J. Fluid Mech., vol.104, pp.1-43.
3) Naot, D. and Nezu, I. (1993); Proc. of 5th Int. Symp. on Refined Flow Modelling and Turbulence Measurements, Paris, pp.687-693.
4) Naot,D., Nezu, I. and Nakagawa, H.(1993a); J. Hydraulic Eng., ASCE, vol.119, No.3, pp.390-408.
5) Naot,D., Nezu,I. and Nakagawa,H(1993b); J. Hydraulic Eng., ASCE, vol.119,No.12,pp.1418-1426.
6) Nezu, I. and Nakagawa, H. (1989); *"Turbulent Shear Flows"*, Springer-Verlag, vol.6, pp.313-337.
7) Nezu, I. and Nakagawa, H. (1991); Proc. of Workshop on Instrumentation for Hydraulic Laboratories, CCIW/IAHR, Burlington, Canada, pp.29-44.
8) Nezu, I. and Nakagawa, H. (1993 a); *"Turbulence in Open Channel Flows"*, IAHR-Monograph, Balkema Publishers, Rotterdam.
9) Nezu, I. and Nakagawa, H. (1993 b): Proc. of 5th Int. Symp. on Refined Flow Modelling and Turbulence Measurements, Paris, pp.603-612.
10) Nezu, I. and Nakagawa, H. (1994): an invited paper submitted to J. Flow Measurement and Instrumentation, Butterworth-Heinemann, UK.
11) Nezu, I., Saeki, K. and Nakagawa, H. (1994); Proc. of 9th Congress of APD-IAHR, Singapore, August 24-26, 1994 (*to be published*).
12) Shiono, K. and Knight, D.W. (1989); Proc. of 7th Symp. on Turbulent Shear Flows, Stanford, pp. 28.1.1.-28.1.6.
13) Tamai,N., Asaeda,T. and Ikeda,H. (1986); Proc. of 5th Congress of APD-IAHR, Seoul, pp. 61-74.
14) Tominaga, A. and Nezu, I. (1991); J. Hydraulic Eng., ASCE, vol.117, No.1, pp.21-41.
15) Tominaga, A. and Nezu, I. (1993); *"Flows in Compound Channels* (Chapter 7)", *Research and Practice of Hydraulic Engineeringin Japan,* J. Hydroscience and Hydraulic Eng., JSCE, Special Issues No.2, pp. 121-140.
16) Tsuda, N., Kobayashi, T. and Saga, T. (1991); Proc. of 6th Symp. on Flow Measurements, Osaka, pp.47-52, (*in Japanese*).
17) Tsuda, N., Koseki, T., Kobayashi, T. and Saga, T. (1993); Proc. of 7th Symp. on Flow Measurements, Osaka, pp.83-87, (*in Japanese*).
18) Utami, T. and Ueno, T. (1987); J. Fluid Mech., vol.174, pp.399-440.

Strain Measurement on the Runner of a Hydroelectric Turbine

by K. Warren Frizell, A.M. ASCE[1]

Abstract: Twelve strain gages were installed on the 10-m-diameter runner of Unit G-23 (700 MW) at Grand Coulee Dam, Third Powerhouse. The gages were located on both the pressure and suction sides of blades 8, 9 and 10 and were positioned in order to measure strains near the trailing edge to crown transition of the runner. This is an area where cracks have previously been discovered on Units G-22 and G-23. Four channels of RF (radio frequency) transmitters were used to record the response of four strain gages at a time while the unit was in various operating conditions. A fixed antenna was placed around the turbine shaft and connected to a receiver. The receiver processed the incoming strain gage signals and they were recorded with a digital audio tape recorder along with the unit speed, wicket gate position and load condition.

Strain gage installation techniques are discussed, especially the waterproofing and securing of the lead wires. Each RF transmitter was enclosed within a custom-made box which included a bridge completion circuit, adjustable balancing resistors, and the batteries which powered the transmitter. A sample of the results will be presented for illustrative purposes.

Background

Cracking of runner blades is not an uncommon occurrence in Francis turbines [Parmakian and Jacobson, 1952]. Cracks are generally found in crown-to-bucket and band-to-bucket transitions and are caused by periodic stresses that fatigue the material [Powell, 1958], defects in the original castings and welds, or sudden loading from debris or misoperations. In most cases, periodic weld repair, or stop drilling at the end of the crack to arrest its growth is sufficient to extend the life of the runner. Changing the blade geometry or adding stiffeners between blades may have to be used if the cracking is a result of excitation of the natural frequency of the blade.

---

[1] Research Hydraulic Engineer, Bureau of Reclamation, Denver, CO 80225.

Measurement of actual stress levels [Haslinger and Quinn, 1987] can be an important factor in formulating a solution to a blade cracking problem. Recent trends have used finite element modelling [Degnan, et al. 1986] to analyze this complex loading case; however, without verification from actual measurements, some judgement in the interpretation of the results needs to be used.

Introduction

The Grand Coulee Third Powerhouse features six vertical-shaft Francis turbine-generators outputing a total of 3,900 MW. Unit G-23 is one of three units rated at 700 MW @ 87-m head. The runners have 15 blades and operate at a rotational speed of 1.428 Hz. In August 1990, project personnel identified 19 cracks on Unit G-23 and a similar number on Unit G-22. Due to the relatively sudden development of some of the cracks and the consequences of losing a piece of a bucket on the 10-m-diameter runners, the decision was made to attempt to measure actual stresses on the runner during different operating conditions.

Operational History

Unit G-23 went into operation in 1979. Operationally, the turbine has been relatively free of problems. Periodic weld repair of cavitation damage on the runner has been needed. Inspection in 1990 revealed many cracks, and these were weld repaired. Subsequent inspections have revealed new cracks. To date the unit has operated more than 60,000 hours. Prior to 1988, there were problems in maintaining tailwater depression during the synchronous condensing mode. Having the tailwater come back up on the motoring runner causes a large instantaneous increase in loading and torque. This problem has been fixed and yet cracks still continue to appear and grow.

Approach

Instrumentation: We decided to directly measure strains present at the areas of interest under a variety of operating conditions. The measurement scheme included mounting four strain gages on three adjacent blades. On each blade, two gages were mounted on the pressure side and two on the suction side. They were located near the trailing edge to crown transition, just at the point of tangency of the weld between the crown and blade. The uni-axial gages were mounted with their sensing direction perpendicular to the crown of the runner and centered on lines 6.35 mm and 101.5 mm from the trailing edge, figure 1. Prior to installation, the runner was sandblasted to remove paint in all areas where gages and wiring were to be attached. We mounted the gages on the prepared runner surfaces using a low-power spot welder. The gages had an active length of 12.7 mm and were 6.35 mm wide.

From the beginning, we realized that it would be a tremendous challenge to strain gage this runner and collect data during full load and flow conditions (800 MW, 85-m head and 850 m^3/s discharge). The gages were waterproofed in the following

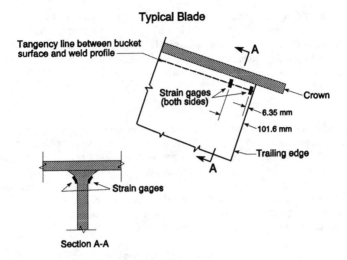

Figure 1: Location of strain gages.

manner. A layer of Teflon film was placed over the solder connections, then a patch of bituthane covered the entire gage. The lead wires were worked into the patch material. Aluminum tape was placed over the entire patch and lead wires and the edges of the tape were sealed with two coats of brushable nitrile rubber. In the gage area, we then applied a coat of aluminum putty. This putty is a two-part, epoxy-based coating which can be machined when dry. The aluminum putty was ground and sanded to a smooth surface, figure 2.

Connecting, routing, and waterproofing the signal wires was the next step. The wiring was passed from the runner into the center of the hollow 3.35-m-diameter shaft through a heater hole in one of the coupling studs. A half coupling was welded to the shaft side of the heater hole and a pressure-tested, pass-through connector was tightened in place. The wires going to the transmitters were then passed through a 19-mm-diameter hole which had been drilled through the shaft into the turbine pit area. At the strain gage end, wires were secured initially with thin stainless steel straps tacked down with the spot-welder. The final solder connections were made and the wires were taped down using aluminum tape. The tape edges were sealed with brushable nitrile rubber. At this point, all the gages were tested at the termination point in the turbine pit, and all were in working condition.

The final treatment to the wires and gages consisted of applying a two-part, epoxy-based brushable ceramic compound. This coating has a relatively long curing time; however, it dries to a very smooth finish. This brushable ceramic was used to coat all of the wiring and gage areas, figure 3.

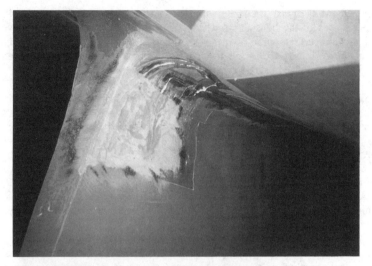

Figure 2: Gages with aluminum putty coating in place.

Figure 3: Final coating of brushable ceramic epoxy-based coating.

The telemetry system we used was manufactured by ACUREX (now Wireless Data Corporation). It contained four channels of signal conditioning. The receivers can be tuned to signals anywhere in the FM radio band between 88 and 108 MHz. The signal conditioning cards process strain gage signals from d.c. to 1000 Hz. The output level is adjustable from 1.0- to 3.0-V peak-to-peak full scale. The transmitters are pretuned and designed to operate with sensors in a Wheatstone bridge circuit. The input ranges of the transmitters when using a bridge circuit with one active element are jumper selectable in ranges of 0-500, 0-1500, or 0-5000 $\mu\epsilon$. The minimum bridge impedance is 350 Ω. The transmitters can be battery powered (2.7V d.c.) or energized using a separate induction coil with a voltage regulator.

We constructed four enclosures for the transmitters which included the batteries for excitation and the bridge completion circuitry. The boxes were bolted to brackets which had been tack-welded to the shaft. The front cover had a terminal strip where the strain gage lead wires were attached to the bridge completion circuit. A variable resistor allowed us to balance the bridge circuit easily. Once powered up, the bridge circuit output is transmitted by RF over a small antenna attached to the transmitter. The receiving antenna was mounted around the perimeter of the shaft, at the same level as the transmitter boxes. The receiving antenna was configured in a "folded tee" and connected to the receiver module which housed the four signal conditioning cards.

The ACUREX telemetry system is significantly different from conventional telemetry in the method of signal transmission. Standard telemetry systems use relatively high power and a high gain antenna system (allowing long distance transmission). This system uses very low power and is generally operated in mechanical situations where high gain antennas are not appropriate. Usually the operating distances are small enough that capacitive or magnetic coupling is used rather than the typical electromagnetic carrier coupling. We used capacitive coupling between the small antenna on the transmitter and the "folded tee" antenna. At the rotating speeds of most hydropower units, continuous data transmission (no signal drop out) is possible.

In addition to the four channels of telemetry, we also recorded unit speed, wicket gate position, and electrical load. The feeds for these inputs were provided by permanent sensors at the powerplant. All signals were recorded on a TEAC RT-111 Digital Audio Tape recorder. Figure 4 shows a cutaway of the turbine, showing the general strain gage and transmitter locations.

Testing: Approximately 2 weeks after the installation was completed, we returned to Grand Coulee to perform the testing. The test plan included collecting data from four strain gages simultaneously, i.e., blades 8, 9, and 10. We wanted to record strain levels for several different operating conditions: speed-no-load, 120 percent speed-no-load, loads up to 800 MW, a load rejection, and synchronous condensing mode.

Results and Discussion

The testing took place over 2 days. Since the unit had been out of service for several weeks for scheduled maintenance, the standard checkout testing was completed prior to the beginning of our tests. These tests included a slow-roll with

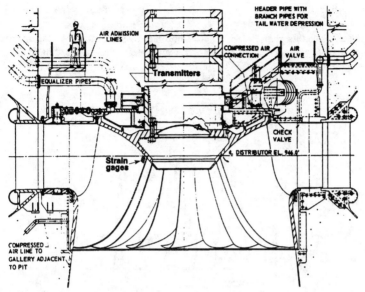

Figure 4: Cutaway of Unit G-23, showing general arrangement.

personnel located inside the air housing of the generator to observe the unit rotation. After some minor adjustments, the unit was brought up to speed in 25-percent increments. During these checkout tests, our instruments were also checked out and were operating well. The transmitters had been set to the 500 $\mu\epsilon$ range with full scale bridge output set at 2.5 V using shunt calibration.

The first day of testing resulted in recording speed-no-load and 120-percent speed conditions for gages 1-4 on bucket No. 8. The day was not without some excitement as a secondary relay was tripped by the overspeed and initiated an emergency closure of the penstock coaster gate. While waiting for the penstock to refill, we changed the gain to the 1500 $\mu\epsilon$ range as the speed-no-load strain levels were higher than we expected. We then tried to begin the load portion of our test but could not successfully load the unit. A governor problem was discovered and the remainder of our tests was postponed until the following day.

Upon arriving at the powerhouse the next morning, we noticed that gage No. 1 was no longer functioning. We decided to switch to blade No. 9 (gages 5-8) so that an entire instrumented bucket would be used during the test. We connected gages 5-8 to the transmitters and balanced the bridge circuits. We were not able to balance gage 5. We balanced the remainder of the gages and proceeded. The unit was brought up to speed and then load was applied. Gage output was inconsistent which indicated that there probably was damage occurring to the gages or wiring.

After the first loading test up to 800 MW, only one strain gage remained in operating condition. This was gage No. 10 (pressure side) on blade No. 10. The test sequence for this gage involved speed-no-load, synchronous condensing, loads up to 800 MW and a load rejection. The gage remained in working condition throughout the entire test. Samples of test data are shown in figure 5a and b.

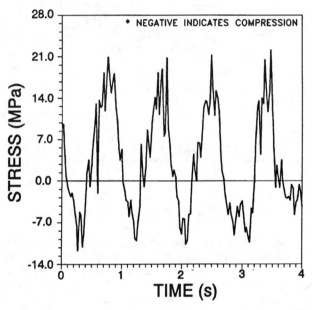

Figure 5a: Time series of stress levels on gage 10 at a 40-percent wicket gate position (note periodic draft tube surge).

Following testing, an inspection of the runner revealed that almost all signal wires had been ripped off by the flowing water. The gages themselves were all intact as the aluminum putty remained in place, figure 6. Although only one gage survived the entire test, critical data were obtained to allow prediction of crack formation and propagation.

The collection of data from rotating machinery is becoming a more straightforward exercise with the use of small commercially available telemetry systems. There is no longer the need for slip rings or expensive induction systems. However, many of the same old problems remain in the installation of the sensors, especially in a submerged environment. The dynamic forces induced on the sensors and associated wiring are substantial. Care needs to be taken in the protection of these features by coatings or mechanical means which will survive the dynamic forces.

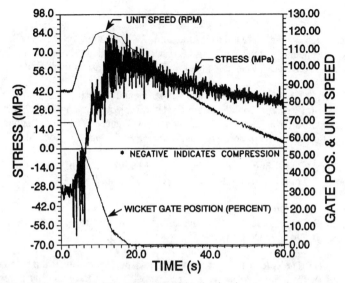

Figure 5b: Output of gage No. 10 during a load rejection.

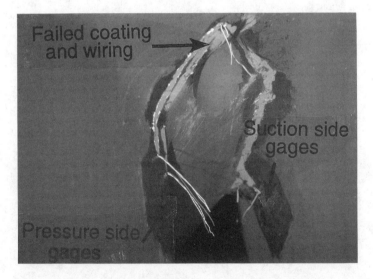

Figure 6: Strain gage installation after the test. Note gages are still intact; however, the ceramic coating failed causing failure of the lead wires.

References

Degnan, J.R., et al., "Modernization and Rehabilitation of Aswan High Dam Francis Runners," ASME Fluids Engineering Divisions Fourth International Symposium on Hydro Plant Machinery, FED vol. 43, December 1986.

Haslinger, K.H. and J.W. Quinn, "Experimental Investigation into the Stress Behavior of a Reactor Coolant Pump Impeller Subjected to a Wide Range of Flow Conditions," International Conference on Flow Induced Vibrations, sponsored by BHRA, Bowness-on-Windermere, England, Paper D2 pp. 133-146, May 1987.

Parmakian, John and R.S. Jocobson, "Measurement of Hydraulic-Turbine Vibration," Transactions of the ASME, July 1952, pp. 733-741.

Powell, Alan, "On the Fatigue Failure of Structures due to Vibration Excited by Random Pressure Fields," The Journal of the Acoustical Society of America, Volume 30, No. 12, December 1958, pp. 1130-1135.

Disclaimer: The information contained in this publication regarding commercial products or firms may not be used for advertising or promotional purposes and is not to be construed as an endorsement of any product or firm by the Bureau of Reclamation.

Testing Turbine Aeration for Dissolved Oxygen Enhancement

Tony L. Wahl[1] A.M. ASCE, Jerry Miller[2], Doug Young[3]

Abstract

During August of 1993 the Bureau of Reclamation tested turbine aeration for dissolved oxygen (DO) enhancement at Deer Creek Powerplant near Provo, Utah. This test required a variety of instrumentation and equipment, assembled during a short time period prior to the tests. Objectives of the testing included determining the effectiveness of aeration, evaluating the impact on power output and mechanical behavior of the turbines, and obtaining data needed to design a permanent turbine aeration system. Variables of interest included standard powerplant parameters (head, discharge and power output), airflow parameters (pressure, temperature, and flowrate), water quality parameters (DO concentration and temperature), and mechanical parameters (shaft runout and bearing temperature). This paper will discuss the design of the tests and the instrumentation involved, as well as plans for additional testing during the implementation of turbine aeration at the site in the summer of 1994.

Introduction

Deer Creek Reservoir is located about 24 km (15 miles) upstream of the city of Provo, Utah and receives inflow from a watershed with extensive agricultural development and increasing commercial and urban development. During late summer months (July-October), the releases from the

---

[1]Hydraulic Engineer, U.S. Bureau of Reclamation, Hydraulics Branch, D-3752, P.O. Box 25007, Denver, Colorado 80225-0007

[2]Water Quality Hydrologist, U.S. Bureau of Reclamation UC-722, Upper Colorado Regional Office, P.O. Box 11568, Salt Lake City, UT 84147-0568

[3]Fisheries Biologist, U.S. Bureau of Reclamation UC-773, Upper Colorado Regional Office, P.O. Box 11568, Salt Lake City, UT 84147-0568

reservoir are made entirely through the powerplant and often have DO concentrations ranging from 0-2 mg/L. Past studies indicate that this has impacted about 3-5 km (2-3 miles) of heavily used blue ribbon trout fishery downstream of the dam.

The powerplant, constructed in 1958, contains a pair of Francis-type turbines and air-cooled generators, rated at 2475 kW each. The two units each discharge about 8.5 m^3/s (300 ft^3/s) at full load. The rated head on the powerplant is 36.6 m (120 ft) and the maximum head is 42.7 m (140 ft). Water levels in the tailrace below the powerplant are controlled by a 3-bay gate structure. The draft tubes on both units are simple conical diffusers discharging into prismatic chambers leading to the tailrace. Figure 1 shows a cross-section through the powerhouse and tailrace pool.

To mitigate a low-flow event on the Provo River during the winter of 1992-93, Reclamation proposed using turbine aeration to raise dissolved oxygen concentrations in the river downstream of the powerplant. Air would be injected into the turbine draft tube through existing passages to produce a mixed air-water flow that would raise DO concentrations. This concept has been tested in both model and prototype situations by many researchers, and in particular by the Tennessee Valley Authority (March, Cybularz and Ragsdale, 1991; Jones and March, 1991; Bohac and Ruane, 1990). However, substantial differences between this site and those tested in the body of past research (especially in the draft tube configuration) made it difficult to predict the effectiveness of the concept for this site. We concluded that a field test was necessary.

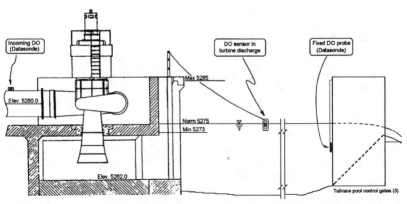

Figure 1. - Deer Creek Powerplant. The turbines installed in 1958 have simple conical diffuser draft tubes rather than formed elbow-type draft tubes for which most turbine aeration research has been done. DO measurements were taken upstream and downstream of the turbines, at the downstream end of the tailrace pool (about 200 ft downstream of the powerplant), and at several downriver locations.

Experimental Design

The test was initially planned for September of 1993, but on short notice was moved forward to August to avoid impacting planned switchyard maintenance. This also ensured that the test could be conducted during the low-DO season, since 1993 was a wet water year with a shorter and less severe low-DO season than normal. The acceleration of the test schedule required that several compromises be made in the instrumentation package for the test. Objectives of the test were to:

- Determine the effectiveness of turbine aeration for DO enhancement;
- Determine impacts of aeration on power output and mechanical behavior of the turbines;
- Collect data necessary for the design of a permanent aeration system for the powerplant;
- Determine effectiveness of aeration obtained by creating a drop across the control gates at the downstream end of the tailrace pool.

Air was supplied to the vacuum breaker systems and snorkel tubes of the two turbines using two diesel-engine-powered air compressors. Dissolved oxygen measurements were made upstream and downstream of the turbines and at the downstream end of the tailrace pool, about 200 ft downstream of the powerplant. Data collected in the powerplant were used to determine the effect of aeration on turbine performance.

The primary test was planned for the vacuum breaker system because air passages were much larger than in the snorkel tube system. Although preliminary estimates showed that axial blowers could supply the necessary flowrates through the vacuum breaker system, we were restricted to equipment that was available in the local area. The two compressors were rated to deliver 0.434 kg/s (0.957 lb/s) of air at pressures as high as 551 kPa (80 lb/in^2). This is equivalent to a volumetric flowrate of 0.354 m^3/s (750 ft^3/min) at standard temperature and pressure (101.3 kPa, 15°C [14.7 lb/in^2, 59°F]). These compressors could supply a 5-6 percent airflow rate (volumetric airflow at standard temperature and pressure compared to volumetric waterflow through the turbines).

Figure 2 shows a schematic diagram of one turbine unit with the two paths of airflow indicated. Air entering the vacuum breaker system travels between the headcover and the runner crown and enters the draft tube through seven holes in the crown of the turbine runner. Air supplied to the snorkel tube system travels through the turbine shaft and then enters the draft tube through the snorkel tube below the turbine runner.

The air supply to each unit was initially adjusted with regulators supplied on the air compressors. Flowrates were measured with two different flowmeters, described in detail in the instrumentation section. Airflow was further controlled by gate valves installed downstream of the flow mea-

428 HYDRAULIC MEASUREMENTS AND EXPERIMENTATION

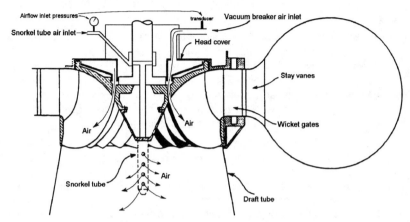

Figure 2. - Air was supplied to the draft tube through the vacuum breaker system (air flows under the headcover and through existing holes in the runner crown) and through the snorkel tube (air flows through a passageway in the shaft).

surement equipment. A check valve was installed on each line between the gate valve and the turbine to prevent possible backflow of water into the airflow meters or into the compressors.

We conducted the testing in two phases. Following installation and shakedown tests on the equipment and instrumentation, we began the balanced load portion of the tests. The two turbines were operated nearly identically, with wicket gate openings of 55-60 percent, and power outputs of about 1700 kW each. We tested aeration rates of 0 to 5.8 percent. Aeration on unit 1 was through the snorkel tube (we found the vacuum breaker system to be partially clogged for unknown reasons), and through the vacuum breaker system on unit 2. The second phase of testing was performed with unbalanced operation of the two turbines to study turbine aeration at high loads and low loads. Unit 1 was operated at a wicket gate opening of 35 percent and power output of 600 kW, and unit 2 was run at 77 percent wicket gate opening and 2750 kW power output. During two days of testing we recorded data for about 15 different combinations of operating conditions and airflow rates.

Instrumentation

Powerplant Performance Data. — Several parameters were required to calculate the combined efficiency of the turbines and generators. Discharge through the turbines was measured by ultrasonic flowmeters permanently installed on the two 1.83-m (6-ft) diameter penstocks. These meters were installed downstream of about 180 m (600 ft) of straight pipe. Head across the turbine was determined from reservoir and tailrace water surface measurements, so friction losses in the penstocks were charged against the turbines in efficiency calculations. Reservoir elevation was determined from the water level recorder in the powerhouse. Tailwater elevations were determined by direct measurement from the downstream deck of the powerhouse.

Power output from the generators was recorded from the analog gages in the powerplant control room. Analyzing the data after the tests showed that these gages were somewhat erratic and lacked precision necessary for this application. This led to significant scatter in the calculated efficiencies, although we were still able to identify significant trends in power output. With better foresight the scatter could have been reduced by making repetitive observations to obtain average readings. With more time before the test, we would have installed a watt transducer.

Vibration and Bearing Temperature Monitoring. — To ensure that turbine aeration would not create future maintenance problems, we monitored shaft runout and bearing temperatures during the tests. Shaft runouts were measured using two proximity probes located just above the turbine guide bearing, oriented at right angles to one another. Outputs from these sensors were recorded with a portable spectrum analyzer and data recorder. We recorded data at several operating conditions with and without aeration.

Both units had a prior history of cooling problems for the turbine guide bearings. This had been corrected with the addition of a cooling system using water withdrawn from the penstocks upstream of the powerhouse. Still, we were wary that any increase in vibration caused by aeration could cause problems. We monitored bearing temperatures throughout the tests, and saw no significant changes in temperature due to aeration.

Airflow and Air Pressure Measurements. — To measure the driving pressure required to inject air into the turbines, we installed a 172 kPa (25 lb/in^2) absolute pressure transducer on the air piping just upstream of our connection to the vacuum breaker system on unit 2. On unit 1, where we were forced to supply air through the snorkel tube due to the partially clogged vacuum breaker system, we installed a 689 kPa (100 lb/in^2) bourdon tube pressure gage (fig. 2). Subatmospheric pressures were not a problem on this unit during aeration due to the small size of the air piping leading to the snorkel tube. We also measured the static vacuum at the vacuum breaker and snorkel tube injection locations under each different turbine operating condition tested. Over the range tested (35 to 77 percent wicket gate setting) we found the static vacuum to be essentially constant and identical at both locations on both turbine units.

Airflow measurements were made with two different instruments. For unit 2 we rented a Hedland variable area pneumatic flowmeter rated for 0.47 m^3/s (1000 ft^3/min). On unit 1 we used a 2.54-cm (1-inch) diameter orifice plate installed in a 4.09-cm (1.61-inch) diameter pipe. This device was a last-minute substitution when our efforts to obtain a second Hedland flowmeter failed just before the test. Both of these meters were installed in-line, between the air compressor hoses and the gate valve we used to control the airflow.

Variable Area Flowmeter. — Airflow through the Hedland meter moves a magnetized, spring-loaded piston to increase the size of an annular orifice formed between the piston and a tapered metering cone in the center of the meter. An external flow indicator magnetically coupled with the piston indicates the flowrate on a pressure-compensated scale. Temperature corrections were made based on temperature measurements made on the piping near the meter with an RTD temperature sensor. The flowmeter can be operated in any orientation and does not require flow straighteners or special piping arrangements upstream of the meter. This meter is advertised to have an accuracy of ±4 percent of full-scale and repeatability within 1 percent. No exhaustive tests of accuracy or precision were performed, but we did find the meter to be easy to use and readings were quite stable. At one point in the test we swapped this meter with the orifice plate meter described below and found that for similar inlet pressures to the turbines we measured essentially the same airflows with both meters.

Orifice Plate Flowmeter. — The orifice plate flowmeter used on unit 1 required measuring several parameters to calculate the flowrate. We measured the pressure differential across the orifice plate with a 172 kPa (25 lb/in^2) Pace differential transducer, and the upstream pressure with a bourdon-tube pressure gage. The barometric pressure was also measured to convert the upstream pressure reading to absolute pressure. Finally, after allowing time for equilibrium to be established, we measured the pipe wall temperature with the RTD temperature sensor. With these measurements, the Reynolds number, discharge coefficient, and flowrate could be determined through an iterative calculation (Fluid Meters, 1959). This arrangement was effective, but more complex and time-consuming than the flowmeter used on unit 2. We found it more difficult to set desired flowrates due to the number of parameters varying as we adjusted the flow.

Dissolved Oxygen Measurements. — DO measurements were made with Hydrolab Datasondes and multi-parameter probes. These probes determine the DO concentration using an electrode assembly and a selective membrane separating the electrodes from the test sample. The electrodes consume oxygen, which depletes the DO at the interface between the sample and the membrane. Thus, to obtain accurate readings, continuous flow must be maintained past the membrane. Use of the membrane electrode probes permitted continuous (timed interval) monitoring, and avoided potential field problems with iodometric methods in which unsaturated samples are exposed to air during the sampling and handling process.

The Datasonde continuous recording probes were installed in the powerhouse to sample inflow to the powerplant prior to aeration, and at the downstream end of the tailrace pool (fig. 1). These units have internal memory, and were preset to record DO and temperature at 15 minute intervals. The instruments were checked and the data downloaded to a portable computer 1-2 times daily. Portable multi-parameter probes were used to make measurements immediately downstream of the powerplant in the flow discharging from each turbine unit (fig. 1), and at several downriver stations. The portable probes were also used as a check for the continuous probe installed at the end of the tailrace.

The probes were calibrated daily or twice daily during the tests according to manufacturer's instructions. This required recording the barometric pressure with a field meter, allowing the probe to come to equilibrium temperature, and then creating a saturated, non-pressurized air pocket above the membrane. A thin film of oxygen-saturated water remains on the probe long enough for equilibrium to be established and the probe is then calibrated against the saturation concentration at the known temperature and barometric pressure.

DO Probe Installations. — It was important to maintain flow past the membranes at the two continuous recording locations. The probe in the powerhouse sampled water drawn from the penstock for cooling of the generator bearings. A line was plumbed from the cooling water supply line to the bottom of a cylindrical well and the DO probe was installed at the bottom of the well, near the point of inflow. Flow from the cooling water line entered the bottom of the well and flowed upwards to an overflow. Although the surface of the well is exposed to air, the overflow continuously discharges the aerated water at the surface. Thus, no aeration exposure occurs at the bottom of the chamber and equilibrium measurement is possible. To confirm the measurements made at this station, the portable probes were used to measure DO in the flow exiting the powerhouse without turbine aeration. These measurements were in close agreement.

The Datasonde at the downstream end of the tailrace was installed at the bottom of a PVC pipe attached to the wall of the gate structure. The pipe was perforated at the bottom around the sensor location. Continuous downstream flow was observed past this location, but separation of the flow from the pipe created a recirculating eddy in the wake of the pipe. This eddy may have increased the exchange time required to reach equilibrium at the probe. To check the validity of data recorded at this station, measurements were made periodically at mid-channel (the middle gate opening) with the portable probes. The readings from the Datasonde in the PVC pipe did lag the mid-channel conditions and were generally lower. This indicated that the installation in the pipe may have limited the flow past the membrane and allowed the instrument to deplete the DO at the membrane interface.

DO Measurements in Bubble Plume. — Turbine aeration created a large bubble plume in the tailrace, so DO measurements made with the multi-parameter probes in the flow exiting each turbine were necessarily made in a portion of the bubble plume. There was some concern that this would produce inflated DO measurements and thus cause some error. This was checked by taking additional manual measurements with the same instrument at the downstream end of the tailrace pool, adjacent to the PVC pipe used for the Datasonde installation. After sufficient exchange time had been allowed in the tailrace, the measurements at this location were in close agreement with the measurements taken in the bubble plume. This indicated that the bubble plume had little or no influence on the DO measurements (assuming DO did not change appreciably between the two stations), or that the additional gas transfer between the two locations was about the same as any distortion of the measurements made in the bubble plume. In addition, the measurements in the bubble plume seemed as stable as those made outside of the bubble plume.

Results

Turbine aeration was very effective, and we achieved DO improvements of up to 3.5 mg/L from an initial deficit from saturation (at the water surface) of about 7 mg/L. For each test an aeration efficiency was calculated from equation 1 as follows:

$$E = \frac{C_d - C_o}{C_s - C_o} \tag{1}$$

where: C_o = incoming DO concentration
C_s = saturation DO concentration at water surface
C_d = downstream DO concentration after aeration

Figure 3 shows the aeration efficiencies plotted against the airflow expressed in percent. The majority of testing was performed with airflow rates of 4 percent or less, and in this range aeration efficiency increased about 10 percent for each 1 percent additional airflow.

In the range of 55-77 percent wicket gate setting, power losses due to aeration were about 0.5 percent for each 1 percent airflow. For the tests at 35 percent wicket gate setting there was an increase in power output of about 1-3 percent. This increase was due to the aeration alleviating draft tube surging at this gate setting.

The vibration monitoring revealed no adverse effects of aeration. The tests also confirmed that axial blowers with a maximum supply pressure of 70-100 kPa (10-15 lb/in^2) would be suitable for a permanent installation. In addition, the natural vacuum in the draft tube was sufficient to draw significant quantities of air into the turbines without blowers. Such a passive aeration system would be most effective with the turbines operating at low discharges when airflow rates could be as high as 2-3 percent.

Testing of the overflow gates at the downstream end of the tailrace pool also showed that approximately 20 percent aeration efficiency (about 1.5 mg/L increase) could be obtained by raising the gates to create a 3 ft drop, although power losses would be about 2.5-3 percent due to reduced head on the powerplant.

Implementation and Further Testing

Based on the results of the 1993 test, passive turbine aeration and manipulation of the tailrace gate will be used throughout the summer of 1994. Active aeration has been shelved indefinitely due to the high cost of

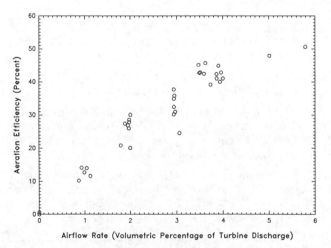

Figure 3. - Aeration efficiency achieved by turbine aeration as a function of airflow rate. This figure includes data collected in the outflow from each turbine, and in the combined powerplant flow at the downstream end of the tailrace pool (recorded with the roving multi-parameter probes). It does not include data from the Datasonde installed in the pipe at the end of the tailrace pool.

upgrading electrical equipment in the plant to provide power to the blower motors. Reclamation's Salt Lake City, Provo, and Denver Offices are now working with the Provo River Water Users Association, the Utah Division of Wildlife Resources, and the National Biological Survey to design monitoring programs for fish and aquatic invertebrates that will evaluate the biological benefits of the improved water quality. Additional hydraulic measurements will be made to measure static vacuum pressures and determine aeration-caused power losses at operating points not tested in 1993.

Acknowledgments

We would like to thank all those who helped arrange and conduct the test under severe time constraints, including Walter Payne (Reclamation, Salt Lake City), Dave Frandsen (Reclamation, Provo), and the powerplant operators, Harold Ford, Jack Powers, Frank Severson and Earl Laycock (Provo River Water Users Association). Reed Oberndorfer (Central Utah Water Conservancy District) assisted with the DO instrumentation and measurements, and Lee Elgin (Reclamation, Denver) assisted with the powerplant and airflow instrumentation and measurements. Bill Duncan (Reclamation, Denver) assisted with the mechanical monitoring of the turbines and bearings. Perry Johnson (Reclamation, Denver) assisted with the evaluation of aeration alternatives and reviewed our test plans.

References

Fluid Meters: Their Theory and Application, Report of the ASME Research Committee on Fluid Meters, 5th ed., American Society of Mechanical Engineers, New York, NY, 1959.

Bohac, C.E., and R.J. Ruane, "Solving the Dissolved Oxygen Problem," *Hydro Review*, Vol. IX, No. 1, February 1990.

Jones, R.K., and P.A. March, "Efficiency and Cavitation Effects of Hydroturbine Venting," *Hydraulic Engineering,* Proceedings of the 1991 ASCE National Conference on Hydraulic Engineering, Nashville, TN, July 29-August 2, 1991.

March, P.A., J. Cybularz, and B.G. Ragsdale, "Model Tests for the Evaluation of Auto-Venting Turbines," *Hydraulic Engineering,* Proceedings of the 1991 ASCE National Conference on Hydraulic Engineering, Nashville, TN, July 29-August 2, 1991.

The information contained in this paper regarding commercial products or firms may not be used for advertising or promotional purposes and is not to be construed as an endorsement of any product or firm by the Bureau of Reclamation.

Eliminating Water Column Separation and Limiting Backspin at a 12,000-Horsepower Pumping Plant

Paul Otter[1], David Hoisington[2], David Raffel[3]

Abstract

Milwaukee's Water Pollution Abatement Program includes a deep tunnel system which collects combined sewage and street runoff during storms for later treatment when capacity is available in the wastewater treatment plants. The Inline Pump Station contains three pumps, each driven by a 4,000-horsepower electric motor. The plant is capable of lifting 150 million gallons per day (mgd) (6.6 m^3/s) 350 feet (107 m) from the tunnels to the ground surface treatment plants.

Prior to system startup, while the pump station was being used to remove tunnel infiltration, water hammer became increasingly severe whenever a pump shut down. Two conflicting problems required solution before continuous pumping could begin. First, the severe water hammer had to be eliminated, and second, pump speed in the reverse direction after shutdown (backspin) had to be limited to 125 percent of the forward speed.

A testing program was undertaken to measure the water hammer wave pressures and to quantify the effectiveness of the proposed solution. Very accurate measuring equipment was necessary because of the small time interval between events.

---

[1] Hydraulic Engineer with CH2M Hill, Milwaukee, WI
[2] President, DATASYST Engineering and Testing Services Inc., Delafield, WI
[3] Director, Protrans Consultants, El Paso, TX

Milwaukee Inline Storage System

The Milwaukee Metropolitan Sewerage District (MMSD) together with its team of consulting engineering firms led by CH2M HILL is in the process of starting up its 2 billion dollar Milwaukee Water Pollution Abatement Program (MWPAP). The MWPAP includes an Inline Storage System and massive expansion and upgrades at two wastewater treatment plants. Prior to construction of the MWPAP, combined sewage and street runoff overflowed into rivers and Lake Michigan whenever sewerage or treatment capacity was exceeded. The MWPAP will decrease the frequency of such overflows from about 50 times to about twice per year.

The centerpiece of the MWPAP is the Inline Storage System (ISS). The ISS consists of a system of deep tunnels which collect combined sewage and street runoff during storms, when near-surface sewerage capability is exceeded, for later treatment when capacity is available in the wastewater treatment plants. The ISS consists of about 20 miles (32 km) of conveyance/storage tunnels which range in diameter from 17 to 32 feet (5.2 to 9.8 m). The tunnels are located about 300 feet (90 m) below the City of Milwaukee and serve two functions: (1) conveyance to the treatment plants and (2) storage until treatment plant capacity is available. The total storage capacity of the ISS is about 380 million gallons (1,440,000 m^3). Sewage is conveyed to the ISS from the near-surface sewers by 300-foot (90 m) vertical dropshafts and is pumped to the treatment plants by the Inline Pump Station (IPS).

Inline Pump Station

The Inline Pump Station (IPS) is the key to operation of the entire ISS. It is located in an excavated rock chamber 350 feet (107 m) below ground and houses three pumps. Each pump is driven by a 4,000-hp electric motor and is capable of lifting 50 million gallons per day (mgd) (2.2 m^3/s) 350 feet (107 m) from the tunnels to either of two ground surface treatment plants.

A 54-inch-diameter (137 cm) intake pipe 40 feet (12 m) long conveys water from the ISS to each pump. Each discharge line has four components:

- Immediately downstream of the pump is a 42-inch (107 cm) diameter horizontal reach 79 feet (24 m) long followed by
- A 42-inch (107 cm) diameter vertical section 344 feet (105 m) long rising to the ground surface elevation
- A 36-inch (91 cm) horizontal section 58 feet (18 m) long at the ground surface elevation
- A 42-inch (107 cm) vertical riser either 18 or 70 feet (5.5 or 21 m) high depending upon which treatment plant the pump serves

Pump No. 1 serves the South Shore Treatment Plant and has a maximum static lift of 386 feet (118 m). Pump No. 3 serves the Jones Island Plant and has a maximum static lift of 334 feet (102 m). Discharge from Pump No. 2 can be directed to either plant. Figure 1 shows an isometric view of the Inline Pump Station and piping.

Water Hammer Source

Initial testing under low discharge head and flow showed that reverse flow of water when a pump was shut down would cause reverse rotational speed exceeding the manufacturers warrantee. Consequently, check valves were installed in the 42-inch (107 cm) discharge lines immediately downstream of the pumps. Knowing that uncontrolled closure of check valves would cause water hammer, spring-loaded valves were selected which were supposed to close an instant before reversal of the water column. In theory, water velocity would be zero and no water hammer could occur. Immediately after valve installation check valve closing did not produce significant water hammer and did control backspin.

In September 1990, the pump station was placed in operation for the purpose of pumping out tunnel infiltration. During late 1991 it was reported that when the pumps shut down, the check valve closed with a loud noise accompanied by apparent movement of pipes on concrete saddles at the surface. The noise and movement were especially noticeable when an unexpected shutdown occurred: typically when power was tripped by a vibration sensor or other protective device. Deliberate shutdown included ramping down pump flow, and appeared to reduce the noise and movement. These reports eventually resulted in the tasking of the engineering group with the responsibility for determining the extent of the water hammer and resulting pipe movement and proposing a method for controlling it. David Raffel, Director of Protrans Consultants was contracted to act as advisor.

Problem Magnitude

A testing program was developed to measure and record pressures and pipe movement during operation and shutdown of the pumps. Datasyst Engineering and Testing Services, Inc., was contracted to install instrumentation and record water pressures, pipe movement and tie rod loads at several locations along the discharge pipes. The problems appeared most severe in Pump No. 1 and its discharge line. This pump operates at a higher head than the other two pumps and was also experiencing vibration trips. Thus it was considered inadvisable to operate Pump No. 1. The vibration was caused by pieces of construction timber lodging in the impeller and was not related to the check valve. However, testing was confined to Pump Nos. 2 and 3. Testing was accomplished in October 1991.

From the tests, it appeared that the source of the problem was a water hammer pressure wave propagating from the check valve closure. During the period of check valve operation (approximately 1 year), the valve closure spring had apparently become unreliable. Water hammer will accompany closure if closure occurs after the pipeline flow has reversed direction and if closure occurs too quickly. In this particular hydraulic system, prevention of severe water hammer requires very precise control of the check valve. If the spring becomes loose or weakens, water hammer pressure will increase. Internal pipe pressures exceeding 500 psi (350,000 kgf/m^2) were recorded as the initial pressure wave passed the upper elbow (at ground level), and some of the second-wave pressure spikes were above 1,000 psi (700,000 kgf/m^2). Using the published yield point of 30,000 psi (21 x 10^6 kgf/m^2) and tensile strength of 60,000 psi (42 x 10^6 kgf/m^2) with a joint efficiency of 0.8, allowable hoop stress in the discharge piping at yield is 24,000 psi (17 x 10^6 kgf/m^2) and at tensile strength is 48,000 psi (34 x 10^6 kgf/m^2). A first-wave internal pressure spike of 500 psi (350,000 kgf/m^2) corresponds to a hoop stress of 28,800 psi (20 x 10^6 kgf/m^2) (120 percent of yield point). The highest recorded spike reached 1,097 psi (770,000 kgf/m^2) during a second wave. This hoop stress was 63,000 psi (44 x 10^6 kgf/m^2) or more than the calculated tensile strength of the steel pipe. Fortunately, the extreme high pressure lasted such a short time that the energy was absorbed by vibration of molecules in the steel without rupturing the pipe. The extremely short duration high pressure spikes might not have been detected with less sophisticated measuring equipment. Low pressures, below atmospheric, were also recorded. Figure 2 shows a typical pressure versus time plot after pump shutdown and check valve closure.

At this point there were two separate but interrelated problems. First, the backspin had to be limited to 750 rpm and second the internal pressures and pipe movements had to be controlled. The pump/motor contract had required pumps and motors which could withstand backspin of 125 percent of "full load drive speed." Motors were required to withstand 125 percent of "nominal motor speed." Nominal motor speed was interpreted as synchronous speed or 600 rpm. Therefore, the motors should withstand instantaneous peaks of 1.25 x 600 or 750 rpm in the reverse direction.

Eliminating Sub-Atmospheric Pressure With Vacuum Relief Valves

Testing of the piping system with the check valves showed that negative pressure could develop at the upper elbow during shutdown. Negative pressures can be dangerous because they may cause separation of the water column and formation of vapor bubbles. Rejoining of the separated water column and/or collapse of these bubbles can produce a violent implosion and very high pressure. This phenomenon could explain the very high second-wave pressure spikes. Consequently, it was determined that vacuum relief valves should be installed at the upper elbows. This was done before further testing.

Eliminating Excessive Backspin With Orifices
General

To limit the pump backspin speed to less than 750 rpm while reducing the high and low pressure spikes, it was proposed that the check valve be replaced by a sharp-edged orifice plate. This could be accomplished by removing a check valve, replacing its disc with an orifice plate welded to the valve seat, and then replacing the check valve body/orifice into the pipeline. The effectiveness of the orifice and vacuum relief valve could then be tested.

During previous testing, the pump suction head could be limited to avoid excessive pump backspin. During orifice testing we no longer had that ability because the construction bulkhead had been removed and the pumping plant was directly connected to the tunnel system. Therefore, the orifice tests would have to be conducted under worst case conditions (very low suction head) and the risk of excessive pump backspin due to an oversized orifice opening was high. Consequently, computer modeling was attempted to predict a safe orifice size for testing.

Computer Modeling

Computer modeling was considered highly desirable in order to size orifices without undue risk of excessive backspin causing damage to the pump motors or bearings. The plan consisted of calibrating an existing computer model to reproduce a previously measured shutdown event with the check valve. Then replace the check valve in the model with various sizes of orifices to develop the relationship between orifice size and backspin speed. This would reduce the risk of field testing too large an orifice and exceeding safe backspin speed.

The system proved extremely difficult to model because of the sensitivity to timing and some important unknowns. First, the pump manufacturer did not provide hydraulic model test results, rotational inertia, or the efficiency of converting water column inertia to rotational inertia during reverse rotation. Second, a relationship between time and velocity of the falling water column could not be established with the flow measuring equipment then available. Third, timing of check valve closure was not consistent.

Two different models were tried without satisfactory results. Despite considerable effort to calibrate the two computer models, the models neither agreed with each other nor reproduced field measurements satisfactorily.

Field Tests

Without useful computer analyses, field testing had to be done to size the orifices. The project team agreed that testing could start with an 18-inch-diameter (46 cm) orifice with minimum risk of overspeed. Datasyst was again contracted to provide instrumentation and record data, and orifice testing was accomplished during January 1992.

The check valve was removed from Pump No. 2 and its disk removed. An 18-inch-diameter (46 cm), 2-inch-thick (5 cm) orifice plate was welded into the check valve body and the body/orifice was replaced into the 42-inch-diameter (107 cm) discharge line.

Because backspin could not be predicted, testing proceeded very cautiously starting with pumping rates as low as were stable and gradually increasing. Fourteen test runs were made with the 18-inch (46 cm) orifice in the Pump No. 2 discharge pipe. Nine tests were made pumping to the Jones Island head tank with discharges ranging from 10,500 to 34,500 gpm (0.7 to 2.2 m³/s). The other five tests involved pumping to the South Shore head tank with discharges ranging from 12,500 to 25,200 gpm (0.8 to 1.6 m³/s).

Similarly, a 20-inch-diameter (51 cm) orifice was installed in the 42-inch (107 cm) discharge line from Pump No. 3. Six tests were run with discharges ranging from 8,100 to 32,800 gpm (0.5 to 2.1 m³/s).

Next, the 18-inch (46 cm) orifice was removed from Pump No. 2, enlarged to 20 inches (51 cm) by onsite flame cutting and grinding, and reinstalled in Pump No. 2. A single test to the Jones Island head tank confirmed that this orifice gave the same results as the factory cut 20-inch (51 cm) orifice had in Pump No. 3. Then two tests were conducted pumping to the South Shore head tank at discharges of 20,300 and 27,300 gpm (1.3 and 1.7 m³/s). Since backspin speeds exceeded 750 rpm (by 5 to 10 rpm) no further testing was done with this configuration. The exceedance was experienced for no longer than a fraction of a second and no harm was done. Figure 3 is plot of backspin speed vs time after pump shutdown.

For the last series of tests the 20-inch (51 cm) orifice was removed from Pump No. 3, enlarged to 22 inches (56 cm) and reinstalled in the Pump No. 3 discharge line. Four more tests were run with discharges ranging from 12,500 to 34,800 gpm (0.8 to 2.2 m³/s).

During all of the tests, pressures never exceeded 200 psi (140,000 kgf/m²) and observed operation was smooth and quiet. No loud noise nor visible pipe movement occurred. Consequently, it was concluded that a properly sized orifice in each pump discharge line, together with a vacuum relief valve, could eliminate the excessive water hammer pressures while effectively limiting the backspin speed.

Field Testing Instrumentation

The test measurement instrumentation was selected to obtain high frequency, high dynamic range test data. The measurement transducers and circuits were calibrated for each test parameter with an accuracy established at 0.5 percent of full-scale range for each transducer channel.

Water pressure sensing was accomplished using strain gage style transducers ranged to monitor potentially high pressure transients at high frequency or fast rise time conditions.

The pipe coupling tie rod loads were monitored utilizing bonded strain gages in a full-bridge, direct-load configuration. Pipe displacements were monitored using potentiometric "string-pot" transducers.

The motor/pump shaft rotational speed was monitored using an electro-magnetic pickup indicating the shaft keyway. The motor contactor was monitored for being energized or de-energized through available motor control relay contacts. All transducers and sensing circuits were hard-wired from the transducer location to the signal conditioning/recording instrumentation. Signal filtering was minimal with 2,000 hertz (0.5 msec.) or better in overall frequency response on each channel.

The transient events during pumping system shutdown were recorded on a fiber-optic CRT light-beam oscillograph in amplitude vs time recordings for immediate, on-site decisions at each planned test sequence. The transient data were also recorded simultaneously on a fourteen channel magnetic tape recorder with analog, frequency modulated (F.M.) electronics as a permanent record of events should further analysis of data be required.

Test data were tabulated from amplitude/time measurements from the oscillograph chart.

Orifice Size

In order to minimize head loss during pumping it is desirable to use the largest size of orifice that will limit the backspin speed to 750 rpm. Backspin speed versus orifice diameter after shutdown was plotted to either head tank. On the basis of these plots, the maximum orifice sizes were selected as 19-1/2-inch (50 cm) for the pumps which can pump to South Shore (Pumps No. 1 and 2) and 23-1/2-inch (60 cm) for Pump No. 3 which can only pump to Jones Island.

Effect of Orifices on Pumping

It was understood that installation of orifices would reduce pumping plant capacity; however, the amount of reduction was not known. Consequently, a computer model was developed to predict pumping plant capacity with different sizes of orifices. Since the required calculations were steady state, rather than transient, it was possible to calibrate the model to reproduce test results.

The most significant calibration requirement was derivation of orifice head loss coefficients that would match measured results. During testing instruments measured steady state pressure on each side of the orifice plate. Late in the testing a differential pressure sensor was added to measure differential pressure across the orifice. Trial and error curve fitting was used to derive the orifice coefficient which would reproduce the measured total dynamic head at the pump when combined with friction and other minor hydraulic head losses.

All of these losses depend upon discharge. Generally, friction and minor losses combined amount to 3.5 to 7 feet (1.1 to 2.1 m) while orifice losses were in the range of 20 to 35 feet (6 to 11 m). The effect of the orifices was to raise the system total dynamic head by between 6 and 9 percent, but the plant capacity remained above the planned 150 mgd (6.6 m³/s).

Conclusions

Testing proved that the pumping plant would operate without excessive water hammer or the accompanying pressure surges and pipe movements if the check valves were replaced by vacuum relief valves and orifice plates. Testing also revealed that with vacuum relief valves in place, an orifice size of 19-1/2 inches (50 cm) in Pump Nos. 1 and 2 and 23-1/2 inches (60 cm) in Pump No. 3 would limit backspin speed to below 750 rpm. Accordingly the 2-inch-thick (5 cm) orifice plates were fabricated and installed during 1992 and the pumps have operated smoothly since then. Because the orifice plates have no mechanical devices, such as springs, or moving parts, the authors consider this a permanent solution.

ELIMINATING WATER COLUMN SEPARATION

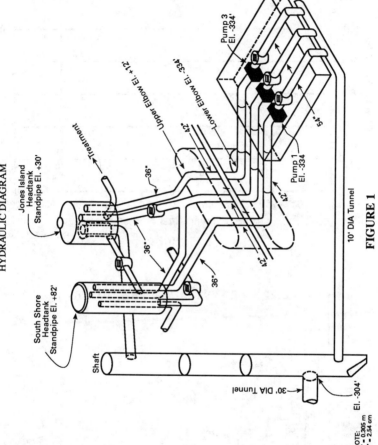

FIGURE 1

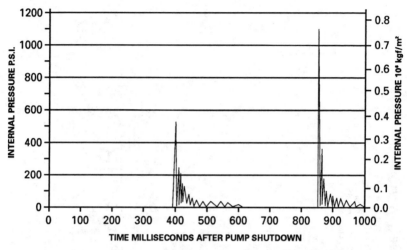

FIGURE 2

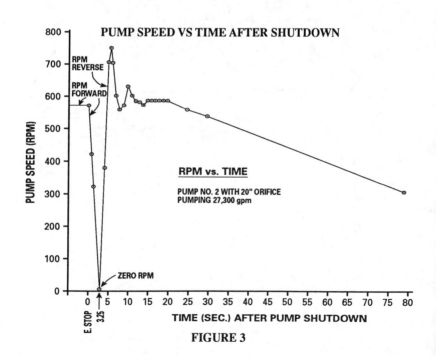

FIGURE 3

Pressure-Time Flow Rate in Low Head Hydro Plants

by

Charles W. Almquist, M. ASCE[1]
David B. Hansen[2]
Gerald A. Schohl, M. ASCE[2]
and
Patrick A. March[2]

Abstract

Very few methods are accepted for measuring the flow in the large intakes typical of low head, run-of-the-river hydro plants. These intakes are characteristically short, non-uniform, non-prismatic, converging multi-bay designs. A previous numerical and theoretical analysis showed that the pressure-time (Gibson) method of flow measurement could be applied to this type of intake, even though such application is not currently approved by relevant test codes. This paper presents the results of a small-scale model study of a low head hydro plant intake which demonstrates the applicability of the method in this situation. The difficulties in dealing with the extremely low level signals involved are described, as are some of the techniques developed for dealing with them. The experiments show that flow measurement in this type of intake using the pressure-time method is possible.

Introduction

The ability to measure flow through hydroelectric turbines is an important component in an electric utility's efforts to operate and manage its hydrogeneration and water resources as effectively and efficiently as possible. Flow measurement though the large intakes and conduits typical of hydroelectric facilities has never been an easy job, and the industry and related test code committees have long been in the forefront of developing, improving, and applying new techniques for the measurement of flow in large conduits. In spite of the variety of flow measurement techniques currently available, which include point-velocity integration, salt-velocity, ultrasonic time-of-

---

[1] Principia Research Corporation, 811 Whirlaway Circle, Knoxville, TN 37923

[2] Engineering Laboratory, Tennessee Valley Authority, Norris, TN 37828

travel, dye dilution, and pressure-time methods, many large plants do not lend themselves to flow measurement by any currently accepted method of flow measurement, particularly in situations where a turbine is being tested to ensure that it performs up to levels guaranteed by the manufacturer. A "Code" test is usually conducted under the auspices of the American Society of Mechanical Engineers Performance Test Code 18 (1992) or the International Electrotechnical Commission Test Code 41 (1991). The situation has been particularly difficult for low-head hydro plants, which usually employ propeller-type turbines, and have short, non-uniform intakes. Because flow measurement techniques typically require a well-established flow resulting from long reaches of uniform conduit, most of the techniques mentioned above cannot be applied to this type of plant. Flow measurement is further complicated by the fact that most low-head intakes have more than one intake passage preceding the scroll case.

One key advantage of the pressure-time (or Gibson) method of flow measurement is that, in principle, it does not require the long reaches of uniform conduit mandated by other methods. In spite of this, test codes have not allowed the use of the method for acceptance testing in the short, rectangular intakes typical of a low-head hydro plant, and as a result the method has not been extensively used or demonstrated for this type of plant. The Tennessee Valley Authority (TVA) has recently been involved in a program of turbine characterization and upgrading, and consequently has pursued an active program of extending the applicability of flow measurement technologies. This paper reports on the results of numerical and experimental analyses aimed at demonstrating the applicability of the pressure-time method to low-head hydro plants.

Review of Theory

The basis for the pressure-time method of flow measurement was described by Gibson in a classic ASCE paper published in 1923 (Gibson, 1923). The method is based on Newton's second law of motion. In brief, the pressure differential between two sections of a conduit carrying a flow is measured as the flow is brought to a halt by closing the turbine wicket gates. As the flow decelerates, Newton's law dictates that the pressure at the downstream section will rise relative to that of the upstream section. When friction losses and leakage flow are properly accounted for, the integral of the pressure difference with respect to time (the total impulse) is directly proportional to the total change in momentum flux during the gate closure. The pre-closure steady-state flow rate can be obtained by solving the impulse equation:

$$Q_i - Q_f = \frac{g}{F}\int_{t_i}^{t_f}\left(h - \left(\frac{Q}{Q_i}\right)^2 h_i\right)dt \qquad (1)$$

where Q_i is the steady-state flow being measured, Q_f is the final (or leakage) flow, Q is the instantaneous flow at any point during the gate closure, g is the acceleration of gravity, h is the pressure difference between the two pressure measurement sections,

h_i is the initial pressure difference, t is time, and t_i and t_f are the times over which the integration is carried out. The factor F is the "pipe factor" defined by

$$F = \int_{x_u}^{x_d} \frac{dx}{A} \qquad (2)$$

where x_u is the location along the conduit axis of the upstream pressure measurement plane, x_d is the location of the downstream measurement plane, and A is the local conduit cross-sectional area. Schohl and March (1991) describe in detail the development and application of Equations 1 and 2 in the traditional pressure-time flow measurement method.

Application to Non-Uniform Intakes

Schohl and March (1991), in a detailed theoretical and numerical investigation, also show that Equations 1 and 2 apply to non-uniform conduits of arbitrary cross-sectional shape. Further, they demonstrate that the friction recovery term (second term in the integrand of Equation 1) can properly account for any pressure difference which is proportional to the square of the velocity head. In the very high Reynolds number flow existing in a hydroturbine intake, this would include virtually any flow-induced pressure difference, including flow acceleration and streamline curvature. This observation is significant for low head intakes, which typically are bounded by converging curved surfaces. Another factor which has historically prevented application of the pressure-time method to short low head intakes is the fact that the measured pressure differences were too small to be accurately recorded and analyzed by traditional methods, including the ingenious Gibson apparatus and Gibson's phenomenally meticulous analysis procedure. Modern instrumentation techniques, including sensitive differential pressure cells and computer-aided data acquisition and analysis, ameliorate the limitations imposed by Gibson's mercury manometers, photographic plates, and planimeters. Improved understanding of the limitations of the method and improved technology allow further extension of the pressure-time method to situations previously considered unmeasurable.

Physical Model Description

The experimental investigation of the applicability of the pressure-time method to low-head hydro plants was conducted in a 1:60 scale model of the intake to one of the units at TVA's Watts Bar Hydro Plant. Plan and profile views of the intake model are shown in Figure 1.

The model reproduced the three bays of the intake, and included the basic shape of the semi-scroll case, although details of the scrollcase, stay vanes, wicket gates, and the turbine were not modeled. Water was delivered to the model via a constant-head tank. The flow through the intake section was manually controlled by a butterfly valve located in the six-inch line just downstream of the scrollcase drain, which approximately modeled the size and location of the turbine discharge ring. An air vent downstream of the butterfly valve served as a vacuum breaker to prevent vaporous water column separation after the butterfly valve was rapidly closed during testing.

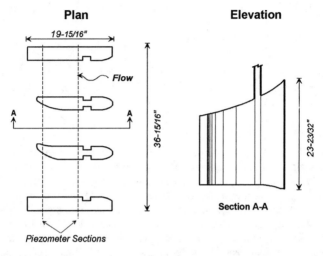

Figure 1. Geometry of Model Low Head Intake

The intake is a three-bay design, as shown in Figure 1, with vertical piers separating the bays. The locations of the piezometer taps are also shown on Figure 1. Each bay was equipped with piezometer taps on an upstream and a downstream plane, with six piezometer taps on each plane. The separation of the tap planes would correspond to about 43 feet in the prototype, which is only about 1/2 of the minimum separation which would be required by the test codes for a circular conduit of equivalent hydraulic diameter.

The six taps on each plane were connected by equal lengths of hard polyethylene tubing to a cylindrical manifold. The upstream and downstream manifolds for each bay were connected to a low-range differential pressure cell, so that pressure-time traces could be recorded independently for each bay. A linear potentiometer was used to record the position of the butterfly valve. A clamp-on time-of-travel ultrasonic flow meter on the discharge pipe was used to measure flow rate. This flowmeter was calibrated in-place against one of the Laboratory's magnetic flowmeters. The pressure, valve position, and flow data were recorded using a PC-based data acquisition system running a Windows-based pressure-time acquisition and analysis software package developed by TVA's Engineering Laboratory.

Test Runs and Data Analysis

The initial series of test runs, which are reported in this paper, were all conducted at the same nominal flow rate of 1150 gpm, the maximum flow for this open-tank, constant-head model. A test run consisted of establishing a steady-state flow in the model, starting the data acquisition system to record the steady state conditions, and then closing the butterfly valve. Recording of the data continued during the valve closure and continued for a period of about 15 seconds after the closure. The valve closure time ranged from about 1.2 to 3.5 seconds. The data acquisition sample rate was about 330 samples per second.

A typical pressure-time trace for the one of the three inlet bays is shown in Figure 2.

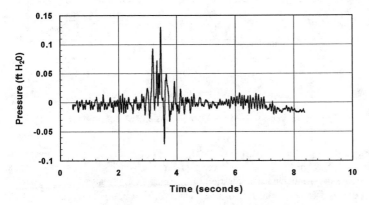

Figure 2. Pressure-Time Trace From Model

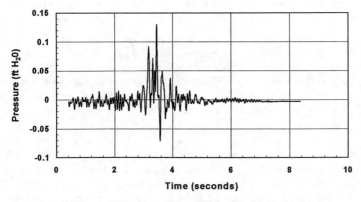

Figure 3. Pressure Time Trace with Digital Damping

Several features of this trace are worth noting, particularly in contrast to a typical prototype pressure-time record taken using the Gibson apparatus as shown in Figure 4.

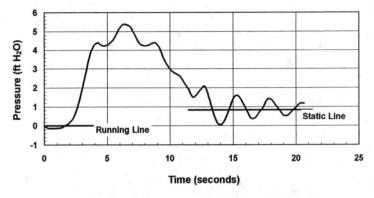

Figure 4. Pressure-Time Trace From Gibson Apparatus

First, the model signal is extremely low-level, with peak pressure of only a few inches of water. Second, the level of noise in the signal is relatively high. Most of the noise in the trace is due to mechanical vibration of the sense lines, because the pressure transducers used were extremely sensitive. This mechanically-induced noise is particularly evident in the trace after valve closure, because the rapid valve closure caused partial water column separation in the discharge pipe, resulting in strong shocks to the model. Finally, it is difficult to visually discern any net area under the pressure-time signal. The analyses of this type of signal proved challenging.

The record of the valve position was used to determine the point at which the signal integration of the pressure-time record would start. The test codes specify that integration should end near one of the peaks of the afterwaves present on the photographic record produced by the Gibson apparatus. Because such a series of after-peaks does not always exist in electronic implementations of the pressure-time method, most analyses carry the integration to a point at which the signal fluctuations have decayed to a small fraction of the peak amplitude of the pressure-time signal. However, such a relatively small amplitude signal was not observed in the model tests until quite some time after the closure was complete and the entire model had "settled down." To handle this situation, a digital damping function was applied to the signal after the point at which valve closure was complete. The damping function was implemented as a first-order decay, with a time constant ranging from 1 - 5 seconds. In principle, the use of the damping function will not affect the results, since the integral of the noise would be zero and the damping function will not introduce a DC component into the signal. The effect of the damping function on the pressure-time trace is shown in Figure 3.

Results

Table 1 summarizes the results for the series of runs made at the nominal flow rate of 1150 gpm. The flows determined by the pressure-time method (P-T) for each of the three bays and the total flow is shown along with the flow measured by the ultrasonic flowmeter for several different valve closure times.

Table 1. Experimental Results

Test	Flowmeter (gpm)	P-T Flow As-Measured (gpm)				P-T Flow With Damping (gpm)				Ratio U/S to P-T
		Left	Middle	Right	Total	Left	Middle	Right	Total	
a	1150	395	459	365	1219	375	460	362	1197	1.041
b	1150	451	353	352	1156	448	348	346	1142	0.993
c	1150	453	430	354	1237	453	435	359	1247	1.084
d	1150	423	392	339	1154	423	393	342	1158	1.007
e	1155	481	445	338	1264	498	447	350	1295	1.121
f	1150	399	376	381	1156	387	371	371	1129	0.982
g	1142	446	373	351	1170	462	376	353	1191	1.043
h	1149	427	378	303	1108	417	405	332	1154	1.004
i	1140	437	394	375	1206	425	395	366	1186	1.040
j	1141	473	451	373	1297	477	453	378	1308	1.146
k	1143	481	397	311	1189	476	397	323	1196	1.046
l	1150	518	395	362	1275	524	388	353	1265	1.100
m	1151	427	411	292	1130	422	411	308	1141	0.991
n	1149	530	396	369	1295	544	399	379	1322	1.151

Figure 5 shows the measured flow ratio as a function of valve closure time. The flow ratio is defined as Q_p/Q_m, where Q_p is the total flow rate measured by the pressure-time method, and Q_m is the flow measured by the ultrasonic flowmeter. The total flow rate shows an overall bias of about 5% as compared to the ultrasonic flowmeter measurement, and a scatter around this value of about ± 10%. While the overall scatter in the total flow rate determined by the pressure-time method would not be acceptable for a Code test of a prototype hydroturbine, it is quite encouraging in the present case where the impulse part of the pressure-time trace is not even visually discernible in the record, and the model-scale induced noise tends to dominate the signal. Valve closure time appears to have little effect on the computed pressure-time flow rate, in keeping with the theory upon which the method is based.

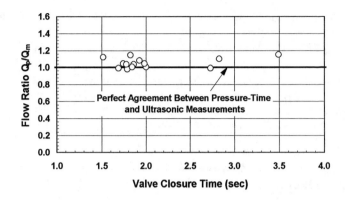

Figure 5. Ratio of Flows Measured by Ultrasonic and Pressure-Time Methods

Figure 6 shows the measured distribution of flow among the three inlet bays. Again, although some scatter from run to run is evident, the relative distribution of flow is consistent, particularly in light of the difficulties involved in signal analysis for these low-differential pressure-time traces.

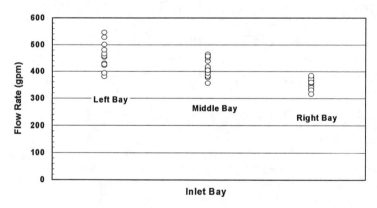

Figure 6. Distribution of Flow Among Intake Bays

Summary and Conclusions

The objective of the analyses and experiments described herein was to demonstrate the applicability of the pressure-time method to the relatively short, non-uniform, non-circular, multibay intakes typical of low-head hydro plants. In spite of considerable experimental difficulties caused by the small scale of the experimental facility and the resultant low-level, high noise pressure-time signals, it was possible to determine the flow rate in the model with reasonable accuracy using the pressure-time method. A key to the analysis was the use of a digital damping function in the analysis of the pressure-time signal, which helped to minimize the effects of noise on the integration of the signal. Based in part on this favorable experience in the laboratory facility, a field demonstration of the pressure-time method is underway at TVA's Chickamauga Hydro Plant, where the method will be compared against ultrasonic and current meter flow measurement methods. Application of the digital damping technique developed for this study will also be investigated.

Acknowledgments

The research reported in this paper was sponsored by TVA's Technology Advancements Group, whose support is gratefully acknowledged. At the Engineering Laboratory, the design of the experimental facility was principally the work of Jerry D. Hubble. The experimental program was conducted with the diligent assistance of Daniel S. McBee.

References

1. American Society of Mechanical Engineers, *Hydraulic Turbines*, Performance Test Code 18, New York, New York, 1992.

2. Gibson, N.R., "The Gibson Method and Apparatus for Measuring the Flow of Water in Closed Conduits", *Transactions of the ASME*, Vol. 45, 1923.

3. International Electrotechnical Commission, *Field Acceptance Tests to Determine the Hydraulic Performance of Hydraulic Turbines, Storage Pumps, and Pump Turbines*, Code 41, Geneva, Switzerland, 1991.

4. Schohl, G.A., and P.A. March, "Theoretical and Numerical Analysis of Pressure-Time Flow Measurement Method", Report No. WR28-1-900-244, Tennessee Valley Authority, Engineering Laboratory, Norris, TN, March 1991.

THE USE OF PIEZOELECTRIC FILM IN CAVITATION RESEARCH

Saurav Paul[1]
Christopher R. Ellis[2]
Roger E. A. Arndt[3]

Abstract: A research program was conducted at the St. Anthony Falls Hydraulic Laboratory to investigate impulse pressures caused by the collapse of transient cavitation bubbles. The impulse pressures associated with vibratory as well as hydrodynamic cavitation were measured. A stationary specimen subjected to vibratory cavitation and a hydrofoil of NACA 0015 section were instrumented with custom-made pressure transducers using polyvinylidene fluoride (PVDF) piezoelectric polymer film. These transducers provided *in situ* measurement of impulse pressures. These measurements made possible the assessment of power associated with cavitation erosion intensities. A practical application of this measurement technique was demonstrated in testing the viability of air injection as a means of mitigating cavitation erosion. The results from the impulse pressure measurements were compared with those obtained from hydrophone and accelerometer measurements. All the three measurement techniques showed similar results. The high sensitivity, minimal thickness, high frequency response, high spatial resolution and low cost of the piezoelectric transducers point to their applicability to wide range of dynamic pressure field measurements.

1. INTRODUCTION

Cavitation is a physical phenomenon associated with the formation, growth and collapse of bubbles within the body of a liquid or at a solid-liquid interface due to the variations of local static pressure. The bubbles form and grow in the regions of low pressure and collapse in the regions of high pressure. This collapse is accompanied by the sudden flow of liquid, which imposes stress pulses capable of causing plastic deformation and eventual erosion of neighboring solid surfaces.

---

[1]Graduate Student, [2]Research Fellow, and [3]Professor, University of Minnesota, St. Anthony Falls Hydraulic Laboratory, Mississippi River @ 3rd Ave. S.E., Minneapolis, Minnesota 55414.

Current research indicates that such impulse pressure is due to the impingement on the surface of a microjet formed during the final stages of the collapse of cavities close to the solid boundary. Prediction of individual pit formation based on any material property, therefore, requires knowledge of the impulse pressure incident on the surface.

A primary objective of this study was to relate the strength of pressure transients striking the solid boundary during the collapse of a cavitation bubble with the resulting erosion. Such pressure pulses are characterized essentially by very high amplitudes (of the order of 10^{11} Pa) lasting for extremely short duration of time (of the order of 10 ns). Various techniques have been explored to measure the pressure transients both spatially and temporally.

Recently, the piezoelectric polyvinylidene fluoride (PVDF) polymer film is being used extensively in various applications of dynamic measurement and control. The use of this film for the measurement of impulse pressures in the study of cavitation erosion is being reported here.

2. DESIGN OF PVDF TRANSDUCERS

The transducers were constructed in-house from piezoelectric film. The integral component of the film is a polarized polyvinylidene fluoride (PVDF) polymer. This homopolymer known for its high degree of piezoelectric activity, is sold under the trade name KYNAR. The film used in this study was a 28 μm thick Nickel metalized film. A sheet of this film was procured from Atochem Sensors, Inc., [1992].

According to the manufacturer, the film has a dynamic range that covers from 100 μPa to 10^{12} Pa and a frequency response that ranges from 0.005 Hz to 10^9 Hz. Estimates of transient stresses reported in the literature are as high as 10^{11} Pa. During initial development, the duration of such transient stresses was estimated to be about 2 μs. In light of these observations the PVDF film was considered to be specially suitable for recording the impulse pressures due to the cavitation events on the hydrofoil. This is probably too optimistic in light of the much shorter durations reported by Vogel and Lauterborn [1988] of 10 to 40 ns. Recent estimates of response time requirements indicate that obtaining an accurate description of cavitation impulses is a formidable task, Vogel et al [1989].

2.1 Layout
In the incipient stage of development, the film was mounted on a PVC button which formed the stationary specimen in the ASTM vibratory cavitation setup. At a standoff distance of 1.5 mm from the horn tip, an output signal in

the range of 3 to 4 Volts could be obtained with a unimorph shielded PVDF film element of 30 mm x 12.5 mm active area. The excellent response of the film and the very clean signal precluded requirements of any amplifier and noise filter.

The signal from the PVDF film is conducted through leads attached to both the metalized surfaces. Simultaneous monitoring of cavitation signals at closely juxtaposed target areas on the hydrofoil posed a major problem in terms of spatial management and architectural layout. This problem was handled by using a common ground for all the individual PVDF probes. Exploratory runs with this concept of common ground were made on the vibratory cavitation setup. The acquired temporal signal showed no evidence of any appreciable cross-talk amongst the signals of the different probes. Also, in terms of the propagation modes of the mechanical waves, this implied that the transmission of the transient stress waves from the exposed surface of the protective elastomeric shield to the PVDF film was predominantly longitudnal in nature. This had an important bearing on the calibration of these PVDF transducers which is discussed in a later section.

This proven technique of fabricating a matrix of target sensors was subsequently used to instrument an interchangeable plug on the hydrofoil. A grid of 14 piezoelectric pressure transducers were attached to a removable section of the suction side of the foil as shown in Fig. 1. These custom made transducers, 6.4 mm in diameter and 150 μm thick, were bonded to the surface of the insert.

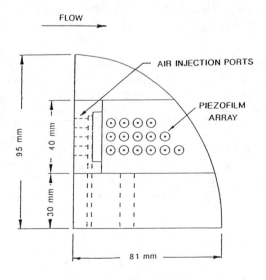

Fig. 1. Details of the hydrofoil.

The construction details, calibration and characteristics of these transducers are discussed in a later section. All but the tip and near-base region of the foil was covered with a polyurethane protective tape (3M Scotch tape #8672). This adhesive tape has an overall thickness of 200 μm with a 50 μm thick layer of pressure sensitive acrylic adhesive. The tape served both to protect and waterproof the transducers and smooth the joints in the surface of the foil. A considerable effort was expended in developing an adequate method of layout, mounting and architecture of the transducer matrix finally used. However, this resulted in a reduction of the sensitivity of the sensor.

2.2 Architecture

The constructional details of the PVDF piezoelectric sensors is shown in a sectional view of Fig. 2. The transducer assembly is shown as a multilayered mechanical structure consisting of a backing block, a unimorph thickness-mode PVDF piezoelectric element, a front face common ground electrode and a protective thermoplastic elastomeric tape.

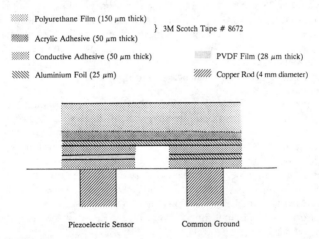

Fig. 2. A sectional view of the developed PVDF piezoelectric pressure sensors.

The backing block consists of 4 mm diameter copper rod threaded into the PVC plastic housing base and machined flush with the surface contour of the hydrofoil. Normally in the design of ultrasonic transducers, the dimensions and stiffness of the backing block plays a critical role in bandwidth control, Hutchins and Hayward [1990], Kuttruff [1991]. However, since the aim of the present study was to measure the pressure pulse height due to the individual cavitation events, and not the temporal pressure profile of the individual events, the backing block essentially formed the electrical conductor in this construction. A 25 μm thick sheet of aluminum served as the common ground electrode for all the PVDF pressure transducers. The various mechanical sections that formed the electrical components of the transducer system were connected by means of 50 μm thick conductive adhesive transfer tape (3M Scotch tape # 9703). This tape has silver-coated nickel conductive particles in a pressure sensitive A-40 acrylic adhesive base and conducts only through the adhesive thickness.

3. CALIBRATION

In the impact sensing mode the essential transduction property of the film is given in terms of the voltage or the stress coefficient g_{ij}. This is defined as the ratio of the open circuit electric field to the applied mechanical stress. Output voltage is obtained by multiplying the calculated electric field by the thickness of the piezoelectric polymer between the electrodes. In our experiment the film was mounted in the "33" mode. This implies that the electric field and the mechanical stress are both along the polarization axis. Thus, for the PVDF film used in this study (thickness 28 μm and g_{33} of -339 x 10^{-3} V/m per Pa) the applied stress calibration is then 1.05 x 10^5 Pa/V. It is noteworthy that this calibration estimate is for ideal mechanical and electrical boundary conditions for the entire metalized surface area of the film that form its electrode being subjected to a uniform stress distribution. The stress calibration should, therefore, be expected to change appreciably under practical construction and operating conditions.

In the acoustic emission sensor constructed for this study, the PVDF piezoelectric film with the aluminum sheet as the front face electrode and the copper rod as the rear electrode and backing block, may be treated as a single unit of the transduction system. For the purpose of calibration, this then is considered to be coupled to the elastomeric shield. Referring to Fig. 2, the elastomeric layer thus becomes the structure medium or the test material in analogous ultrasonic nondestructive testing (NDT) procedures. The studies on the characterization and calibration of acoustic emission sensors have shown that the ball drop is an adequate impulse source for the calibration of the PVDF sensors developed for this study.

Calibration using a dropping ball technique indicated that the calculated sensitivity was very optimistic. Simplifying assumptions of the sensor transduction process that was used to facilitate calibration include one dimensional transduction, singleness of mode, system linearity and fixed input and output impedances. Additional assumptions include (*i*) the signal is gated in time domain and band limited in frequency domain, (*ii*) the phase spectra are disregarded, (*iii*) the transfer function of the structure medium is known to be flat or diffused, and (*iv*) the detection mode is primarily longitudinal. Using plastic deformation theory, this was estimated as 1.1 x 10^6 Pa/V for a permanent indentation radius of 1.5 mm. The details of calibration of the PVDF pressure transducers is presented in Paul [1994]. The significant reduction in sensitivity is attributed to the particular method of mounting and coating with a protective film.

3.1 Amplitude Response

The amplitude calibration of the PVDF piezoelectric pressure sensors on the instrumented hydrofoil was performed by dropping a stainless steel ball of 4.76 mm diameter directly on the center of each transducers. The drop height was kept constant at 82 cm. The average rebound height h_2 was measured to be 4 cm.

The case of a hard spherical indenter impinging on a horizontal flat surface of a softer metal has been studied for calculating dynamic hardness of metals, Tabor [1951]. If the impact is such that the spatial-mean pressure exceeds about $1.1Y$, where Y is the uniaxial yield stress or the elastic limit of the metal, a slight amount of plastic deformation will occur and the collision will no longer be truly elastic. A release of elastic stresses takes place in the indenter and the indentation, as a result of which rebound occurs. Thus, the indenter strikes the surface and rebounds, leaving an indentation in the surface.

The use of full-scale plastic deformation requires a detailed study and understanding of the related basic principles that are involved in high velocity impact dynamics typical in cavitation erosion, Blazynski [1987]. The basic principles that pertain include the conservation laws, the role of wave propagation, the influence of inertia, and an understanding of the material behavior under high rate of loading. However, for the problem at hand of calibrating the pressure transducers using a ball drop technique, the analysis may be limited to the stages of partial plastic deformation. The temporal-maximum spatial-mean dynamic pressure P_d that occurs at the end of such deforming process has been shown by Tabor [1951] to assume the form,

$$P_d^5 = \frac{h_2^4}{(h_1 - \frac{3}{8}h_2)^3} \frac{10^4}{\pi^5 4^3 3^4} \frac{mg}{r_1^3} \left(\frac{1}{f(E)}\right)^4, \tag{1}$$

where $f(E)$ is a function of the elastic constants of the sphere and target materials, given by

$$f(E) = \frac{(1 - v_1^2)}{E_1} + \frac{(1 - v_2^2)}{E_2}. \tag{2}$$

Here E_1 and E_2 are the Young's moduli for the indenter and the surface respectively, and v_1, v_2 are Poisson's ratios; m is the mass of the spherical indenter of radius r_1.

It is pertinent to note that the dynamic yield pressure P_d is not necessarily the same as the static pressure P_s required to cause plastic flow. If the dynamic yield pressure is calculated from the energy required to produce an indentation of a given volume, it is larger than the static yield pressure and increases with the velocity of impact.

The dynamic yield pressure tends to increase during the impact process and also varies across the cross-sectional area of the indentation. At any distance x from the center of the indentation, the pressure distribution over the circle of contact is given by,

$$P_d(x) = P_0 \left(1 - \frac{x^2}{a^2}\right)^{1/2}, \qquad (3)$$

where P_0 is the pressure at the center of the circle of contact. It then follows that the maximum dynamic yield pressure P_0 is related to the spatial-average yield pressure P_d of Eq. (1) as,

$$P_0 = 1.5 P_d. \qquad (4)$$

The time of impact for plastic collisions still remains a formidable task. Hutchings [1977] studied the strain rate effects of impact velocities in microparticle impact and concluded that the orders of magnitude of the impact duration do not depend critically on whether plastic flow occurs during impact. For elastic collisions the impact time t_e is given accurately by Hertz's equation as

$$t_e = 2.94 \frac{z_0}{u_1}, \qquad (5)$$

where z_0 is the maximum distance the indenter sinks into the surface during collision and may be written as,

$$z_0 = \frac{3\pi a P_d}{4} f(E).$$

The chordal radius a of the circle of contact at the end of the loading time can be written as

$$a = \left[\frac{10}{3} \frac{mgh_2}{\pi^2 P_d^2 f(E)}\right]^{1/3}. \qquad (7)$$

The impact velocity u_1 due to free fall is given as,

$$u_1 = \sqrt{2gh_1}. \qquad (8)$$

In the following, use was made of Eqs. (1), (4), (5) and (7) in the amplitude calibration process.

The pressure signal from the dropping ball calibration test for each of the individual PVDF pressure sensors on the instrumented hydrofoil was recorded on a Philips 50 MHz digital storage oscilloscope. The time trace on the photograph negative was digitized for input to a computer through a Nikon LS-3510AF 35 mm film scanner. Fig. 3 shows the computer output of one such digitized image. The rise time and the corresponding peak-positive voltage due to pressure pulse for the array of PVDF sensors was measured using commercially available software (Aldus Photostyler version 1.1). It can be seen in the time trace of the pressure pulses that the voltage signal during the rise time can be closely approximated as a linear function of time. The average value of the slope was calculated to be about 0.1 Volts per μs.

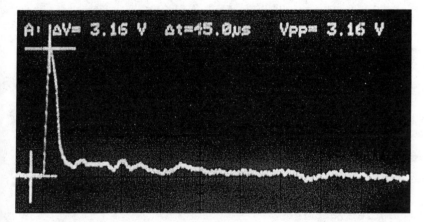

Fig. 3. Computer output of the time trace of the pressure signal acquired through the film scanner.

The pressure amplitude calibration of the PVDF transducer was done in terms of the temporal-peak spatial-peak dynamic yield pressure P_0 given by Eq. (4). The collision time was estimated using the principles of energy conservation during elastic collisions as expressed in Eq. (5). For the purpose of stress analysis and calibration, the elastic properties of only the elastomeric layer was taken into consideration. For a rebound height of 4 cm, a was calculated as 1.5 mm, and t_e as 24.5 μs. The corresponding value of P_0 was estimated at 2.73 MPa. Comparison of the estimated impact time with the measured rise time of the generated voltage signal, shows that the plastic deformation theory can be used to a good approximation in estimating the amplitude calibration of the PVDF pressure sensors. The amplitude calibration in terms of Pascal per Volt was obtained by dividing the product of the estimated P_0 and $(1/t_e)$ by 0.1 Volts per μs. This was calculated as 1.1×10^6 Pa per Volt.

3.2 Frequency Response

The basic half wavelength resonance frequency f_r of the piezoelectric film can be expressed as,

$$f_r = \frac{c_0}{2t}, \qquad (9)$$

where c_0 is the speed of sound in the PVDF polymer, and t is the nominal thickness of the film. With $c_0 = 2.2 \times 10^3$ m s^{-1}, and $t = 28$ μm, the resonance frequency of the transducer is about 40 MHz. In terms of the Nyquist sampling criteria in digital signal processing, this corresponds to a time resolution of 50 ns. However, due to the low quality factor Q (≈ 10) of the PVDF film, the pressure transducer is expected to have a highly damped resonance frequency. The complete frequency response of the developed PVDF pressure sensors can be obtained from the spectrum of the time trace of the calibrated voltage signal during the ball drop calibration test. In digital data acquisition, the frequency resolution was limited by the record length of the acquired signal. This was typically 2^{11} samples at a minimum sampling interval of 10 μs which gave a frequency resolution of 48.8 Hz.

4. UNCERTAINTIES

The use of target specific piezoelectric pressure transducers in the experimental determination of acoustic transients due the collapse of a cavitation bubble raises a few basic difficulties. These are related to (*i*) time resolution, (*ii*) spatial resolution, (*iii*) mechanical damage of the transducer, and (*iv*) transducer coupling.

4.1 Time Resolution

The duration of the pressure pulse was shown to range between 10 and 40 ns. Since these pulse durations are much smaller than the rise time of typical pressure sensing devices, significant measurement errors can occur. For this reason, the pressure transducer must have a very high natural frequency to reproduce as reliably as possible the sudden rise in pressure. If not, the signal is underestimated. In addition, signal processing must be suited to such high frequencies. The time correction can be based on the measured pulse duration of Vogel and Lauterborn [1988]. They suggest that the peak amplitude of an exponential pulse is too low by a factor of

$$m = \frac{\tau}{\tau_a \ln 2} \qquad (10)$$

where τ is the pulse time which is much shorter than the rise time τ_a of the pressure transducer. For typical transducer rise times, this factor can be of the order of 0.01 to 0.05 which explains the unreasonably low values of collapse pressure measured in most experiments.

4.2 Spatial Resolution

The sensitive surface should be smaller than the size of the impact, to avoid once more, an underestimation of the pulse height. If not, the measured quantity is actually the force or the equivalent pressure which would give the same output if it were uniformly applied to the entire sensitive area of the transducer. The above calibration estimate of the pressure sensors corresponds to an active region of the sensor given by the indentation diameter.

The integral property of a piezoelectric material, such as the PVDF film, is its capacity to develop charge when being stressed mechanically. In open circuit configuration of transduction the charge Q developed on its metalized surface that form the electrodes is then related to the developed voltage V through the capacitance C of the transducer as,

$$V = \frac{Q}{C}. \qquad (11)$$

Here the capacitance of the transducer using PVDF film of uniform thickness t is given by,

$$C = \frac{\epsilon A}{t}, \qquad (12)$$

where ϵ is the permittivity of the PVDF piezo film, and A is the metalized area of the film of thickness t. In the present construction, the metalized area A is same as the area of the individual probe A_{probe}. For the 6.4 mm used in this study, A_{probe} is 31.7×10^{-6} m^2. The permittivity ϵ of piezoelectric polymer film at 10 kHz is 106×10^{-12} F m^{-1}. The capacitance C of this probe is 120×10^{-12} F.

When only a fraction of the metalized surface of the film is excited by the stress field, the remaining metalized film acts as a passive capacitor in parallel with the active capacitor due to the active film. The total capacitance C_{total} is then obtained from

$$C_{total} = C_{active} + C_{passive}, \qquad (13)$$

where C_{active} and $C_{passive}$ can be obtained from Eq. (12) for the corresponding areas.

In such situations, the charge is then distributed over the entire area. The recorded voltage $V_{recorded}$ is then related to the generated charge Q as,

$$V_{recorded} = \frac{Q}{C_{total}}. \tag{14}$$

Noting that the sum of the active and passive areas is the area of the probe A_{probe}, the equation for the voltage ratio assumes the form

$$\frac{V_{true}}{V_{recorded}} = \frac{A_{probe}}{A_{active}}. \tag{15}$$

The above equation shows that in situations where the excited area is smaller than the size of the transducer, the recorded voltage involves a correction in terms area. If the excitation area is a circular section of diameter d_{active}, Eq. (15) can be written for the circular probe diameter d_{probe} as,

$$\frac{V_{true}}{V_{recorded}} = \left[\frac{d_{probe}}{d_{active}}\right]^2. \tag{16}$$

In the ball drop calibration test d_{active} is twice the chordal radius a of the contact circle. Referring to Sec. 3.1, a was estimated as 1.5 mm. From Eq. (16) this gives the voltage as 4.6 times the recorded voltage. The pressure amplitude calibration of the temporal-peak spatial-peak dynamic yield pressure P_0 in terms of the true voltage is then 2.4 x 10^5 Pa per Volt.

If it is assumed that the diameter of a cavitation pit is the measure of the corresponding excitation area, then the voltage recorded for a 10 μm diameter pit should be multiplied 2^{12} x 10^2 to obtain the true voltage. The peak pressure P_0 can then be estimated from the pressure amplitude calibration in terms of true voltage as mentioned above.

4.3 Mechanical Damage

The transducer must be sufficiently resistant not to be damaged. Coleman and Saunders [1989] reported the occurrence of transient cavitation during measurements of the pressure amplitude and waveforms in the focus of extracorporeal shock wave lithotriptors (ESWL). They had used PVDF film sensors developed by Preston et al [1983]. They observed that the damage of the PVDF film due to the cavitation microjet impact on the sensors did not alter the calibration of the film by any appreciable amount. Calibration of the film before and after exposure to cavitation showed a difference of amplitude of about 5%. Similar effects have been observed by Simpson and Tussolini [1985] with the calibration of the PVDF film damaged due to high velocity impact of microparticles in space studies of dust particles in Haley's comet.

4.4 Transducer Coupling

The source of much of the variability and many of the limitations associated with the piezoelectric transducers is their coupling to a test medium. The amplitude and time characteristics of ultrasonic signals can be affected by phase shift, interference, and attenuation effects associated with wave propagation through the couplant layer, Truell et al [1969], Krautkrämer and Krautkrämer [1990].

The conductive adhesive tape and the adhesive layer of the polyurethane protective tape were expected to exert a negligible influence on the amplitude response in the reproducability of the individual sensors and their relative variability. This follows from the observations of Ono et al [1984] that for a particular couplant material, the mounting pressure do not have significant effect on the amplitude calibration. However, detailed studies of the stress wave interaction and transducer characterization needs to be done to prove that such expectations were adequately met.

5. RESULTS AND DISCUSSION

5.1 ASTM Vibratory Apparatus Tests

Several different types of tests were carried out with the vibratory apparatus in the course of developing the PVDF transducers. Fig. 4 shows the power spectrum measured beneath the horn with 6.4 mm diameter transducers

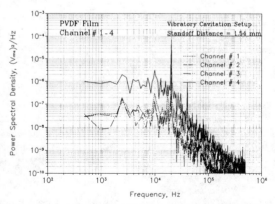

Fig. 4. Measured pressure spectra beneath the horn in the ASTM vibratory cavitation apparatus.

at a stand-off distance of 1.54 mm. Channel 4 is from a transducer positioned at centerline of the horn. Channels 1, 2, and 3 are from transducers each positioned at a radial offset distance of 7.9 mm and angles of 0°, 90°, and 180°. A common ground was positioned at 270°. As expected, a strong peak is noted at a vibration frequency of 20 kHz with another peak at the first harmonic of 40 kHz. The signal from the centerline transducer is higher, as expected. Some discrepancies are noted in the signals from the other three transducers, which should be identical.

A second test was carried out to determine the effect of transducer size. A circular shaped transducer was split into 3 pie shaped pieces having included angles of 60°, 120°, and 180° respectively. The pressure spectra of the signals obtained by placing this configuration under the horn is plotted in Fig. 5. The signal from each pie shaped transducer here can be seen to be essentially the same, indicating that effect of size was minimal.

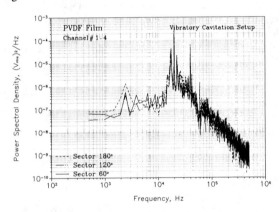

Fig. 5. Measured pressure spectra of the signal from three pie-shaped piezoelectric sensors beneath the horn in the ASTM vibratory cavitation apparatus.

5.2 Pressure Pulse Height Spectra

Since it was expected that the pressure signal would be dominated by pressure impulses of high amplitude due to cavitation, pulse height spectra were compared at different velocities. A typical comparison is shown in Fig. 6. Scaling of pressure amplitude with velocity depends on the frequency of occurrence, as expected.

High frequency of occurrence corresponds to turbulent pressure fluctuations which scale like U_0^2. High amplitude, low frequency of occurrence pulses scale with a much higher power of velocity. Presumably high amplitude pulses are due to cavitation events. The technique is limited by a sampling rate of only 100 kHz, which appeared adequate for measurement of spectra. However, measurements of mean square pressure and pulse height spectra are consistent. The mean square pressure for a series of pulses of amplitude p_m and duration τ is given by

$$\overline{p^2} = N p_m^2 \tau \qquad (17)$$

Typical values for N and τ at a velocity of 17.5 m s^{-1} are 195 s^{-1} and 5.3 μs. At this flow velocity, the pits formed due to cavitation were found to be predominantly of 10 μm in diameter. The values of p_m for this sized pits could then be estimated from the amplitude calibration as 8,650 MPa for recorded voltage of 0.088 V. This agrees with a measured value of the mean squared signal of 8 x 10^{-6} V^2 which corresponds to $\overline{p^2}$ = 7.73 x 10^{16} Pa^2.

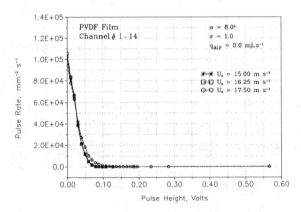

Fig. 6. Comparison of pressure pulse height spectra at different velocities.

Pulse height spectra were expected to be a good indicator of cavitation erosion intensity after further development. However, our initial findings suggest that comparisons of the mean square signal in the frequency band 10 kHz to 30 kHz offer the best indication of erosion rate in the present study.

5.3 Effects of Air Injection

A comparison of relative spectral power versus normalized air flow rate is shown in Fig. 7. In this figure, erosive power is considered to be proportional to either: (*i*) the mean square of the hydrophone signal in the 10 kHz to 30 kHz band, (*ii*) the average of the three most intense mean square pressure signals in the 10 kHz to 30 kHz band or, (*iii*) the mean square modulation acceleration defined by Abbot et al (1993). The air flow rate per unit width, q, in the normalized air flow rate, $q/U_0 c$, is obtained by dividing the total air flow rate applied by the number of holes (5) and the spacing between holes; c refers to the chord length of 68.9 mm at the position of the center air injection port.

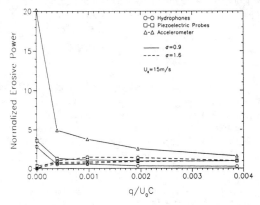

Fig. 7. Comparison of normalized erosion power versus normalized air flow rate at $U_0 = 15$ m s^{-1} and $\sigma = 0.9$.

The favorable influence of air injection is clearly evident. Although the modulation method appears to be the most sensitive technique, the effects of air on the pressure pulses are also evident. This is illustrated in Figs. 8a and 8b. These are photographs of the sheet cavitation without and with a relative amount

Fig. 8a. Photograph of sheet cavitation at $U_0 = 17.5$ m s^{-1} and $\sigma = 0.9$. Superimposed on the picture are isobars of mean square pressure fluctuations.

of air injection of 8.3 x 10^{-4}. Superimposed on the pictures are isobars of the mean square pressure. The favorable influence of air injection over the entire surface of the foil is evident.

Fig. 8b. This is the same situation as Fig. 8a except for an air injection rate q/U_0c of 8.3 x 10^{-4}. Note the dramatic reduction in pressure intensity over the entire surface of the foil.

5. CONCLUSIONS

In situ measurements of pressure pulses encountered in transient cavitation were made with acoustic emission sensors designed and developed at the St. Anthony Falls Hydraulic Laboratory with a PVDF piezoelectric polymer film. Different configurations of the dynamic pressure sensor using the PVDF film were tried with the ASTM cavitation erosion setup in the stationary specimen mode. The techniques developed thereby were used to instrument the hydrofoil for instantaneous *in situ* measurements of the pressure pulses due to transient cavitation.

Pressure amplitudes of the order of 10,000 MPa were estimated for impulses sufficient to cause pits of 10 μm diameter. In the light of the small duration of these transients, these estimates are expected to be even higher by about a factor of 10. However, further correction of the amplitude estimates is limited at the present time due to the uncertainties in the measurements of the impulse duration and the characteristic frequency of the transducer.

Data analysis of the acquired signal included mean-squared pressure measurements in terms of the power spectral density and the strength of the individual events in terms of the pulse height spectrum. The results obtained from these analyses were compared with other diagnostic techniques. These included cavitation noise measurements with hydrophones and structural vibration measurements using accelerometers.

A practical application of the erosion monitoring technique developed in this study was demonstrated in testing the viability of using air injection as means of mitigating cavitation erosion in hydropower units. Dynamic pressure measurements using the PVDF film have shown air injection to be effective in minimizing erosion. This has been confirmed with other methods of detection and data analysis techniques.

ACKNOWLEDGEMENTS

This research was funded by the Legislative Commission on Minnesota Resources under the Stripper Well Alternate Energy Program. The studies on air injection were done in a collaborative effort with Accusonic Division, ORE International, Inc. and was funded by Pacific Gas and Electric Company, San Francisco, California. The authors would like to thank the technical support staff of Atochem Sensors, Inc. for their helpful disposition and consultation during the development of the PVDF sensors. The various adhesive tapes that were tried in the construction of the piezoelectric sensors were provided gratis by Dr. Richard Hartman of 3M Industrial Tape and Specialties Division. This is sincerely appreciated.

REFERENCES

1. Abbot, P. A., Arndt, R. E. A. and Shanahan, T. B. [1993], "Modulation Noise Analysis of Cavitation Hydrofoils", *Proc. ASME Intl. Symp. on Bubble Noise and Cavitation Erosion in Fluid System*, Winter Annual Meeting, December.

2. Atochem Sensors, Inc. [1992], P.O. Box 799, Valley Forge, PA 19482, USA.

3. Blazynski, T. Z. [1987], *Materials at High Strain Rates*, Elsevier Applied Science, New York.

4. Coleman, A. J. and Saunders, J. E. [1989], "A Survey of the Acoustic Output of Commercial Extracorporeal Shock Wave Lithotripters", *Ultrasound in Med. & Biol.*, Vol. 15, No. 3, pp. 213-227

5. Hutchins, D.A. and Hayward, G. [1990], "Radiated Fields of Ultrasonic

Transducers", *Physical Acoustics*, Academic Press, Vol. **XIX**, pp. 1-80.

6. Hutchings, I. M. [1977], "Strain Rate Effects of Microparticle Impact", *J. Phys. D: Appl. Phys.*, Vol. **10**, L179.

7. Krautkrämer, J. and Krautkrämer, H. [1990], *Ultrasonic Testing of Materials*, Springer-Verlag, New York.

8. Kuttruff, H. [1991], *Ultrasonics: Fundamentals and Applications*, Elsevier Applied Science, New York.

9. Ono, M., Higo, Y., et al [1984], *Prog. in AE*, **2**, pp. 343-350.

10. Paul, S. [1994], "Towards Laboratory Prediction of Cavitation Damage Intensities in Prototype Applications", *MS Thesis*, University of Minnesota, Minneapolis.

11. Preston, R. C., Bacon, D. R., Livett, A. J. and Rajendran, K. [1983], "PVDF Membrane Hydrophone Performance Properties and Their Relevance to the Measurement of Acoustic Output of Medical Ultrasonic Equipment", *J. Phys. E. Sci. Instrum.*, Vol. **16**, pp. 786-796.

12. Simpson, J. A. and Tussolino, A. J. [1985], "Polarized Polymer Films as Electronic Pulse Detectors of Cosmic Dust Particles", *Nuclear Instr. and Methods in Phys. Res.*, **A236**, pp. 187-202.

13. Tabor, D. [1951], *The Hardness of Metals*, Oxford University Press, London.

14. Truell, R., Elbaum, C. and Chick, B. B. [1969], *Ultrasonic Methods in Solid State Physics*, Academic Press, New York.

15. Vogel, A. and Lauterborn, W. [1988], "Acoustic Transient Generation by Laser-Produced Cavitation Bubbles Near Solid Boundaries", *J. Acoust. Soc. Am.*, Vol. **84**, No. 2, pp.719-731.

16. Vogel, A., Lauterborn, W. and Timm, R. [1989], "Optical and Acoustic Investigations of the Dynamics of Laser-Produced Cavitation Bubbles Near a Solid Boundary", *J. Fluid Mech.*, Vol. **206**., pp. 299-338.

Design and Operation of a System to Monitor Sediment Deposition for Protection of an Endangered Mussel

Michael S. Griffin[1] and David S. Mueller[1] M.ASCE

ABSTRACT

The U.S. Army Corps of Engineers (COE), Louisville District is replacing Ohio River Locks and Dams 52 and 53 with a single facility located near Olmsted, Ill. The endangered orange-footed pearly mussel (Plethobasus cooperianus) is found in two mussel beds in the lower Ohio River, near the site of the new locks and dam. The construction and operation of the Olmsted Locks and Dam could change the current sediment deposition and erosion patterns. If sediments were to be deposited on the mussel beds, the viability of the beds and the survival of the orange-footed pearly mussel could be threaten. The guidelines in the biological opinion, set by the U.S. Fish and Wildlife Service (USFWS) to ensure protection of the mussels, state that sediment accumulations on the mussel beds cannot exceed 2 centimeters above baseline levels. According to the biological opinion, if USFWS guidelines are not met, construction activities will cease and corrective actions will be taken.

The U.S. Geological Survey (USGS), Kentucky District, in cooperation with the Louisville District COE, has designed and implemented a multi-transducer acoustic ranging and telemetry system to continuously monitor the changes in elevation of the river bed over the mussel bed, located downstream of the construction project. This system will be operational during construction and normal lock and dam operation until such time that it is determined that the new locks and dam are not adversely affecting the mussel bed by sediment deposition or erosion.

---

[1]Hydrologist, U.S. Geological Survey, Water Resources Division, Kentucky District, 2301 Bradley Avenue, Louisville, KY 40217

INTRODUCTION

The Ohio River flows 1578 kilometers (km) from the confluence of the Allegheny and Monogahela Rivers at Pittsburgh, Pa. to the Mississippi River near Cairo, Ill. The entire river, except for the last 30.6 km (from Lock and Dam 53 to the Mississippi River), has been altered by construction of new locks and dams to provide an improved navigational channel. Over the past 50 years, all but 2 of the original 53 wicket dams and locks have been replaced with high-lift dams (fig. 1).

The Ohio River and its tributaries historically supported a multitude of aquatic species, including fish and mussels. Construction of navigation facilities on the river significantly altered the river into a series of lake-like pools. Extensive artificial impoundments of the river slowed current velocities, and subsequent accumulation of silt resulted in reductions in mussel faunae (Thorp and Covich, 1991). The lower Ohio River, from Lock and Dam No. 53 to its confluence with the Mississippi River, contains the only remaining free-flowing riverine habitat in the entire main stem of the river and supports the largest populations of pre-impoundment fish and mussel species (U.S. Fish and Wildlife Service, 1993).

On the basis of investigations of the lower Ohio River study area, with respect to improving navigational conditions, the U.S. Army Corps of Engineers (COE), Louisville District was authorized to replace Locks and Dams 52 and 53 with a single structure. This new facility will be located at Ohio River Mile 964.4, near Olmsted, Ill., and will be known as Olmsted Locks and Dam. The facility, located approximately 2.9 km downstream of Lock and Dam No. 53, will consist of two 33.5 meters (m) by 366 m lock chambers constructed on the Illinois side of the river, a 671 m navigable pass section controlled by 220 remotely operated hydraulic wickets, and a 130 m section of fixed weir on the Kentucky side of the river. The navigational pool created by the new dam will be operated to maximize use of the navigable pass and minimize frequency of lockage. Upon completion of the Olmsted Locks and Dam, the existing Locks and Dams 52 and 53 will be destroyed and removed from the Ohio River.

Problem

The endangered orange-footed pearly mussel (Plethobasus cooperianus), is an Interior Basin species usually found in medium to large rivers at depths of 3 to 8.8 m (U.S. Fish and Wildlife Service, 1993). The mussel buries itself into sand and gravel leaving only part of its shell and feeding siphon projecting above the river bed. The Ohio, Cumberland, and Tennessee River drainages contain the only three known populations remaining anywhere in the historic range of the species. The only population that is known to be reproducing is located in the Tennessee River (U.S. Fish and Wildlife Service, 1993). The two mussel beds (one population) located in the lower Ohio River near the site of the Olmsted Locks and Dam project are probably reproducing, so loss or significant adverse impact to the population could threaten the survival of the species. One mussel bed is located upstream of the project, and the other is located downstream of the project. Bed material lost during dredging for cofferdam construction, creation of a temporary navigational channel, changes in the water current direction and magnitude, associated river traffic, and riverside development during the construction and operation of the new locks and dam may change the sediment deposition and erosion patterns that existed prior to construction. Changes to the sediment regime could result in water-quality degradation and habitat alteration. The guidelines in the biological opinion, set by the U.S. Fish and Wildlife Service (USFWS) to ensure protection of the mussel bed, state that sediment accumulations cannot exceed 2 centimeters above baseline levels, and if these guidelines are not met, construction activities must cease and corrective actions taken (U.S. Fish and Wildlife Service, 1993). The U.S. Geological Survey (USGS), Kentucky District, in cooperation with the Louisville District COE, has designed and implemented a multi-transducer acoustic ranging and telemetry system to continuously monitor the changes in elevation of the river bed over the mussel bed located downstream of the construction project.

Purpose and Scope

This paper describes a unique application of a multi-transducer acoustic ranging system interfaced with radio and satellite telemetry. The system determines river bed elevations collected at a remote site and sends warnings, based on predefined criteria, to multiple offices.

SYSTEM DESIGN

To ensure continued viability of the mussel beds, the COE implemented a monitoring program designed to monitor the mussel populations and record effects of project construction. The U.S. Geological Survey (USGS), Kentucky District, in cooperation with the COE, designed and implemented a system to continuously monitor the changes in elevation of the river bed over the mussel bed, located downstream of the project. The system is currently collecting baseline data. System operation is planned to continue during construction and normal lock and dam operation until such time that it is determined that the new locks and dam is not adversely affecting the mussel bed by sediment deposition and erosion.

Functional Requirements

To meet the guidelines of the biological opinion issued by the USFWS, regarding protection of the endangered mussel species (U.S. Fish and Wildlife Service, 1993), the monitoring system has to provide real-time river bed elevation measurements capable of detecting 2 cm of sediment deposition at remote locations on the Ohio River, and provide these data to personnel located in various offices. To avoid continuous manual interpretation of the data, the system should have the capability of monitoring the river bed elevation, comparing the data to criteria established by USFWS, and issuing of alarms in the event that the criteria were in jeopardy of being exceeded. The alarm feature should include the ability to call multiple phone numbers in priority order and to issue verbal warnings of the conditions detected. Dial-in capabilities are also required to allow personnel to check data and collect additional data in the event of an alarm.

Physical Configuration

Six transducers positioned in a "T" configuration (in plan view) with the base of the "T" pointing upstream are located at four sites over the mussel bed. Each leg of the "T" has two transducers equally spaced over a distance of 107 m (fig. 2). A 0.8 centimeter (cm) diameter steel cable is laid between a mooring pile at the center of the cluster and each of the transducer stands. The steel cable helps secure the transducer cable and is used as a guide for divers to locate the transducer stands. A transducer stand consists of a 15.2 cm H-pile, driven into the river bed with a 0.9 m long I-beam bolted to the H-pile, in a horizontal position about 1.2 m above the river bed and pointing upstream. The transducers are mounted to a stainless steel bracket attached to the upstream end of the I-beam. Kevlar[2] reinforced transducer cable, with mating underwater pluggable connectors, is used to connect the transducer to the surface electronics, and is attached to the 0.8 cm stainless steel cable (fig. 3). Each site has a 3 m diameter by 0.6 m high mooring buoy, with a 1 m by 1 m steel box, used to house the acoustic ranging system and telemetry equipment. Each mooring buoy is attached by a 4.4 cm steel cable to a 25.4 cm H-pile that is driven into the river bed at the center of each cluster. Each buoy is equipped with a U.S. Coast Guard approved navigation light.

Acoustic Ranging System

Monitoring changes in river bed elevation requires elevation data to be collected at numerous points over the mussel bed. Several technologies were evaluated and the acoustic ranging systems were found to be the most cost effective method of monitoring the change in river bed elevation. The equipment must be rugged and capable of providing the required measurement accuracy over a broad range of environmental conditions with minimal manual adjustment, and operate with low power consumption. To minimize damage caused by submerged debris, all electronics are located above the water surface, and only the transducers beneath the water surface. This design limited the length of cable between the transducer and electronics to a maximum of 152.4 m. The cable was reinforced with Kevlar[2] to provide strength while maintaining cable flexibility. Each transducer has a 2 m pigtail, with underwater pluggable connector, to

---

[2]The use of trade, product, industry, or firm names in this report is for descriptive or location purposes only and does not constitute endorsement of products by the U.S. Government nor impute responsibility for any present or potential effects on the natural resources.

allow underwater connection during installation and maintenance of the transducers. The connector serves as a weak link in the event that debris catches on the transducer cable. To minimize data loss, the system continues to collect data on all undamaged transducers. The minimum and maximum range of the system is important because this dictates how far above the river bed the transducers have to be mounted. The higher the transducer is mounted the more likely it is to be damaged; however, if it is mounted too low, the transducer could be buried during extreme sediment deposition and data would be lost. The transducers were mounted about 1.2 m above the river bed.

Accurate detection of 2 cm of sediment accumulation requires the acoustic ranging system to have an accuracy of plus or minus 1 cm in environmental conditions typical of the Ohio River. Most acoustic ranging systems measure the depth to the nearest object that reflects sufficient acoustic energy to exceed a predetermined energy threshold. To minimize the influence of false bottom readings resulting from fish and submerged debris, the acoustic ranging system measures the river bed elevation every 2 minutes, ranks the data from minimum to maximum for each hour, and determines the hourly minimum, maximum, and median elevations. The hourly data are transmitted to a base station for storage and retransmission; however, in the event of radio failure, the acoustic ranging system provides internal data storage for more than 3 months to ensure no data loss. A single RS-232 interface is used to download data and to change setup parameters for the instrument.

The Model PSA-902 Multi-Channel High Resolution Acoustic Ranging System by Datasonics[2], Inc., was selected for this application. The PSA-902[2] provides internal data storage using a 128K Byte memory card and includes an internal battery backup in the event of power failure. The transducers operate at a frequency of 200 kilohertz (kHz), and the system has a stated resolution of 1 cm and a range of 0.3 m to 30 m. The PSA-902[2] operates on 24 plus or minus 2 volts direct-current (VDC) at 100 milliamps and provides 12 hours of battery backup power. Communication with the PSA-902[2] is through a PC compatible RS-232 port. Custom firmware was provided to facilitate the measurement scheme required by this application.

Remote Communication and Warning System

The remote communication and warning system interfaces with the acoustic ranging system, provides redundant data storage, allows access to the data and to the acoustic ranging system by telephone, calls a prioritized list of telephone numbers in the event of an alarm, and provides routine data to various offices by Geostationary Operational Environmental Satellite (GOES). The remote units located on each buoy communicate with the acoustic ranging system through the RS-232 interface, provide backup storage of the hourly data, measure battery voltage of the power system, and communicate with the base station. The base station provides secondary backup data storage and serves as the hub for all on site communications with the monitoring system. Communication between the base station and remote units is accomplished by ultra high frequency (UHF) radios. Regular transmission of the hourly data and battery voltages to the various offices are accomplished with the GOES system. Continuous remote access to the data is provided through one telephone line while a second telephone line gives priority to outgoing messages sent by the alarm system.

Access to the system by telephone modem is controlled by three levels of access, each with a unique password. Interaction with the base station at all three levels is menu driven. The first level of access is utilized by the USGS and provides unrestricted access to all components of the system to allow remote troubleshooting and configuration. The second level is utilized by the COE and allows the user to evaluate the alarm state, view existing data, collect new data, and respond to the alarm as appropriate. The third level of access is utilized by personnel not associated with the operation and maintenance of the system and allows viewing and downloading of the stored data.

The alarm system provides 24 hour monitoring of data and alerts appropriate personnel if the data indicate that the criteria established by the USFWS is in jeopardy of being violated. In the event of an alarm, an automatic calling procedure is initiated. Telephone numbers are dialed in a priority order until a call is properly acknowledged. The acknowledgment scheme requires a specific code from the person being called. This scheme helps to alleviate problems with answering machines or other than appropriate personnel responding to the telephone alert. If false alarms are repeated, an option to temporarily turn off the alarm is provided.

The Sutron 9000[2] is an expandable multitasking Remote Terminal Unit (RTU) system designed for remote data acquisition and control and is the system selected for this application. Plug-in modules provide data collection, process control, and data communication between remote RTU's and a central site. The hardware modules have a low power consumption and are designed to operate at temperature extremes typical of the site location. The real-time multitasking operating system is programmable with a Pascal-like programming language, Sutron Data Language (SDL). Communication between the base station and remote units is provided by a half-duplex (two-way) UHF radio. An omni-directional antenna at the remote sites prevents loss of communications due to rotation and movement of the buoy. The base station is similar to the remote units except for the two telephone modem modules that provides dial-in and dial-out capabilities, and the UHF satellite radio module that provides concurrent synthesized communication capabilities with the GOES system

Power System

The sediment deposition monitoring system is powered by the combination of batteries, solar panels, voltage regulators, and direct current (DC) to DC converters. Each buoy has three 12-volt, 60 amp hour sealed lead-acid batteries connected in parallel. The batteries are recharged by two 50-watt solar panels. To ensure durability in the remote environment, solar panels constructed without glass are used. The solar panels are connected to a voltage regulator that has a blocking diode to prevent battery discharge through the solar panels. The DC to DC converter is used to convert the

12 volts from the voltage regulator to 24 volts that are needed to power the PSA-902[2]. The power system should provide 14 days of power with the batteries operating at 75 percent capacity.

SYSTEM OPERATION

The PSA-902[2] Acoustic Ranging System was installed in July 1993. Approximately 2 weeks, using a 4-person dive team, were required to install the 24 transducers and lay approximately 2,000 m of transducer cable. A Datasonics[2] field engineer was on site to verify adequate installation, program, and calibrate the PSA-902[2]. Divers verified the PSA-902[2] readings by physically measuring the distance between the transducer and the river bed. Initial software and procedural problems were quickly corrected with little loss of data. Steel plates mounted on the river bed under one transducer in each cluster are being used as quality checks on the acoustic ranging system. The Datasonics[2] field engineer and a 4-person dive team were deployed to the site in November 1993 to verify transducers yielding questionable data and to recalibrate each of the PSA-902's[2] after the transducers had become acclimated to the water. No problems attributable to the transducers or system design were evident, but four transducers yielding erratic data were found to be located nearer to the river bed than allowed by the acoustic ranging system operating specifications. This problem may result from the H-piles being driven too far into the river bed and some change to the river bed geometry. The data collected and physical measurements made by the divers indicate that the PSA-902's[2] have met the specified resolution of 1 cm (fig. 4). The only problem which has not been resolved, to date, is the four transducers mounted less than 0.3 m from the bottom. Of the 24 transducers, four are providing erratic data and four are mounted over steel plates for quality assurance verification; therefore, 16 transducers are providing good and reliable data which are being used to monitor the sedimentation conditions over the mussel bed. The internal data storage provided by the memory card has been extremely valuable and has prevented any loss of data despite radio link problems.

A Sutron 9000[2] RTU was installed at each of the four buoy sites and at the base station located at Lock and Dam No. 53 in September 1993. Although problems with the radio link, software, and satellite link have been common, the conceptual design of the system appears to provide the features required for this application. The problems identified during operation of the system are being addressed by the manufacturer, and at the writing of this paper, they have not been resolved. At this time, the PSA-902's[2] are collecting baseline data during phase I construction of the cofferdams, and the alarm system is not needed. For this reason, the alarm system software has not been loaded into the system and no information on its operational characteristics is available at this time.

SUMMARY AND CONCLUSIONS

The construction and operation of the Olmsted Locks and Dam may have an impact on two mussel beds located near the site. These mussel beds contain the endangered mussel species (Plethobasus cooperianus) or orange-footed pearly mussel. Guidelines set by the USFWS, to ensure protection of the mussel beds, state that sediment accumulations cannot exceed 2 centimeters above baseline levels. The USGS, in cooperation with the COE, has designed and installed a system to continuously monitor changes in elevation of the river bed over the mussel bed located downstream of the Olmsted Locks and Dam project. The installed system utilizes a multi-transducer acoustic ranging system to provide river bed elevation measurements accurate to plus or minus 1 cm. Data from the acoustic ranging system are transmitted to a base station by UHF radios. The base station retransmits the data every two hours on the GOES system and provides continuous access to the data by telephone modem. The base station has the capability of evaluating the data based on predefined criteria and initiating warnings based on the conditions. The system has been in operation since July 1993. Although the communication system is not working as planned, the acoustic ranging system is providing data adequate for monitoring a 2 cm change in river bed elevation. The system is currently collecting data to assess the baseline sediment deposition and erosion characteristics of the mussel bed.

ACKNOWLEDGMENTS

The authors would like to acknowledge the U.S. Army Corps of Engineers, Louisville District for the financial support of this effort, and Michael Turner, Robert Biel, Donald Griffin, and Susan Snyder of that office for their cooperation and assistance in the design and installation of this system.

REFERENCES CITED

Thorp, J.H., and Covich, A.P., eds., 1991, Ecology and classification of north american freshwater invertebrates: Academic Press, Inc., New York, 911 p.

U.S. Fish and Wildlife Service, 1993, Supplemental biological opinion for the proposed Olmsted Lock and Dam Ballard County, Kentucky: Cookeville, Tenn., 15 p.

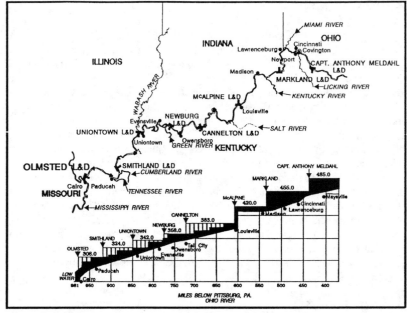

Figure 1. New high lift dams on the Ohio River.

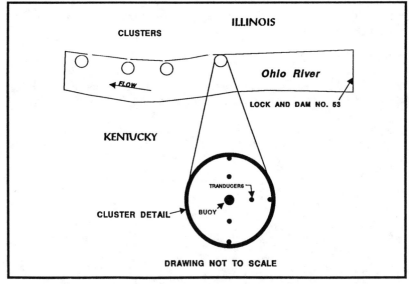

Figure 2. Physical configuration--Plan view.

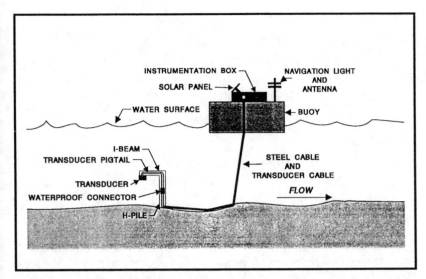

Figure 3. Physical configuration--Profile view.

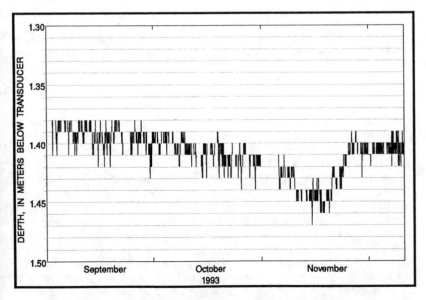

Figure 4. PSA-902 median hourly readings from transducer 2 at buoy 3.

Model-Prototype Conformance of a Submerged Vortex in the Intake of a Vertical Turbine Pump

by K. Warren Frizell A.M. ASCE[1]

Abstract: Submerged vortices are commonly found in sump intakes of vertical turbine pumps. These vortices are known to cause increased vibration and at times degrade pump performance. Approach flow conditions and sump and pump bell geometry are very critical in the formation of submerged vortices. Scale model tests are a common method used to evaluate the presence of free-surface and submerged vortices in pumping plants. While scale effects in reproducing free-surface vortices are fairly well documented, scale effects involving modeling of submerged vortices are somewhat more unknown due to the difficultly in actually observing a prototype vortex.

This paper will discuss the comparison between measurements and observations of a 1:8.74 geometric scale model of Twin Peaks Pumping Plant and the prototype operation.

Introduction

Twin Peaks Pumping Plant is located near Tucson, Arizona and is part of the Central Arizona Project. The plant contains six vertical turbine single-stage pumps which lift project water about 23 m. Rated capacity at rated head is 17.58 m^3/s. The plant contains three 4.25-m^3/s units in individual sumps and three 1.42-m^3/s pumps in a common sump. Upon initial startup of the plant, the larger pumps were undercapacity when compared to the shop tests. One suspected reason for this deficiency was the possibility that submerged vortices were present and entering the pump bells. Initial testing at Twin Peaks confirmed the presence of a submerged vortex attached to the floor and entering the pump bell (Nystrom, 1992). These tests included visualizing the vortex by injecting compressed air near the vortex location and recording it on an underwater video camera. Other instrumentation, accelerometers and proximity sensors, verified that there was a component of vibration at the blade-passing frequency which would be characteristic of a vortex

---

[1] Research Hydraulic Engineer, Bureau of Reclamation, Denver, CO 80225.

entering the pump bell; however, the magnitude was less than 10 percent of the component present at the rotational frequency.

The Model

As a result of these initial observations, a 1:8.74 geometric scale model of the Twin Peaks Pumping Plant was constructed in Reclamation's Denver hydraulic laboratory. The model included a portion of the canal, the inlet transition, and the pump sumps, figure 1. The pump bells and columns were constructed from acrylic and connected to PVC piping. The outlets were manifolded with the pipeline forming a closed loop system. A 11.2-kW centrifugal pump located inline provided the recirculation and

Figure 1.- 1:8.74 scale model of Twin Peaks Pumping Plant constructed in Reclamations' hydraulic laboratory.

an orifice-venturi meter was used to measure discharges. Each pump column was equipped with an elbow meter so that individual unit flows could be adjusted when more than one unit was operating simultaneously.

Model similitude was based on equal Froude numbers in the model and prototype. The model scale was chosen to ensure that viscous and surface tension effects were kept to a minimum. The critical Reynolds number, based on the velocity of approach and the incoming depth, suggested by Knauss (1987) was matched.

Measurements

The Model: In addition to making visual observations with air bubbles and dyes as tracers, we made several different measurements to help us evaluate the sump performance. These measurements included velocity profiles in the sumps approaching the pumps, pre-rotation in the pump column, and velocity profiles in the pump column. The sump velocity profiles were taken with either an Ott propeller-type meter or a two-component Marsh-McBirney electromagnetic velocity meter. The pre-rotation measurements were taken with a vortimeter. This meter measured tangential velocity in the pump column, assuming solid body rotation (Lee and Durgin, 1980). The velocity profiles in the pump columns at the impeller location were measured using a standard pitot-static tube, traversed across the column on lines at 45° increments.

The Prototype: The main measurements used in evaluating the prototype performance were underwater video recordings. These video images were taken with a Deepsea Power & Light FM-1000 CCD camera which had a fixed focus and remotely controlled panning features. Compressed air was injected from a manifold mounted below the pump bell in order to visualize the vortex. In addition to these observations, we also measured runout at the lower bowl bearing with proximity sensors and pressure pulsations in the discharge line with a pressure transducer. Pump discharges were verified with a permanently installed ultrasonic flowmeter.

Results

Submerged vortices were present in all model sumps with every operating sequence tested. These vortices were only visualized with the aid of tracers (air bubble and dye). The vortex was not strong enough to pull vapor bubbles out of solution. The velocity measurements taken in the sumps indicated that the approach flows to the pump bells were skewed to one side, figure 2. The resulting velocity profile in the pump column showed this skewness also, figure 3. Swirl measurements showed inconsequential amounts of swirl (solid body rotation) present in the pump column. This corresponds to the thin vortex (3mm- to 5-mm diameter) present in the model, figure 4.

Results from the prototype are mostly in the form of underwater video footage. As in the model, we had to use a tracer (compressed air) in order to visualize the vortex in the prototype. This would indicate that the vortex strength is similar to that observed in the model. With air injected into the vortex core, it appeared to be between 25-mm and 75-mm in diameter. Shaft runout at the lower pump bowl bearing and pressure pulsations in the discharge line did not indicate significant vibration due to the presence of the submerged vortex.

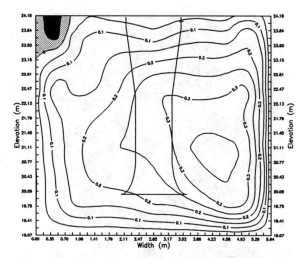

Figure 2: Elevation view of velocity profile in sump No. 1, one pump bell diameter in front of the pump bell, minimum water surface elevation, Q=4.39 m³/s, velocity in m/s.

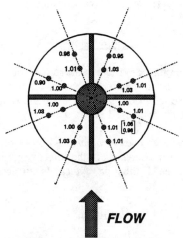

Figure 3: Normalized velocity profile in pump column, original configuration, Q=4.39 m³/s, values in parentheses show range of data due to fluctuations caused by the submerged vortex.

Figure 4: Vortex attached to floor cone. Air injected to enable visualization.

Modifications

After verification of the prototype vortices in the scale model - including similar size and strength - we proceeded to modify the model sump to eliminate the vortex. After many iterations, we installed a series of floor and wall splitters which were successful in eliminating all submerged vortex action, figure 5.

This arrangement is similar to ones recommended by other researchers (Nakato, 1989; Prosser, 1977). The measurement of the velocity profile in the pump column showed a much more even distribution, typical of an acceptable radial inflow pattern, figure 6.

Verification

The sump modifications developed in the model were installed in sump No. 1 at Twin Peaks Pumping Plant. The video taping and additional measurements were repeated. No visual evidence of a submerged vortex was observed with the underwater video, with or without the compressed air tracer. Runouts and pressure pulsations were essentially unchanged from the previous tests. The discharge capacity and efficiency did not change from the values measured during the previous test with the original sump design.

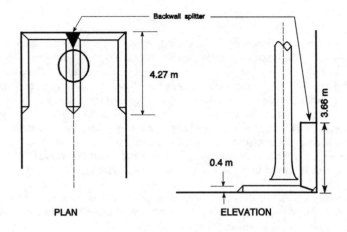

Figure 5: Final sump modification which eliminated all submerged vortices.

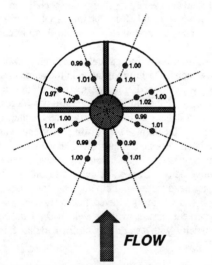

Figure 6: Normalized velocity profile in pump column. Normalizing velocity is the average of all point measurements.

Discussion

Submerged vortices (floor-attached), directly beneath the pump bell were verified in all sumps. This type of vortex is usually generated by local flow conditions near the pump bell. Considering the uneven velocity distributions approaching the pump bell and that several of the sump dimensions (based on the supplied pump bell diameter) were outside of the recommended guidelines, the conditions were prime for submerged vortex formation (Triplett,1988; Sweeney, 1979). Very little information is available about scale effects in modeling submerged vortices because of the difficultly in actually visualizing prototype vortices. In this case our ability to observe the prototype vortex through underwater video and actually compare it to the vortex observed in the model was especially meaningful.

In the past, many techniques have been used to assist in the modelling of vortices. In general, the current feeling is that a model should be of a large enough scale that the Reynolds numbers remain large enough ($Re > 10^5$) that viscous effects are negligible. Many researchers distorted the discharges up to 1.5 times the scaled discharge to aid in the production of vortices in models. However, the majority of present-day studies have abandoned this technique, especially on larger scale models, because of the distortion which results from the increased flows incorrectly modeling the streamlines and shear zones around and near the sump boundaries.

Perhaps the most meaningful of the model measurements in evaluating sump performance were the velocity distributions taken in the pump column at the impeller location. As previously mentioned, the pre-rotation at this location was small, having a swirl angle of only 2° maximum. Observations showed that the submerged vortex does not appear to have a significant influence on the large-scale swirl (or pre-rotation) at the impeller location. Drop off in performance or insufficient pump capacity is generally related to large scale swirl. The velocity distribution measured for the original design indicated that point velocity measurements deviated up to 10 percent from the mean velocity. Uneven velocity distributions have been shown to be consistent with higher head losses in the pump bell and the presence of vorticity. The final configuration (no vortices) showed that all velocity measurements were between -3 and +2 percent of the average velocity. This uniform distribution indicates a lack of vorticity with essentially uniform radial flow approaching the impeller location.

The data collected during the prototype testing proved to be very valuable. In general, pump shaft runout at the lower bowl bearing and pressure pulsations in the discharge line did not show any sensitivity to the presence of a submerged vortex. In addition, pump discharge capacity also remains unchanged. However, the video tape revealed that, indeed, the sump modifications were successful in eliminating all vortex activity. In addition, size and strength of the vortex (prior to the sump modifications) seemed to scale quite well. The sizes were estimated from the video footage and the prototype vortex appeared to be 8 to 10 times larger (core diameter) than what we observed in the model. The only true measure of vortex strength is a measure of the minimum pressure in the core. While we were not able to measure this, neither the model or the prototype vortex were strong enough to pull air/vapor out of solution. In both cases we had to use a tracer (air bubbles or dyes) in order to visualize the submerged vortex.

Through our ability to compare relative size and strength of submerged vortices from a model to a prototype facility, we can have increased confidence that if current state-of-the-art practices in modeling pump/sump intakes are followed, the results will be representative and free from scale effects.

References

Lee, H.L. and W.W. Durgin, "The Performance of Crossed-vane Swirl Meters," *Vortex Flows*, The American Society of Mechanical Engineers, edited by W.L. Swift, P.S. Barna, and C. Dalton, 1980.

Nakato, Tatsuaki, "A Hydraulic Model Study of the Circulating-water Pump-intake Structure: Laguna Verde Nuclear Power Station Unit No. 1, Comision Federal de Electricidad (CFE)," IIHR Report No. 330, Iowa Institute of Hydraulic Research, The University of Iowa, May 1989.

Nystrom, James B., "Tracer Dilution Flow, Sound, Vibration, and Video Measurements at Twin Peaks Pumping Plant," 151-92/M198F, Alden Research Laboratory, Inc., September 1992.

Prosser, M.J., "The Hydraulic Design of Pump Sumps and Intakes," British Hydromechanics Research Association and Construction Industry Research and Information Association, July 1977.

Sweeney, C.E., R.A. Elder and D. Hay, "Pump Sump Design Experience: Summary," *Journal of the Hydraulics Division, ASCE*, Vol. 105, No. HY9, September, 1979, pp. 1053-1063.

Swirling Flow Problems at Intakes, IAHR Hydraulic Structures Design Manual No. 1, Coordinator-editor Jost Knauss, A.A. Balkema, Rotterdam, 1987.

Triplett, G.R. et al., "Pumping Station Inflow-Discharge Hydraulics, Generalized Pump Sump Research Study," Technical Report HL-88-2, Department of the Army, Waterway Experiment Station, February 1988.

Disclaimer: The information contained in this paper regarding commercial products or firms may not be used for advertising or promotional purposes and is not to be construed an an endorsement of any product or firm by the Bureau of Reclamation.

SIMULTANEOUS FLOW VISUALIZATION AND HOT−FILM MEASUREMENTS

Fabián López[1], Yarko Niño[1], and Marcelo García[2], A.M. ASCE

Abstract

High−speed video recording of flow visualizations in the near wall region of a turbulent open channel flow was synchronized with hot−film measurements of flow velocity and bed shear stress probes. Analysis of the video images provided information about coherent flow structures associated with the occurrence of ejections near the wall. These structures consisted mainly of oscillating shear layers that were convected in the downstream direction and lifted away from the wall. A visual detection criterion was used to obtain ensemble averages of the velocity and shear stress data during ejection events. Preliminary results obtained from the application of cross−correlation and VITA analyses of the hot−film data were compared with those obtained from the visual analysis, and a general good agreement was found. Further analysis is required to define better criteria to set the parameters required by the VITA algorithm, and also to obtain a better understanding on how wall shear stress and velocity signals relate to the observed coherent structures in the wall region.

1. Introduction

The general structure of turbulence reflects the local balance between production, transport, and dissipation of turbulent kinetic energy. In turbulent boundary layers the primary role is played by the production term, which is responsible for the maintenance of the turbulent state, counteracting the effects of dissipation. Since the earlier landmark experimental work of Kline et al. (1967), several experimental as well as numerical results have supported the concept of a coherent structure of turbulent boundary layers. The majority of the turbulence production appears to occur in the buffer region during intermittent or quasi periodic events called bursts, which are characterized by violent outwards ejections

---

[1] Research Assistant, [2] Assistant Professor. Department of Civil Engineering, University of Illinois at Urbana Champaign. 205 N. Mathews, Urbana, IL 61801. USA.

of near−wall fluid (associated with negative fluctuations of the wall shear stress) and inrushes of high speed fluid towards the wall (associated with positive fluctuations of wall shear stress). At the same time, the streamwise velocity field in the viscous−dominated wall region appears to be organized into alternating narrow streaks of high− and low−speed fluid, which are quite persistent in time.

The interest in studying the coherent structure of the turbulent boundary does not come only from a purely fluid mechanics point of view. In fact, there is sufficient evidence supporting the idea that turbulent bursting would have a major effect over sediment transport processes, such as entrainment and transport of particles in the near wall region of a turbulent open channel flow (Niño et al, 1994). From this point of view the study of flow ejections in the wall region appears to be an important topic, related to the particle entrainment mechanism, which requires much research, particularly in open−channel flows.

Flow visualizations have played a major role in the study of the coherent structure of wall turbulence. In general, flow visualization methods offer much higher information density than single point probes, however, because of the rapid dispersion of marker patterns at high flow velocities, they are limited to relatively low Reynolds numbers. Besides, three−dimensional vortices are extremely difficult to characterize in the laboratory, which makes the conclusions obtained from visual studies partially ambiguous. An alternative approach has been the use of statistical analysis techniques in order to detect and characterize the structure of organized turbulent motion from time series of measured turbulent quantities. A number of conditional−sampling techniques oriented to detect coherent structures from single probe measurements have been developed with this aim (see Luchik and Tiederman, 1987, for a brief review). However, those techniques always involve the use of certain calibration parameters that need to be set somehow arbitrarily. Because of this, results obtained from the analysis of probe data do not usually agree with those inferred from visual studies.

The experimental work reported herein aims at reconciling results obtained using flow visualization and single probe data acquisition in the study of coherent structures in wall turbulence. Images of a tracer moving in the wall region of an open−channel turbulent flow taken with a high speed video system were synchronized, and simultaneously recorded with the signals from standard hot−film and flush−mounted sensors. This study represents a novel effort which could lead to define better criteria to set the parameters required by structure−detection algorithms, and also to obtain a better understanding on how wall shear stress and velocity signals relate to the observed coherent structures in the wall region.

2. Experiments

Facilities

The experiments were carried out in an open channel, 18.6 m long, 0.297 m wide, and 0.279 m high. The slope of the channel was set to a value of 0.0009. The test section was located about 12 m downstream from the entrance, it was about 0.9 m long and had the right wall made of plexiglass to allow for visualization studies. Two different hot−film probes were used in the study, namely, a ruggedized hemisphere velocity probe and a shear stress probe mounted flush to the bottom surface of the channel. Data acquisition was made using an A/D card connected to a personal computer and appropriate software. The equipment used had the capability of digitizing two channels simultaneously with different triggering options.

A high−speed video recording system Kodak Ektapro TR Motion Analyzer was also used in the study. The system has the capability to record up to 1000 frames per second. This equipment, together with proper strobe−light illumination, was employed in order to visualize coherent structures in the near wall region of the open channel flow in study. The video system outputs a synchronization signal, which was employed to trigger the acquisition of the probe data simultaneously with the video recording. The video images were digitized into a personal computer using a frame grabber, and analyzed with the help of the National Institute of Health's Image public domain software.

Experimental Method

Signals from velocity and shear stress probes were recorded simultaneously with flow visualization images taken with the high−speed video system. The velocity probe, which measured the streamwise velocity component, was located at a height of 2.5 mm from the channel bottom and at a distance of 10 mm downstream from the shear stress probe. Preliminary results indicate that this distance maximizes the cross−correlation between velocity and shear stress signals for the range of Reynolds numbers covered herein. A solution of white clay and water was injected through an orifice in the channel bottom located 30 mm upstream of the shear stress sensor to act as a marker for flow structures developing at the wall. The dye discharge was controlled as to minimize disturbance of the flow, and to allow the tracer to displace attached to the wall before flow ejections lifted dye filaments away from it. This technique allows to visualize flow ejections fairly well, however it does not provide visual information about the flow structure in the region above the ejection, nor it is efficient in marking sweeps events.

Hot−film data were acquired at a rate of 500 Hz. This satisfies Nyquist's criteria, given the maximum frequencies of about 200 Hz and 90 Hz of the velocity and shear stress signals, respectively. Each data series had a total of 15000 data points. Video images were recorded at 500 frames per seconds, hence there is a direct correspondence between data points and video frames. Video recordings had a duration of about 20 sec, which gives a number of about 10000 frames per experiment.

Experimental Conditions

Experiments were carried out under uniform flow conditions, in a channel with hydraulically smooth walls. Data corresponding to two different Reynolds numbers are reported in this paper. The experimental conditions are shown in Table 1, where h denotes flow depth, U_m denotes flow mean velocity, U_* denotes shear velocity, and Re denotes flow Reynolds number defined as

Re = $U_m h/\nu$, with ν denoting water kinematic viscosity. The shear velocity U_* was estimated from a best fit of measured mean velocity profiles to the semi−logarithmic law.

TABLE 1 Experimental Conditions

h (m)	U_m (m/s)	U_* (m/s)	Re
0.023	0.231	0.016	5300
0.036	0.344	0.020	12400

3. Method of analysis

Different analyses of the probe and video data obtained were performed. Standard statistical analysis of the velocity and shear stress time series provided mean, standard deviation, and higher order moments of those series. Also, cross−correlation analyses and the VITA (variable−interval time average) pattern recognition algorithm (Blackwelder and Kaplan, 1976) were applied to velocity and shear stress series with the aim of detecting and characterizing coherent structures present in the near wall region of the flow.

Analysis of the recorded video images provided information about the geometry and convective velocity of the flow structures observed. These structures consisted mainly of shear layers as described by Thomas and Bull (1983), resulting from flow ejections away from the wall. A visual criterion was used to obtain ensemble averages of the velocity and shear stress signals during ejection events observed in the video recordings. Taylor's frozen turbulence hypothesis together with measured convective velocities were used in order to transform temporal data into spatial data, which allowed estimation of the flow velocity structure associated with the shear layers observed.

4. Results

As an example of the results obtained, Fig. 1a shows a sequence of images of a coherent structure, in the form of a shear layer, being convected in the downstream direction and lifted away from the wall. The images are 0.04 sec apart, and Re=5300. The velocity probe (marked P1) can be easily seen at the right end of the images. The shear stress probe can not be seen but its location has been marked P2 in the upper image. Fig. 1b shows digitized shapes of the shear layer as it develops in time and space (structures marked SL1, SL2 and SL3 correspond to those of Fig. 1a), with spatial dimensions, X_+, Y_+, given in wall units (i.e., made dimensionless with the length scale ν/U_*). Values of velocity and shear stress fluctuations (u, τ), measured during the passage of the flow structure, are also shown in Fig. 1b, made dimensionless with corresponding standard deviations (u_{rms} and τ_{rms}), and plotted as a function of t_+, time measured in wall units (i.e., made dimensionless with the time scale ν/U_*^2). Time t_+ and streamwise coordinate X_+ have been scaled in Fig. 1, so as to make them correspond when the structure crosses the velocity probe level (marked P1), by using the observed mean

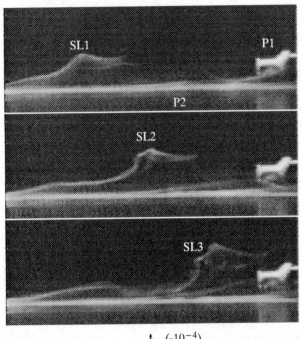

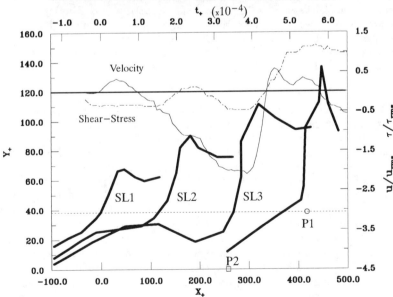

Figure 1. (a) Video Images ; (b) Digitized Images and Acquired Data

convective velocity (U_{mc}) of each flow structure. Mean values of dimensionless convective velocities computed from the digitized structures were $U_{mc}/U_* = 9.7$ for Re = 5300, and $U_{mc}/U_* = 9.1$ for Re = 12400. These values agree fairly well with reported experimental observations as well as with results from numerical simulations for propagation velocities of coherent structures in channel flows (Guezennec et al, 1989).

Figs. 2 and 3 show ensemble averaged values of dimensionless velocity fluctuations during the passage of several different coherent structures, for values of Re of 5300 and 12400, respectively. Upper and lower bounds of the velocity signal are also shown. Temporal probe data was converted to spatial data by introducing the frozen turbulence hypothesis and observed convective velocities of the flow structures. The zero of the X_+ coordinate corresponds to the location where the flow structure intersects the level of the velocity probe. In those figures, the shape of the observed shear layers is also plotted.

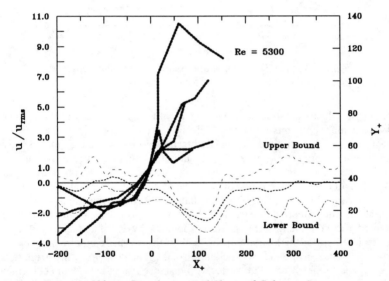

Figure 2. Velocity Distribution and Shape of Coherent Structures
Re = 5300

From the results shown in Figs. 2 and 3, the velocity structure of the flow in the vicinity of the coherent structure can be inferred. An average shear layer is plotted in Fig. 4, together with the ensemble averaged velocities corresponding to the values of Re of 5300 and 12400. As it is observed therein, the streamwise fluctuating velocity has a negative peak of about twice the value of the standard deviation of the signal, which occurs at a distance of about 100 to 200 wall units downstream from the structure. Supposedly, this negative peak of the

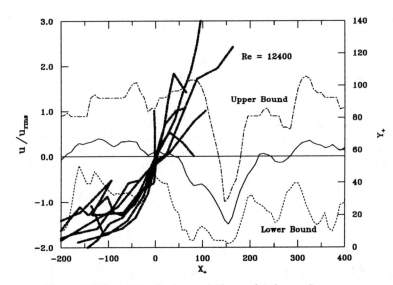

Figure 3. Velocity Distribution and Shape of Coherent Structures
Re = 5300

streamwise velocity component is associated with a positive peak of the vertical velocity component, thus defining an ejection event. Also, if we assume that Reynolds number variations are negligible under proper scaling (in this case scaling with wall units is proposed), the results shown in Fig. 4 would give an idea of the vertical structure of the streamwise velocity field in the vicinity of the shear layer. Apparently, the magnitude of the negative peak of the velocity decreases as Y_+ increases, which seems to indicate that the ejection event tends to be less intense in the upper part of the structure. It is worth to note that similar results were obtained by conditionally averaging results obtained by the VITA technique. Also, the displacement of the location of the peak in the downstream direction as Y_+ increases seems to respond to the inclination of the flow structure. This inclination has an angle to the bottom that varies with distance from the wall and coincides with average reported values between 8° (Rajagopalan and Antonia, 1979) and 18° (Brown and Thomas, 1977).

The inclination of the coherent structures became also evident from space–time correlations between velocity and wall shear–stress fluctuations:

$$R_{u\tau} = \frac{<u(t+\Delta t)\,\tau(t)>}{u_{rms}\,\tau_{rms}} \quad (1)$$

where the angular brackets stand for ensemble averages. Figure 5. illustrates the variation of the cross–correlation with time–lag in wall units, for a separation of 10 mm between both sensors (sensor P2 upstream) and Re = 12400. The negative

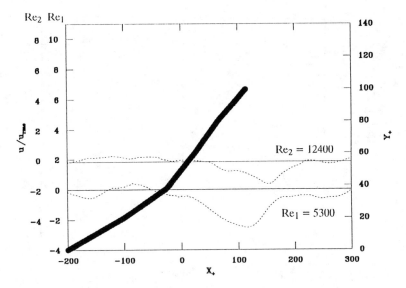

Figure 4. Associated Shear Layer with Average Velocity Distribution

value of Δt corresponding to the peak of $R_{u\tau}$ clearly indicates that, in the average, the sensor P1 detects the occurrence of a coherent structure before it crosses sensor P2, implying that those structures are inclined to the bottom. The numerical value of the peak of $R_{u\tau}$ agrees well with those observed by Rajagopalan and Antonia (1979).

The analysis of the video images indicated that as the value of Re increases, the occurrence of multiple flow structures during one ejection event (or multiple ejections within a burst) becomes more common. This was also observed in the velocity and shear stress data. An example of a multiple event detected in the velocity time series corresponding to Re=12400 is shown in Fig. 6. Three negative peaks are observed therein, which are associated with three different structures being convected one after the other through the probe. The analysis of the video images corresponding to this event revealed that, indeed three different shear layers developed and interacted during this particular event, as they were convected downstream. The interaction between structures was observed to give rise to flow patterns that are much more complicated than the clear shear layers observed at lower values of Re.

5. Conclusions

Simultaneous flow visualization and hot−film data acquisition tecnique allowed to obtain valuable information about the main characteristics of turbulent coherent structures in an open channel flow. Improving the present experimental set up (for example by using a hydrogen−bubble generator) could

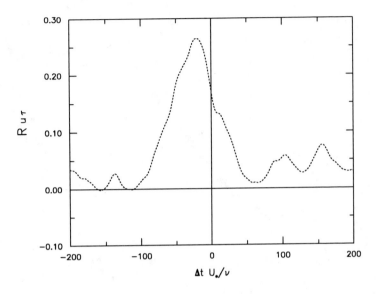

Figure 5. Cross−Correlation between Streamwise Velocity and Bed−Shear Stress Fluctuations − $\Delta X = 10$ mm

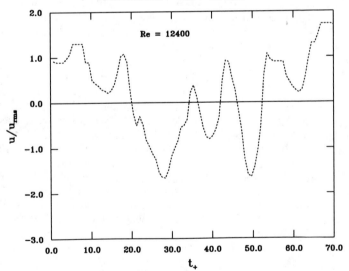

Figure 6. Velocity Distribution in the Presence of Several Ejections

lead to the acquisition of more data concerning the velocity field and the related sweep events. Further analysis is required, which would lead to define better criteria to set the parameters required by the VITA algorithm, and also to obtain a better understanding on how wall shear stress and velocity signals relate to the observed coherent structures in the wall region.

6. Acknowledgments

The support of the Fluid, Hydraulic, and Particulate System Program of the National Science Foundation (Grant CTS−9210211) is gratefully acknowledged

7. References

BLACKWELDER R.F. and R.E. KAPLAN, 1976, *On the wall structure of the turbulent boundary layer.* J. Fluid Mech., 76, pp 89–112

BROWN G.L. and THOMAS A.S.W., 1977, *Large Structure in a Turbulent Boundary Layer.* Phys. Fluids, 20, 10, Pt II, pp S243–S252.

GUEZENNEC Y.G., PIOMELLI U. and KIM J., 1989, *On the Shape and Dynamics of Wall Structures in Turbulent Channel Flow.* Phys. Fluids A 1(4), pp 764–766.

KLINE S.J. et al, 1967, *The Structure of Turbulent Boundary Layers*, J. Fluid Mech. 30, pp 741–773.

LUCHIK, T.S. and TIEDERMAN, W.G., 1987, *Timescale and Structure of Ejections and Bursts in Turbulent Channel Flow.* J. Fluid Mech. 174, pp 529–552.

NIÑO Y. and LOPEZ F., GARCIA, M.H., 1994, *High–Speed Video Analysis of Sediment–Turbulence Interaction.* ASCE Symposium on Fundamentals and Advancements in Hydraulic Measurements and Experimentation. Buffalo, NY.

RAJAGOPALAN S. and ANTONIA R.A., 1979, *Some Properties of the Large Structure in a Fully Developed Turbulent Duct Flow.* Phys. Fluids, 22(4), pp. 614–622.

THOMAS A.S.W. and BULL M.K., 1983, *On the Role of Wall–Pressure Fluctuations in Deterministic Motions in the Turbulent Boundary Layer.* J. Fluid Mech. 128, pp 283–322.

Non-Intrusive Experimental Setup for Inflatable Dam Models

T. A. Economides[1] and D. A. Walker[2]

ABSTRACT

Experimental research on scaled models of inflatable dams has challenges that are traditional to flow-induced vibrations, and other challenges that are related to inflatable structures. A new method for quantifying the dynamic response of inflatable dam models is presented. The method is used for overflow and impounding water conditions. There is no interference to the motion of either structure or fluid. Internal pressure variations are measured instead of membrane stresses or membrane displacements. The response of pressure transducers located inside the dam, is successfully compared with the response of transducers outside the dam connected with a 10cm long tubing of 0.8mm diameter. Pressure variation voltage signals lower than $10\mu V$ are identified based on the instrumentation presented. A specialized timer is also presented, that aids to a non-intrusive validation of the primary oscillations of the structure under overflow conditions, using video images.

Upstream velocity records are based on the channel's flowrate measurements and the upstream water depth readings. Basic principles are employed with the aid of a miniature camera, a TV monitor, and the water's surface tension forces. The result is a 0.013mm (0.0005 inches) accuracy on the upstream water depth, with the advantage of remote operation. Flow visualization at the upstream side of the dam is also presented, based on Helium-Neon laser light sheets and naturally entrapped air-bubbles and water-borne debris. The captured image is digitized and enhanced. Typical response measurements with respect to the upstream water depth and average internal pressure, are also presented.

---

[1] Post-doctoral research associate at the Department of Engineering Science and Mechanics and the Department of Civil Engineering, Virginia Polytechnic Institute and State University, Blacksburg, VA 24061-0219.

[2] Adjunct Professor of the Department of Engineering Science and Mechanics of Virginia Polytechnic Institute and State University, and at Paracelsus Technology Corp., 254 Shakespeare Court, Severna Park, Maryland 21146.

INTRODUCTION

Inflatable dams were invented in 1956 as a flow-control hydraulic structure, to divert the seasonal waters of the Los Angeles River (1957) for groundwater recharging (Firestone 1968). It was necessary that the hydraulic structure used would function as a gate that would divert the water or allow free escape. The dams can be simply described as cylindrical membranes lengthwise attached on a single or double anchorage line (Figure 1). At their early days it was popular to use water for

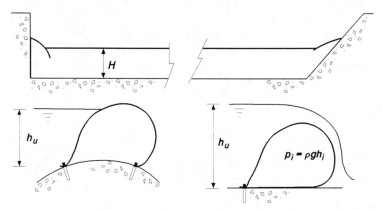

Fig. 1 Typical inflatable dams with vertical or inclined abutments, and double or single anchorage systems.

inflation media, but nowadays air-inflation is predominately used. They can be anchored on a reinforced concrete foundation, or existing concrete structures like concrete dams. They are usually used in irrigation projects as storage dams.

Experimental and analytical investigation of these structures started appearing as refereed publications in the mid-sixties (Baker *et al.* 1965). A considerable volume of references is now available, including some work on their dynamic response. So far, all the previous experimental work done involved instrumentation and sensors that intruded to the structure or the water flow, or both. Examples of such practices include strain gages, point gages, and classical flow velocity measurement instruments.

The response of inflatable dams under overflow conditions, involves the consideration of the *upstream water head*, h_u, the *internal pressure head* of the dam, h_i, and the maximum upstream water head without overflow or the *maximum static head* possible, h_s. It is clear that the maximum static head is the height of the dam, H, thus the two quantities can be used interchangeably. Oscillations resulting in the dams' failure, can occur at the right combination of the parameters mentioned. Membrane characteristics are also very important but not critical.

Flow separation effects at the downstream side of the dam are responsible for the vibration problems of inflatable dams. Whether these effects drive or just perturb the dam to unstable behavior depends on the parameters set earlier. An instrumentation practice that will precondition the flow field due to flow-intrusion, or localized stiffening effects on the dam's membrane, can predetermine the behavior of the models at the onset of their vibrations. It is of interest to map the behavior of inflatable dams with these parameters. In doing so, it is found that significant changes of behavior occur at slight parametric variations (Economides 1993). Therefore, precision in modeling practice could determine the ability to detect certain patterns of behavior.

RESPONSE MEASUREMENTS

Traditionally, the response of a structure is measured in terms of displacement, velocity, acceleration, and strain. Exploring the options available for this problem it is very difficult (or relatively expensive) to find a sensor or measurement method that will be non-intrusive to both flow and structure.

Two non-intrusive instruments were developed, one based on Infra-Red optoelectronic design and the other on Hall-effect sensors (Economides 1993). Both instruments, however, have the traditional problem of any simultaneous multiple-point-measurement technique. It is therefore necessary to provide a measured quantity which is inclusive of all the response characteristics of inflatable dams, without regard to the instrumentation positioning. The internal pressure variations of the dam provide this unique type of measured response.

Internal Pressure Variations

The internal pressure of inflatable structures is identified to be the most important parameter of their structural stiffness. Under an excitation load, the internal pressure variation is representative of this load as well as the response of the structure, similarly to the membrane strains. In addition, the internal pressure response is insensitive to nodes of structural modes of vibration. The result is a continuous spectrum of vibration frequencies limited only by the data sampling rate, as opposed to traditional measurement methods with positional and directional limitations resulting in inhibiting relevant frequency component(s).

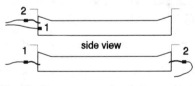

Fig. 2 Pressure transducers.

Two pressure transducers are used in two different formations. The first case has both transducers located at the same end. The one transducer is placed inside the dam and the other outside the dam connected with a 0.8mm by 10cm tubing (Figure 2). This setup is used to compare the two signals for possible phase differences or dampening effects, a possible source of error. In the second setup, each transducer is located at each of the dam's ends, outside the dam, connected with a 0.8mm by 10cm tubing. The two signals are compared for possible phase difference, again, which could be interpreted as an internal pressure-wave propagation along the dam. In this respect, wave surging and longitudinal snaking have been reported on prototype dams (Binnie *et al.* 1973).

Variations of internal pressure are considerably smaller compared with the absolute average. This results in problems related to voltage resolution and signal

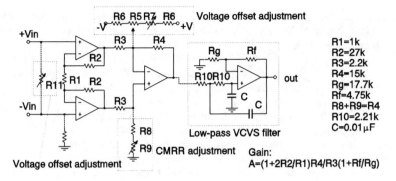

Fig. 3 Circuit schematics of the large-voltage-offset instrumentation amplifier with the a low-pass filter.

amplification. A large-voltage-offset instrumentation amplifier is used for this purpose (Figure 3). The output signal from the pressure transducers is connected at the Voltage-In (±Vin) inputs of the instrumentation amplifier. Then, the signal (approximately amplified by 500) is further conditioned with a dynamically correct low-pass filter, to limit the noise inherently generated by the pressure transducers (5μV typical). This filter is a VCVS (voltage controlled voltage source) Sallen and Key type, with a second order Bessel function approximation. This is a typical

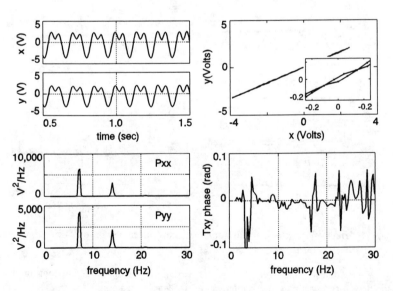

Fig. 4 Internal (x) versus external (y) pressure transducers response of inflatable dam's at high overflow conditions ($h_u/H=1.55$, $h_i/h_u=2.38$). [C57]

instrumentation amplifier based on the TL074 quad op-amps which have comperable characteristics with the LF412. Many other operational amplifiers are suitable for such an instrumentation.

The typical effect of the extension tubing on the pressure transducers output is shown in Figure 4. The case presented in this graph showed very high cross-sectional vibrations of the dam. The *aspect ratio*, L/H, of the model used in Figure 4 was 8.4. There is some phase difference between the two signals (x and y), up to 5 degrees (0.09 rad). However, at the vicinity of the frequency values of interest the phase difference diminishes. The frequency components of the two signals are exactly the same, which is the primary concern of this investigation. The power spectral energy (V^2/Hz) is reduced to 2/3 of the actual signal (x).

Significant phase differences between the dam's ends are shown when longitudinal vibrations occur (Figure 5). Comparing the two graphs (in Figures 4 and 5), it is clear that the phase difference between the two signals in this case is beyond any error introduced by the tubing extension. Displacement levels are small compared with the previous case (Figure 4) as well as the internal pressure variations. Orbital formations of xy plots are typical of 3D oscillations in inflatable dams. These become more distinct when water-filled dams are used instead of the traditional air-inflated.

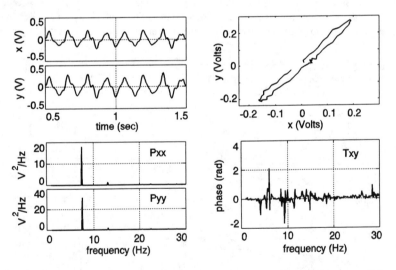

Fig. 5 Effect of longitudinal vibrations (snaking) at moderate overflow conditions ($h_u/H = 1.36$, $h_i/h_u = 1.06$). [C249]

Evidently, the internal pressure variations are an effective measurement method for describing the dynamics of inflatable structures in general. In this problem, the non-intrusive characteristic of the method is the most important feature. With the use of internal pressure variations, conclusive results are relatively straight forward. This is demonstrated in Figure 6 where a critical *pressure ratio*, h_i/h_u, is immediately visible at approximately $h_i/h_u = 1.15$. In addition, it shows the effect of the Reynolds number to the energy of oscillations. The Froude number shows comparable correlation and it should be used for scale-up of the results.

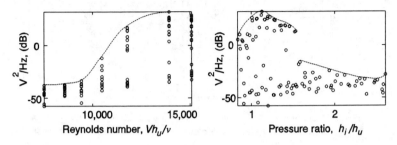

Fig. 6 Collective response characteristics of the power spectral density (Volts²/Hz), to the Reynolds number ($\mathbf{Re} = Vh_u/\nu$), and the Pressure ratio (h_i/h_u). [C151-254]

Video Images

Video movies of the vibrating dam provide useful information about the main vibration frequency of the dam. This process is time-consuming and not very accurate for computing the frequency of oscillation. It is mainly used here to verify the primary vibration frequency detected by other measurement methods. This setup used a regular VHS video camera (29.9 frames per second) with close-up filters when necessary. A video-cassette player with slow motion capability for a noise-free screen frame-by-frame playback is used.

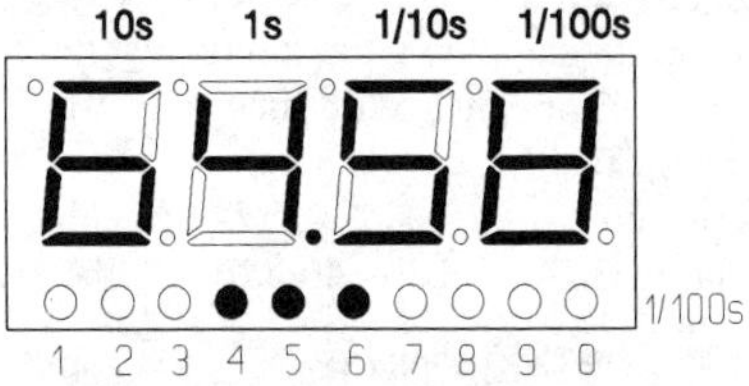

The time accuracy of the events is enhanced by using an independent timer at the scene of the events. This timer features 7-segment LED (light emitting diodes) numbers and a 10-LED bar display. It is designed to have a resolution of 1/100 to 1/200 of a second, with the ability for much higher resolutions with the addition of extra digit modules. The principle of operation is based on the fact that the camera used here, records three consecutive 1/100s digits in a single frame, rendering the display unreadable. This is made clear in Figure 7, showing the effect of the numbers 4, 5, and 6 overlapping on the timer's 1/100s digit. The 10-LED bar display which is synchronized with the 1/100s timer provides this information. On a single frame the timer will show three lighted LEDs on the bar display. The last LED indicates the exact 1/100s number that the frame grabbing process stops. The basic schematics of this timer are shown in Figure 8.

Fig. 7 1/100s timer for a 30 frames/s video camera.

It is emphasized that this is a non-intrusive measurement verification method and not a basic data acquisition process. This principle may be upgraded with the use of high speed cameras and a higher resolution timer. In addition to the dam's

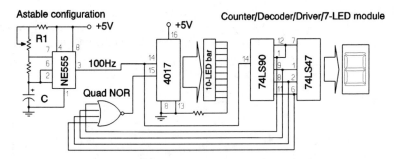

Fig. 8 Basic schematics of the external Video Timer with a time accuracy of 1/100 to 1/200 of a second, for video cameras with only 30 frames/s resolution.

response measurements, the same timer/camera setup may be used for velocity measurements. In this respect, suspended particles or floating bodies were used to successfully verify other velocity measurement methods.

UPSTREAM VELOCITY MEASUREMENTS

Accurate velocity measurements present another challenge as the average upstream velocity, V, ranges only between 8.7cm/s to 29.2cm/s. The method used here combines a good measurement accuracy with minimum expenditures. It requires knowledge of the flow rate in the channel and it measures the height of the water upstream of the dam. A point gage is used with an accuracy of 0.025mm (0.001 inch). What makes this setup different from any previous work, is the use of a camera and monitor to display the point gage readings. This enhances the gage's resolution to 0.00635mm (0.00025 inches) and it also allows for remote operation. This may be motorized with a step-motor or hand operated with mechanical links, according to the arrangement of the experimntal facilities.

The meniscus created by the surface tension at the point gage's tip, behaves as a concentrating lens on incoming light. The result is a visible light-dot at the side or bottom surface of the channel, under normal ambient-light conditions. This helps in identifying the point of water-contact, when remote operation is used. The process is very reliable and provides an accuracy better than 0.013mm (0.0005 inches). The error level may increase in the presence of surface waves, and it is necessary to use a mean value. In this case, the dam's oscillations under overflow conditions do not generate any surface waves, even at very high amplitude oscillations. Surface waves may exist due to insufficient wave dissipation mechanisms of the experimental facilities. For this work, honeycomb filters and fine wire-meshes were used.

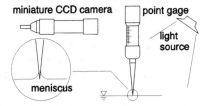

Fig. 9 Upstream water-head, h_u, measurement setup.

FLOW VISUALIZATION

Flow visualization was performed at the upstream side of the dam. Visualization of the downstream side of the dam is not possible at such small scale models, unless very high overflows are studied. A Helium-Neon laser sheet is used with naturally entrapped air-bubbles and water-borne debris. Larger sizes of particles escape through a series of honeycomb filters, and fine-wire meshes. The result is a very clear picture in the test section, but it imposes some difficulties when is transferred on a video tape or 35mm film because of the very small particle sizes.

Fig. 10 Laser-sheet flow visualization of particles at the upstream side of the dam. The curved intense-white line is the upstream face of the model dam ($H=6.7$cm, $L/H=8.4$).

The pictures are enhanced through image processing techniques. A single video frame or a 35mm photograph may be digitized and post processed with a graphics program (like Adobe Photoshop) to enhance clarity. Alternatively, numerical processing on the image may be performed with programs like MATLAB 4.0. The results of the first process is shown in Figure 10. Similar results may be produced with numerical processing, where additional information can be extracted from the RGB (red green blue) image matrices.

CONCLUSIONS

Internal pressure variations are a good response characteristic of inflatable structures and they can successfully describe their dynamic sate. The use of such a measurement method in the case of inflatable dams allows for non-intrusive evaluation of the response of the dams under overflow and impounding water conditions. The instrumentation amplifier necessary is relatively inexpensive. Along these lines an accurate average velocity measurement method was also presented.

These methods were verified with the timer/camera setup which enhances the resolution of any video system, by one more digit. Additional advantages may relate to other applications since the timer is external and not physically connected with the camera. The flow visualization confirms the expected results of the flow-field.

ACKNOWLEDGMENTS

The contribution of the National Science Foundation grant #MSM-9008518, and the help of Professor D. P. Telionis of the Department of Engineering Science and Mechanics and Professor R. H. Plaut in the Department of Civil Engineering in Virginia Tech is acknowledged.

APPENDIX I. REFERENCES

Baker, P. J., Buxton, D. H., and Worster, R. C. (1965). "Model Tests on a Proposed Flexible Fabric Dam for the Mangla Dam Project, Pakistan", *The British Hydromechanics Research Association, R.R. 803 & 827*, February.

Binnie, G. M., Thomas A. R., and Gwyther, J. R. (1973). "Inflatable Weir Used During Construction of Mangla Dam", *Proceedings of the Institution of Civil Engineers, Vol. 54*, pp. 625-639.

Economides, T. A. (1993). "Experimental Investigation on the Dynamics of Inflatable Dams", Ph.D. dissertation, Virginia Polytechnic Institute and State University, Blacksburg, Virginia.

Firestone (1968). "Presenting the Imbertson Fabridam", advertising brochure, Firestone Coated Fabrics Co.

APPENDIX II. NOTATION

C = capacitance (μF, micro-farads)
h_i = inflatable dam's internal pressure water head
h_u = upstream water head
H = height of the inflatable dam
L = length of the inflatable dam
Ri = resistance (Ω, ohms)
V = upstream average velocity
V = voltage differential (V, volts)
ν = kinematic viscosity

A Computer-Controlled, Precision Pressure Standard
Othon K. Rediniotis[1]

Abstract

The operating principles as well as the technical aspects of the implementation of a precision pressure standard are presented here. The instrument can have dual use: either as a pressure source or as a pressure transducer. The device is mostly intended for use in problems where small differential pressures are of interest, i.e. 0-20 Torr and high accuracy is desirable. Such a range, for example, encompasses most of the pressure measurement applications in subsonic wind-tunnel testing. The device interfaces to a PC and is ideal for fully-automated pressure-transducer-calibration applications. The accuracy of the pressures produced or measured by the device is 0.08% F.S., or better.

Introduction

Many differential pressure transducers, especially silicon pressure transducers, used today in basic research facilities require often calibration since their response tends to change with time and temperature. For example, an excellent series of electronically scanned pressure transducers, the ESP series from PSI Inc. have to be calibrated once every hour if an accuracy of 0.1% F.S. should be expected. Such an accuracy, although sounding very high, might be the bare minimum required in several applications such as subsonic aerodynamic research, when for example, a +/- 10 in H_2O F.S. pressure transducer is used. An automated calibration process requires a known, constant and accurate value of pressure, a pressure standard.

In the above sentence, "automated," "known" and "accurate" are the key words. With rather simple hardware, one could generate, with relative ease, a constant value of pressure. The challenge is in whether this pressure value is known and to what degree of accuracy, as well as whether the pressure standard could be interfaced to an automatic calibration process. Moreover, in a calibration process, both positive and negative differential pressures have to be generated if a +/- range of calibration is desired.

---

[1] AEROPROBE Corp., 3138 Indian Meadow Drive, Blacksburg, VA 24060

A very compact, versatile and accurate device, the ACCUPRES, which meets all of the above requirements, is described in the present work.

Other pressure calibrators are currently available in the market. However, there are significant differences between them and the ACCUPRES. The principle of operation of the ACCUPRES, which will be described later is more basic thus resulting in two rather major advantages over the other units: a) the cost of the ACCUPRES is on the order of quarter of their cost, b) pressure ranges of +/- 2 Torr F.S., which are very common in low subsonic wind-tunnel testing, cannot be provided by the other units while maintaining a decent accuracy. However, ACCUPRES units with a full scale pressure range of +/- 2 Torr and a 0.08% F.S. accuracy are common place.

Principle of Operation

First, a summary of the principle of operation is given. In the continuation, each physical principle as well as the technical aspects of the implementation are discussed in more detail.

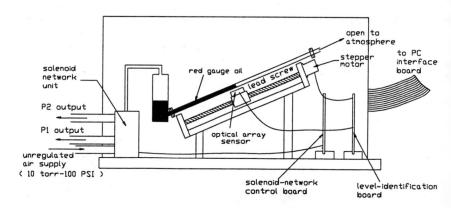

Figure 1. Schematic of the device.

A schematic of the unit is presented in Figure 1. As shown there, an elevation difference is generated between the levels of the liquid in the two members of an inclined manometer, through a network of solenoid valves. The main element of the unit, a 64-pixel CCD sensor, is mounted on a computer-controlled traversing mechanism. This assembly optically searches for the location of the liquid-air interface in the inclined member of the manometer. Once the interface has been identified, mass conservation and liquid incompressibility arguments along with the knowledge of the unit's geometry, accurately yield the generated pressure value.

Optical Sensor Operation

The unit's main element is a very cost-effective linear CCD optical sensor from Texas Instruments (Texas Instruments, 1992). A schematic of it is shown in Figure 2. Its optically sensitive surface has an overall length of 8 mm and is comprised of 64 discrete photosensing areas which will be called pixels. The distance between consecutive pixels is 125 μm. Each pixel's dimensions are 120 μm × 70μm. Light energy striking a pixel generates electron-hole pairs in the region under the pixel. The electrons are collected at the pixel element while the holes are swept in the substrate. The amount of charge accumulated in the element is directly proportional to the intensity of the incident light and the integration time, i.e., the time between consecutive interrogations of the pixel by the sensor-driving circuit. Every time a pixel is addressed and interrogated, its accumulated charge is translated, through internal circuitry, to voltage which appears at the sensor's output pin. Figure 3 provides information on the sensor's sensitivity. It presents a plot of a pixel's voltage output versus integration time with the light irradiance Ee (light energy incident on a unit surface per unit of time) as a parameter. Among the several comments that could be made about this plot, one is of particular interest to us: if the integration time is kept constant and since the pixel's surface area is constant, the pixel's voltage output is proportional to the amount of light incident on it. The above comment along with Figure 4 will explain the functionality of this sensor in the pressure unit.

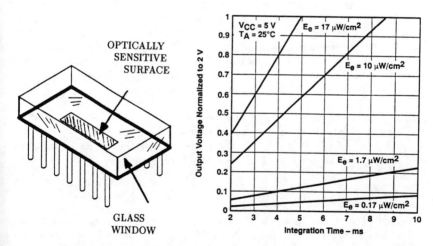

Figure 2. Schematic of the optical array sensor employed in the unit.

Figure 3. Sensor's normalized output voltage versus integration time.

The inside diameter of the inclined manometer member was selected small enough ($\frac{1}{8}$in) so that the mean liquid level is almost perpendicular to the manometer axis (Figure 4). As shown in Figure 4, the light source used is an L.E.D. with

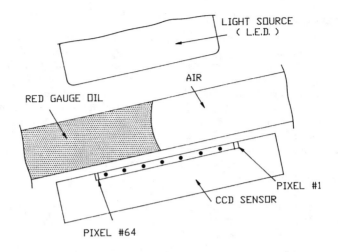

Figure 4. Structural detail in the sensor's neighborhood.

diameter almost equal to the length of the pixel array. The L.E.D. is fixed with respect to the sensor and they are both traversed in unison. This insures that the light irradiance incident on the sensor is constant, regardless of the sensors position along the manometer. The pixels right underneath the liquid-filled part of the manometer collect more light compared to those underneath the air part because the liquid with the glass tube act as a focusing lens to concentrate the L.E.D. light onto the pixels. Figure 5 presents the sensor's voltage output as it appears on the screen of an HP Dynamic Signal Analyzer. The last half of the pixels receives significantly more light than the first third of the pixels causing their voltage output to saturate to approximately 3.7 volts. The intermediate section of steeply rising voltages corresponds to the pixels right underneath the liquid-air interface. This rising part of the curve is further processed to detect the location of the interface. The figure presents several consecutive scans of the sensor.

Liquid-Air Interface Detection

The CCD sensor is driven by a linear frame-grabber circuit which scans through the sensor's pixels, processes its voltage output and interfaces the sensor with the PC. Figure 6 presents a schematic of the rising section of the sensor's voltage output as the driving circuit scans through the pixels. The output Vo is obviously not a continuous function of time. For every positive clock edge the sensor receives, an internal shift register is advanced from the current pixel i to the next pixel $i+1$ and the voltage Vo present at the output pin corresponds to that pixel. These voltages are serially sent to a comparator circuit. This circuit,

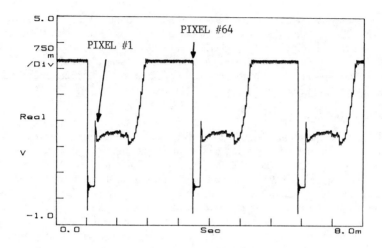

Figure 5. Sensor's voltage output for several consecutive pixel scans.

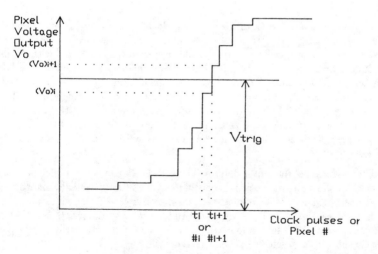

Figure 6. Detail of sensor's voltage output.

when a threshold voltage V_{trig} is exceeded, latches the data from the binary counters of the frame-grabber onto the data bus of the PC. These data are nothing but the binary equivalent of the integer $i + 1$ representing the number of the current pixel, the output of which triggered the comparator circuit. This is the pixel that corresponds to the location of the liquid-air interface. The rate with which the frame grabber circuit scans through the sensor's pixels can be varied up to 0.5 MHz. For the present application a rate of 20 KHz was selected. This means that an entire frame is acquired approximately every $T_{sw} = 3.3 \times 10^{-3}$ sec. This yields an image sweeping frequency $f_{sw} = \frac{1}{T_{sw}} = 300$ Hz.

Device Operation

Before the operation of the device is discussed, the following remarks should be made. Each inclination setting of the unit's manometer is characterized by two figures. The first is the ratio $\Delta P/\Delta \ell$ and it shows how much the pressure output will rise or fall if the liquid level is displaced along the inclined manometer member by a length $\Delta \ell$ upwards or downwards, respectively. This figure is determined upon construction of the unit. It was found that the value of this ratio stays constant for a temperature range 10° - 30° C which covers a wide range of working environments. If a user desires to employ a unit in temperature conditions beyond this range, temperature functions are estimated and incorporated into the unit's software. The correct value of the ratio $\Delta P/\Delta \ell$ is automatically selected by the software upon the user's selection of a working temperature. In an environment where wide temperature variations are common and automation is desired, a thermocouple could be used with its output being interfaced to the PC.

The second characteristic of the unit is the location of the liquid level along the inclined manometer that corresponds to zero differential pressure. The liquid used is red gauge oil with specific gravity of 0.826. Experiments conducted with it show that its evaporation rate is negligible. Therefore, the zero-pressure level needs to be determined only once. This is done upon manufacturing of the unit and the value is incorporated in the software. However, to account for changes in this level because of possible oil leakage due to reasons like mishandling of the unit during transportation from one test site to another, the unit automatically performs a self-detection of the zero-pressure level every two months or at user-specified time intervals.

The operation of the unit is described below. The reader should refer to Figure 7. Let us define a one-dimensional coordinate system $0x$ with is axis aligned with the inclined member of the manometer as shown in Figure 7. The location of the origin 0 is not important. Here, we take it to be located at the junction of the two manometer members. In the following analysis all liquid levels will be specified by their coordinate ℓ_i with respect to system $0x$.

Let us assume that the user desires to generate a differential pressure value P_1 (with respect to atmospheric pressure). The unit's hardware and software achieve that through the following process. If ℓ_0 is the zero-pressure level location, the traversing mechanism moves the CCD sensor such that its pixel #30 is located

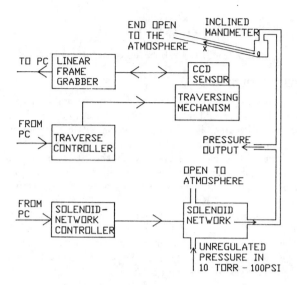

Figure 7. Device's functional block diagram.

at coordinate ℓ_1 so that

$$\frac{P_1}{\ell_1 - \ell_0} = \frac{\Delta P}{\Delta \ell}.$$

This means that if the liquid level were located at the 30th pixel the pressure output would be equal to P_1. Then, the solenoid network allows the unregulated pressure input to slowly increase the pressure at the output through a pressure regulator and a needle valve. The pressure output is assumed to be connected to the instrument that accepts that pressure as an input, for example, a pressure transducer. That increase in pressure in turn causes the liquid level in the inclined manometer to rise. At the same time the sensor is checking for the existence of liquid-air interface above it. When the rising liquid level reaches the 50th sensor pixel, the solenoid traps the generated pressure. It should be noted here that the solenoid valves of the network have a response time of about 20msec. As a result, although the solenoid network receives the signal to trap the pressure when the level is at the 50th pixel, the valves actually engage to do so by the time the liquid level, which has continued to rise, has reached the neighborhood of the 30th pixel. However, if for any reason the level significantly overshoots the 20th pixel, the solenoid network releases some of the trapped pressure in increments of approximately 10 pixels, until the level reaches the neighborhood of the 30th pixel. This neighborhood is defined as the pixels between 20 and 40. Let us assume that the final location of the level is at pixel i, with $20 < i < 40$. The final output pressure will therefore be:

$$P = P_1 + (30 - i) \cdot w \cdot \left(\frac{\Delta P}{\Delta \ell}\right)$$

where $w = 0.125 mm$ is the distance between consecutive pixels.

Application Example

In this section we present an example of implementing an ACCUPRES unit in an automated pressure calibration process. The setup is described in Figure 8. The ESP-16 BP of PSI Inc. is an 8-port electronic differential-pressure scanner. The pressures the user desires to measure connect to any or all of the eight pressure input ports along with a reference pressure. In the "RUN" mode, the ESP samples the pressures and communicates the data with the PC. However, the pressure transducers of the ESP need to be recalibrated once for every hour of operation. The ESP features a "CALIBRATION" mode: if a 100 PSI pressure is applied to its port C1, all eight pressure transducers are manifolded to port CAL, $i.e.$, any pressure applied to port CAL will be felt by all eight transducers, while the pressures applied to the regular eight pressure inputs are blocked off. Similarly, the transducers' reference manifolds to port CAL REF. When the calibration is over, 100 PSI applied to port C2 switches the ESP back to "RUN" mode. The entire calibration process is performed by a modified ACCUPRES unit as shown in Figure 8. The software interrupts the pressure data-acquisition process, the ACCUPRES directs the 100 PSI input to C1 to switch the ESP to "CALIBRATION" mode. The

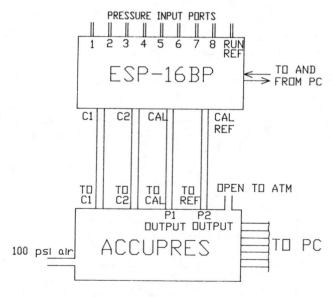

Figure 8. An ACCUPRES application example.

ACCUPRES then produces a series of accurate pressures which are directed to the CAL port of the ESP, with the CAL REF port communicating with the atmosphere. For each of these positive differential pressure values, each transducer's voltage output is sampled by the computer. When this process, which calibrates the ESP for the positive pressure range, is completed, the ACCUPRES sends another series of accurate pressures to port CAL REF, while leaving port CAL open to atmosphere. This calibrates the ESP for the negative pressure range. When the entire calibration process is completed, the ACCUPRES directs the 100 PSI input to port C2, which sets the ESP back to "RUN" mode. The data-acquisition process is resumed with new accurate calibration curves for the ESP.

To give the reader an integrated idea of the usefulness of the above described setup, we present, in Figure 9, an integrated fully-automated system for three-component velocity and pressure measurements in wind-tunnel testing. The seven-hole probe (Rediniotis et al., 1993) is an instrument which, by measuring seven pressure at its tip ports, yields, through extensive calibration routines, all three components of the velocity vector at the location of its tip as well as the local static and dynamic pressures. The PDA-3101 is an interface and data-acquisition board for the ESP. The above setup, through menu-driven software, measures the velocity and pressure field on entire planes with user-defined grid size and spacing and presents the data in terms of velocity vectors, vorticity and pressure contours or outputs the data to ASCII files for further processing by the user.

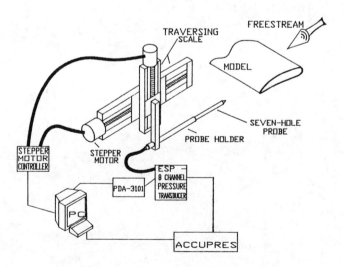

Figure 9. An integrated flow-diagnostics systems employing the setup of Figure 8.

The ACCUPRES as a Pressure Transducer

In a second mode of operation, the ACCUPRES can be used as a very accurate pressure measuring device. Its frequency response is low, therefore it is not suitable for fast varying pressure signals. However, it is ideal for static pressure measurements. Referring to Figure 7, the user has to connect the pressure to be measured to the unit's output. The software takes care of the rest. The unregulated pressure input is no longer required. The accuracy in this mode of operation is maintained at 0.08% F.S.

Conclusions

The operation and use of a very accurate, computer-controlled pressure- calibration/pressure-measurement device was presented. The unit can generate or measure low differential pressures (up to $\pm$ 20 Torr) with an accuracy of 0.08% F.S. It is ideal for the automated calibration of pressure transducers employed in subsonic wind-tunnel testing as well as for static pressure measurements. The simplicity of its principle of operation makes it much more affordable than the other calibrator units available in the market.

References

Rediniotis, O.K., Hoang, N. T. and Telionis, D. P., "The Seven-Hole Probe: Its Calibration and Use," Forum on Instructional Fluid Dynamics Experiments, Vol. 152, pp. 21-26, June 1993.

Texas Instruments, "TSL214, 64×1 Integrated Opto Sensor," May 1992.

Observations on the Growth of an Internal Boundary Layer with a Lidar Technique

Chia R. Chu[1], Marc Parlange[2], William Eichinger[3], Gabriel Katul[4]

Abstract

The distribution of atmospheric water vapor across a step change in surface humidity (dry-wet) was observed by a Lidar measurement technique under neutral atmospheric stability condition. The technique uses multiple elevation scans from a scanning water Raman-Lidar to construct a time-averaged image of the variation in water vapor concentration with height and distance. The measurements were obtained from a bare soil field in California's central valley. The growth of the vapor blanket over the wet surface was identified by the spatial distribution of the specific humidity over the field. The height of the local humidity boundary layer under neutral atmospheric stability condition was observed to increase as $\delta_v \propto x^{0.87}$, where x is the downstream distance from surface humidity discontinuity line. Evaporation rates calculated from the humidity profiles agreed satisfactorily with the fluxes independently measured using a lysimeter.

---

[1] Depart. of Civil Engineering, National Central University, Chung-Li, Taiwan, R.O.C.
[2] Hydrologic Science, University of California, Davis, Davis, CA 95616
[3] Los Alamos National Laboratory, Los Alamos, NM 87545
[4] School of Environment, Duke University, Durham, NC 27708

Introduction

The development of the internal boundary layer associated with the horizontal advection of air across a discontinuity in surface properties has been of interest for a number of years. Various numerical solutions and experimental studies on the turbulent structure of the internal boundary layer have been reported in the literature. Most of these studies are concerned with the response of the internal boundary layer with a sudden change in surface roughness. These studies generally indicate that under neutral atmospheric stability conditions the height δ of the momentum boundary layer increases with fetch x as $\delta \propto x^{0.8}$ (Bradley, 1968; Rao et al., 1974). A comprehensive review of these studies can be found in Garratt (1990).

Relatively few studies have been devoted to the growth of the internal boundary layer across a change in surface humidity or temperature. For the case where a fully-turbulent steady wind crosses a discontinuity in surface humidity, the water vapor can be used as a passive tracer to observe the growth of the internal boundary layer. Making use of the assumptions for two-dimensional boundary layer flow: $U >> W$, $\partial/\partial z >> \partial/\partial x$, and $V = 0$, and steady state conditions, the conservation equation for the mean specific humidity $<q>(x,z)$ can be written as

$$U\frac{\partial <q>}{\partial x} = -\frac{\partial <w'q'>}{\partial z} \tag{1}$$

where $U=U(z)$ is the horizontal wind velocity in the streamwise direction, $<w'q'>$ is the humidity flux in the vertical direction, x and z are the coordinates in streamwise and vertical directions, respectively. Sutton (1934) used a power law velocity profile and a first order closure model

$$<w'q'> = K_v \frac{d<q>}{dz} \tag{2}$$

to solve this advection equation analytically. His solution indicated that the height δ_v of the humidity boundary layer increases with fetch as $\delta_v \propto x^{0.875}$. However, this result has never been confirmed experimentally.

Due to the difficulties in conducting field scale experiments with multiple fixed towers, which interfere with the boundary layer flow, relatively few field observations on the growth of the humidity boundary layer have been reported. Rider et al. (1963) reported a field study concerning the step change of both surface humidity and roughness. In their experiment, air moved from an extensive dry, smooth (z_o = 0.002 cm) surface to a grassy (z_o = 0.14 cm) well-irrigated area. They measured the profiles of wind velocity, air temperature and specific humidity on a series of masts in the downwind direction. Their measurements show the changes of velocity, air temperature and specific humidity over a fetch of 20 m. However, due to the limited fetch and number of data points of their experiments, they did not present the growth rate of internal boundary layer. In this study, we use a Lidar technique to investigate the growth of the humidity boundary layer under both neutral atmospheric stability condition. The Lidar is well suited for this study since the humidity field can be sampled rapidly in space and time over a relative large area without distorting the flow field.

Experimental Setup

The Lidar experiment was carried out at the Campbell Tract facility of the University of California, Davis in August, 1991. The site is a 500 m x 500 m uniform bare soil field with an average momentum roughness height (z_o) of 0.002 m. Within the site, there is a sprinkler irrigation system, which covers a surface area 150 m x 130 m. Figure 1 shows the experimental setup at the Campbell Tract. Accurate measurements of evaporation rate were available on a 20-minute basis from a circular (6.0 m diameter) weighing lysimeter located within the irrigated section of the field. Further details regarding the Campbell Tract facility can be found in Parlange et al. (1992).

The scanning, solar-blind water Raman-Lidar used in this experiment was developed at the Los Alamos National Laboratory. The principle of the Raman-Lidar is based on the technique pioneered by Cooney et al. (1985). A pulse of high energy laser beam is emitted by the Lidar system, and the Raman shift return light

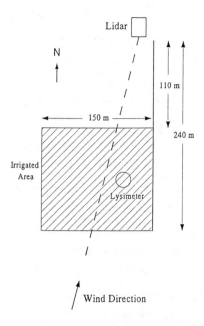

Figure 1. Schematic diagram of the experimental site.

from both atmospheric nitrogen and water vapor is collected using a telescope and then digitized. The ratio of the return water vapor signal to the return nitrogen signal is directly proportional to the absolute water vapor content in the atmosphere. Because atmospheric ozone will attenuate the return signal from nitrogen more strongly than that from water vapor, a correction for the differential attenuation is required. The effect of the correction is a few percent and increases with range.

A Lambda Physics EMG 203 MSC excimer laser with 248 nm wavelength is used as the laser source. The laser beam is aligned so that it is coaxial to the 40.6 cm diameter F/8 Cassegrain telescope. The detector is a Thorne EMI 9813QB photomultiplier, and the digitizer is a Lecroy 8818 signal convertor. Using a periscope mirror assembly mounted at the exit of the telescope, the system is able to scan across 85 degrees in azimuth and up to 35 degrees in elevation. The maximum range of the Lidar is about 450 m in the scanning mode when the output of the photomultipliers is directly digitzed. This truck-mounted Lidar system is

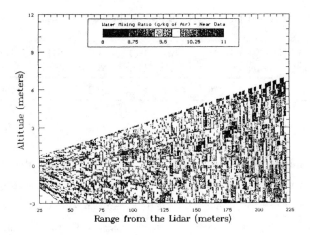

Figure 2. Spatial variation of specific humidity from one Lidar scan.

totally self-contained, including its own generator for electric power, and it has been demonstrated to duplicate the water vapor measurements by the conventional psychrometer/hygrometer to within three to five percent (Eichinger et al., 1993).

The data used in this study are obtained from elevation scan experiments. The laser source was placed 3.0 m above the ground. The scanning system positions the mirror assembly to align at a given azimuth angle (the predominant wind direction), and the vertical scanning angle ranges from -4.0 to 2.0 degree with a resolution 0.1 degree. The water vapor concentration was sampled every 1.5 m along the laser beam at each scanning angle. The scanning time for one elevation scan was about 1 minute. The result is a plot of the water vapor concentration as a function of distance and height above the ground (see Figure 2). Ten of these elevation scans were averaged to obtain the distribution of water vapor concentration along the mean wind direction. Because the intermission time for re-alignment of the telescope between each scan is approximately two minutes, the total sampling duration for the averaged profile is about 25 minutes.

Irrigation was carried out for 6 hours the evening prior to each experimental day to sufficiently saturate the upper soil surface and maintain potential evaporation condition for at least the next 24 hours. The momentum roughness length of the dry

Table 1: Meteorological condition during the Lidar experiment.

Time	V m/s	u. m/s	T_a oC	Rn W/m^2	G W/m^2	L_eE W/m^2	H W/m^2	L m
16:20	6.64	32.5	275	63.3	134	0.38	55.9	-55.6

and wet areas are assumed to be the same (i.e. z_o = 0.002 m), and the surface displacement height d_o is taken to be zero. The mean meteorological condition within the irrigated area was shown in Table 1. The mean wind velocity V was measured by a 3-D ultrasonic anemometer, the net radiation Rn was measured by a Fritchen type net radiometer, the soil heat flux G was measured by a Thornwaite soil heat flux plate, and the air temperature (at 80 cm above the ground) was measured by a Campbell Scientific temperature probe. The friction velocities $u_* = <-u'w'>^{1/2}$ were calculated from direct measurement of Reynolds stress $<-u'w'>$ with a 3-D sonic anemometer. The sensible heat flux H was calculated using the energy budget equation $H = Rn - G - L_eE$, where the latent heat flux L_eE is measured by the weighing lysimeter within the irrigated area, L_e is the latent heat of evaporation. The Obukhov length L is calculated from

$$L = \frac{-\rho u_*^3}{\kappa g \left[\frac{H}{T_a c_p} + 0.61E\right]} \quad (3)$$

where $\kappa = 0.4$ is the von Karman constant, g is the gravitation acceleration, ρ is the air density, T_a is the air temperature, and c_p is the specific heat capacity of dry air. For this case, $|z/L| << 1$, where z is the height of internal boundary layer. This indicates that the atmospheric stability effect is negligible, i.e., this case can be treated as being under neutral stability condition.

Results and Discussion

The spatial distribution of specific humidity q is shown in Figure 3. The

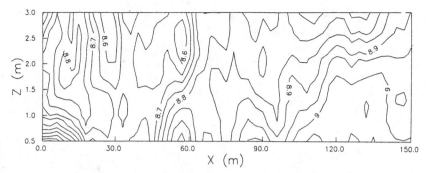

Figure 3. Spatial variation of mean water vapor concentration.

wind is blowing from left to right, and the leading wet edge starts at x = 0 m. The irrigated surface extends from x = 0 m to x = 130 m. Because of the light scatter caused by the surface reflection, the data close to the ground (z = 0.0 - 0.3 m) are generally less reliable and are not plotted in Figure 3. The growth of wet blanket from left to right is clearly seen in the range from x = 50 m to x = 150 m. However, the specific humidity profiles in the transitional region (x = 0 - 50 m) do not show a clear growth pattern.

The values of mean water vapor concentration $<q>(x,z)$ at any position can be linearly interpolated from its four closest adjacent points, which are on the grid points of scanning mesh. Figure 4 shows the verical profile of q(z) from x = 50 m to x = 130 m. Except the first profile (x = 50m), all the profiles were shifted 1 g/Kg respect to the one on its left to allow for comparison. As can be seen, humidity decreases gradually as height increases. The height of humidity boundary layer δ_v is set at the height humidity dose not show significant decrease anymore, and this was done as objectively as possible. The concentration within the vapor blanket is indicated by the solid circles in Figure 4.

The variation of δ_v is plotted as a function of x in Figure 5. This figure suggests that δ_v/x is roughly on the order of 1/100, which is in agreement with the usual rule of thumb for scaling in the surface layer (Brutsaert, 1982). The best linear regression fit of the data gives $\delta_v = 0.0194 x^{0.87}$ with $r^2 = 0.94$. This result compares remarkably well with the Sutton's (1934) analytical solution $\delta_v \propto x^{0.875}$

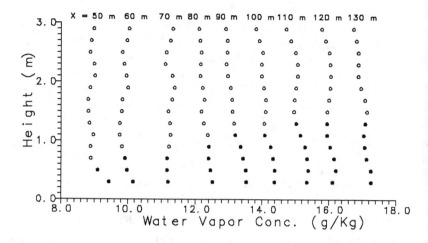

Figure 4. Vertical profiles of water vapor concentration at various downstream distances.

and as well with the growth of momentum boundary layer over a sudden change of surface roughness $\delta \propto x^{0.8}$ (Bradley, 1968; Rao et al., 1974).

The profile derived evaporation rate E_p is calculated using Monin and Obukhov (1954) similarity theory

$$\langle q(z) \rangle = A[\ln(z) - \Psi_v(\xi)] + B \qquad (4)$$

where

$$A = \frac{E_p}{\rho \kappa u_*}, \qquad B = [-\frac{E_p}{\rho \kappa u_*} \ln(z_0) + \langle q_s \rangle]$$

q(z) is the specific humidity at height z above the ground, q_s is the specific humidity at the surface, $\xi = z/L$, and Ψ_v is the stability correction function for specific humidity (Businger et al., 1971). Under neutral stability condition $\Psi_v(\xi) \simeq 0$. Therefore, the value of E_p can be determined from a linear regression between q(z) and ln(z).

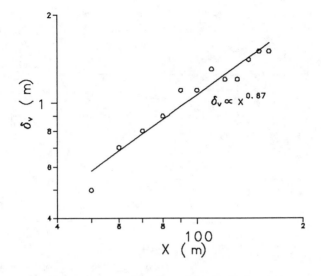

Figure 5. Growth of humidity boundary layer height δ_v.

Figure 6 shows the spatial variation of evaporation rates E_p calculated from the humidity profiles. The values of E_p are normalized by the evaporation rate E_L measured with the weighing lyismeter. The results indicate that after the transition region (roughly 50 m from the leading edge), there is a fairly close agreement between the evaporation rates obtained from these two methods. For the transition region (x = 0 to 50 m), the large scatter of E_p/E_L indicates that the similarity theory, which assumes local equilibrium, can not be used to calculate the humidity flux. The similarity theory yields good results once the internal boundary layer is well developed over the wet surface.

Conclusions

The response of water vapor distribution over a step change in surface humidity has been observed by a Lidar technique. The technique uses multiple elevation scans from a scanning water Raman-Lidar to construct a time-averaged image of the variation in water vapor concentration with height and distance. The

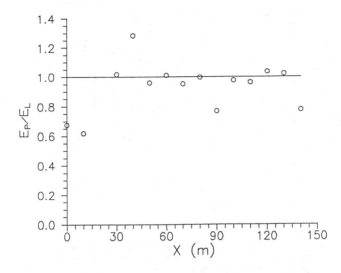

Figure 6. Comparison between the evaporation rate E_p calculated from humidity profiles and the evaporation rate E_L measured by the lyismeter.

growth of the wet blanket is clearly seen in the spatial distribution of specific humidity over the field. The height of the internal boundary layer under neutral stability conditions increases as $\delta_v \propto x^{0.87}$. The fetch requirement of 100:1 for growth of the internal boundary layer appears to be adequate for small roughness elements. The evaporation rates calculated from the humidity profiles are in good agreement with the values measured with the lysimeter. Finally, this study demonstrates the usefulness of the Lidar to study the spatial and temporal variability of water vapor concentration.

Acknowledgements

The authors gratefully acknowledge the support of NIEHS Super Fund Basic Research Program P42 ES04699, INCOR cooperative grant, Kearney Foundation, USGS and California Water Resources Center.

Reference

Bradley, E.F.: 1968, A micrometeorological study of velocity profiles and surface drag in the region modified by a change in surface roughness. *Quart. J. Roy.*

Meteorol. Soc. 94, 361-379

Brutsaert, W.: 1982, *Evaporation Into the Atmosphere: Theory, History and Applications.* D. Reidel Publishing Co. Holland, 299pp.

Businger, J.A., Wyngaard, J.C., Izumi, Y. and Bradley, E.F.: 1971, Flux-profile relationships in the atmospheric surface layer. *J. Atmos. Sci.* 28, 181-189

Cooney, J., Petri, K. and Salik, A.: 1985, Measurement of high resolution atmospheric water vapor profiles by use of a Solar-blind, Raman-Lidar. *Applied Optics*, 24, 104-108

Eichinger, W.E., Cooper, D.I., Hof, D.E., Holtkamp, D.B., Karl, R.R., Quick, C.R., and Tiee, J.J.: 1993, Derivation of water vapor fluxes from Lidar measurements. *Boundary Layer Meteor.* 63, 39-64

Garratt, J.R.: 1990, The internal boundary layer - a review, *Boundary Layer Meteor.* 50, 171-203

Monin, A. S. and Obukhov, A.M.: 1954, Basic Laws of Turbulent Mixing in the Ground Layer of the Atmosphere. *Tr. Geofiz. Instit. Akad. Nauk, S.S.S.R.* No. 24 (151), 163-187

Parlange, M.B., Katul, G.G., Cuenca, R.H., Kavvas, M.L., Nielsen, D.R., and Mata, M.: 1992, Physical Basis for a Time Series Model of Soil Water Content. *Water Resource. Res.* 28, 9, 2437-2446

Rao, K.S., Wyngaard, J.C., Cote, O.R.; 1974, The structure of the Two-dimensional Internal Boundary Layer over a Sudden Change of Surface Roughness. *J. Atmos. Sci.* 31, 738-746

Rider, N.E., Philip, J.R., and Bradley, E.F.: 1963, The horizontal transport of heat and moisture -A micrometeorological study. *Quart. J. Roy. Meteorol. Soc.* 89, 507-531

Sutton, O.G.: 1934, Wind Structure and Evaporation in a Turbulent Atmosphere, *Proc. Roy. Soc. (London) Ser.* A, 146, 701-722

Design and Execution of Hydrodynamic Field Data
Collection using Acoustic Doppler Current Profiling Equipment

Timothy L. Fagerburg[1], M., ASCE and Thad C. Pratt[2]

Abstract

Proposed major changes to the navigation channel depth in the Cape Fear River prompted an investigation of the existing hydrodynamics of the river system. The observed data are to be used in a numerical model study of the proposed improvements. Verification and the subsequent application of a numerical model requires sufficient field data to demonstrate that the model is reproducing the existing hydrodynamics of the system. The U.S. Army Engineer District, Wilmington and the Waterways Experiment Station have recently concluded a detailed data collection program to provide these data on the Cape Fear River system. The data collection effort began in August 1993 and ended in October 1994. Long-term (45 days) and short-term (2 days) data were collected at the boundaries of the proposed model as well as various interior locations. These data consist of water surface elevations, magnitude and direction of currents, and salinity concentrations. Current profiles were obtained using Acoustic Doppler Current Profiler system.

Introduction

The navigation channel of the Cape Fear River (Figure 1) extends 18 kilometers from the mouth of the river to the upper turning basin near the port of Wilmington, NC. The river system serves as major thoroughfare for commercial shipping, commercial fishing, and recreation vessels. In an effort to improve navigation of the river, it has been proposed to deepen the navigation channel from the existing channel depth of 11.6 meters to depths ranging from 13.2 to 14.5 meters

---

[1]. Research Hydraulic Engineer, Hydraulics Laboratory, US Army Waterways Experiment Station, 3909 Halls Ferry Road, Vicksburg, MS 39180

[2]. Research Physicist, Hydraulics Laboratory, US Army Engineer Waterways Experiment Station, 3909 Halls Ferry Road, Vicksburg, MS 39180

is planned without any changes to the existing channel alignment.

Prior to the deepening, feasibility studies are required to determine the potential impacts on the project area. Areas of concern include shoaling volume in the main river channel, river and tidal current magnitude and direction, salinity distribution, and ship maneuvering practices. The US Army Engineer Waterways Experiment Station (WES) was contacted to design and execute a field data collection program that would provide the necessary data for input to the numerical model for the impact analysis.

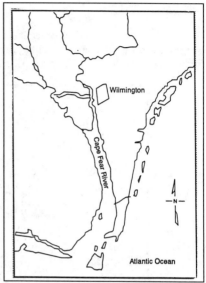

Figure 1. Project location map

Design of Field Data Collection Program

Team leaders from the WES field data collection group, modeling group, and the Wilmington District provided input into the design of the field data collection effort. Model boundaries were determined to identify the upper and lower limits of long-term equipment installations. Decisions were also made as to the approximate location of data collection areas within the model boundaries. Data needs were identified in terms of parameters, sampling intervals, and length of data collection period. The field data collection efforts were designed to obtain the hydrodynamic and salinity conditions of the Cape Fear River extending from the mouth of the river at the Atlantic Ocean to various tributaries above Wilmington, NC. The field efforts consisted of an intensive two day (short-term) exercise and a 45 day (long-term) data collection period for remotely deployed self-recording equipment. The field data provide a unique characterization of the hydrodynamic and salinity changes in a deep, swift river. This paper provides a description of the data collection techniques, preliminary results of the field efforts and some lessons learned on how to best obtain such data.

Field measurements during the intensive short-term effort included current speed and direction, salinities, and water samples for suspended sediment concentrations. The long term data included water levels, current speed and direction, and salinities.

Field Data Collection

The two-day field data collection effort was performed during August 1993 to provide short-term intensive hydrodynamic data for numerical model verification and boundary conditions. The principal monitoring was

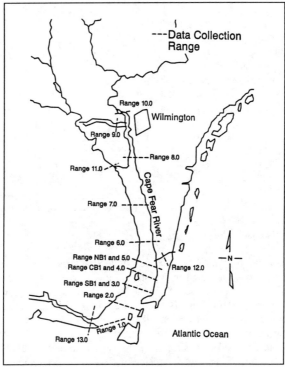

Figure 2. Data collection ranges

performed from four boats which traversed ranges taking discrete water samples and continuous acoustic transects over a two day period. The data collection approach was to use acoustic methods to obtain high-resolution spatial coverage of currents and salinity concentrations over the transects. Sixteen data collection ranges were established at regular intervals along the Cape Fear River, as shown in Figure 2, covering the 18 kilometers from the mouth of the river to Wilmington, NC. One-half of the data collection ranges were monitored each day of the two-day effort. In addition to the data collection ranges, long-term hydrodynamic data were obtained at nine locations along the same 18 kilometer reach of the river plus three additional locations farther up several major tributaries of the Cape Fear River. The parameters monitored at the 9 river locations were water levels, near surface salinities, and temperatures. Currents speed and direction were monitored by moored instrument arrays at ten of the twelve locations as shown in Figure 3. All of the long-term recording instruments were installed and operational prior to the two day intensive data collection effort.

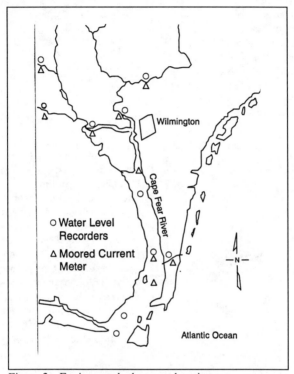

Figure 3. Equipment deployment locations

Boats equipped with an Acoustic Doppler Current Profiler (ADCP), collected cross-sectional current velocity data. Direct-reading broad band ADCP's were mounted on three boats. The ADCP operates by transmitting acoustic pulses (see Figure 4) from four transducers each oriented 30 degrees from the vertical at 90 degree intervals in the horizontal plane. The return signals are gated to resolve up to 128 depth increments. The Doppler principal is applied to resolve current components from backscattered acoustic signals.

Instrumentation and Sampling

The acoustic systems acquired high-density current and acoustic backscatter information which revealed surprising three-dimensional structures. Between acoustic transects, water samples were obtained over depth for salinity and total suspended material (TSM) concentration. The samples were refrigerated and transported to WES. Analysis for salinity concentrations and TSM were performed at WES in-house laboratory facilities.

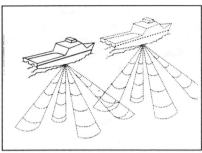

Figure 4. ADCP beam configuration

Almost 200 acoustic transects and five hundred water samples were obtained over the two day period. Continuous recording of current speed, direction and salinity over the 45 day period were obtained using from one to two ENDECO 174 current meters. The meters were deployed at 50% and/or 80% of the depth in a moored array configuration. A typical installation of this type of equipment is shown in Figure 5. Changes in water level and near-surface salinity concentration were obtained with ENDECO Model 1152 recording water level/salinity gages.

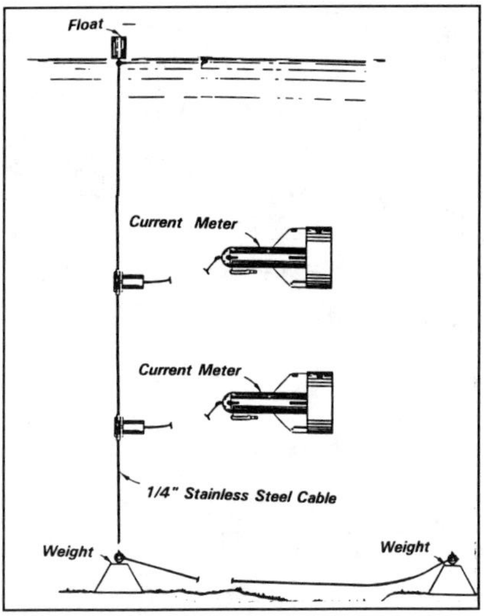

Figure 5. Typical current meter deployment

Data Presentation

The field data collected have been reduced and cast into the proper format to provide information for the numerical model to begin simulation of existing conditions. The ADCP data was taken on hourly intervals across the channel. These data were processed into files for extracting current data for a known depth at a particular time. The data were extracted at five depths; near-surface, one-quarter depth, middepth, three-quarter depth, and near-bottom. The data were then plotted, as shown in Figure 6a, in a time series format for use by the numerical modelers. The current data were then compared to the currents obtained by the moored current meter in the vicinity of the data collection range. Example data from the moored current meters are shown in Figure 6b. These figures illustrate a good correlation of the maximum flood and ebb velocity magnitudes between the ADCP and the fixed-depth moored current meter. The data is currently being used for simulation runs and verification purposes.

Figures 7- 9 are representative examples of the data obtained. These include water-surface elevation, salinity concentrations, and suspended sediment concentrations. The salinity concentrations and suspended sediment concentrations were determined by laboratory analysis of water samples collected during the intensive data collection effort. The concentrations of suspended sediment in the water samples has been successfully correlated to the acoustic backscatter intensity for the purpose of sediment flux calculations and verification of sediment transport models.

Summary

A large data collection program has been conducted on the Cape Fear River for the purpose of verifying a numerical model to be used to address the issue of the impact of channel improvements on salinity intrusion in the Cape Fear River system. Water surface elevations, current speeds, salinity and temperature were collected continuously for the 45 day period. Extensive spatial data were collected at sixteen ranges throughout the system over a two day period. It is apparent that sufficient information were collected to provide an adequate data set for model simulation runs and verification. These data will be made available to the scientific community upon request

Acknowledgements

The field data collection program discussed herein was conducted by the U. S. Army Engineer Waterways Experiment Station and the U. S. Army Corps of Engineer District, Wilmington, NC. Permission to publish this paper was granted by the Chief of Engineers.

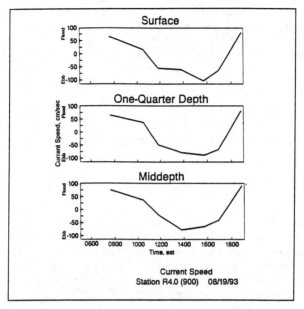

a. ADCP data

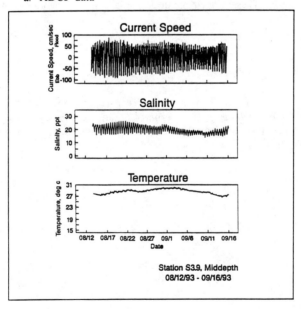

b. Moored current meter data

Figure 6. Current meter data comparison

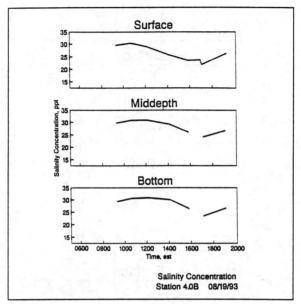

Figure 7. Salinity concentrations at current data collection location

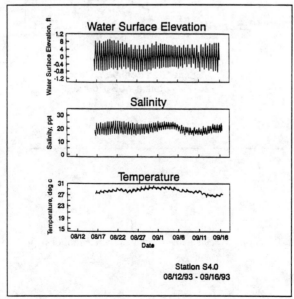

Figure 8. Water surface elevations, salinity, and temperature at Station S4.0

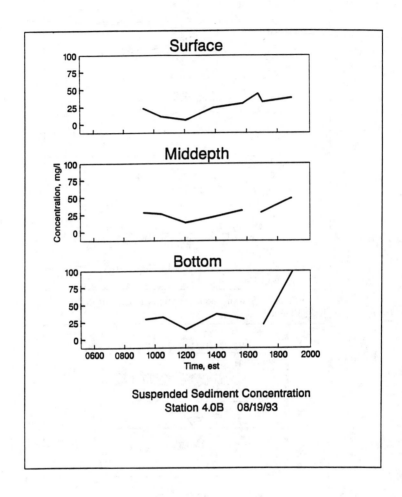

Figure 9. Suspended sediment concentrations

Instrumentation Needs and Possibilities: A Dialogue Between Suppliers and Users

Robert Ettema[1] and Vito Latkovitch[2]

Abstract

The present symposium includes a panel discussion to address and illuminate needs and possibilities in hydraulic instrumentation. The panel discussion is organized with the intent of promoting greater dialogue between suppliers and users of hydraulic instrumentation. The suppliers are drawn from commercial developers and suppliers of instrumentation. The users are drawn from a large governmental agency charged with monitoring much of the nation's waterways, an industry reliant on high accuracy in flow measurement, and a university hydraulic laboratory. Some of the issues for discussion are briefly summarized herein.

Introduction

The substantial advances in instrumentation for laboratory and field measurement of diverse flow and water properties have occurred in recent years. Where until not too long ago hydraulic laboratory experimentation and field measurement programs had acquired a "low tech" image, they increasingly are becoming relatively "high tech." Sophisticated instruments, based on physics of light, sound, electrical response, chemical reaction, and other physics principles, coupled with powerful computational software and portable computers, are facilitating acquisition of flow and fluid-property data that only recently was regarded as unattainable. New instrumentation and data-acquisition methods are on the horizon, and soon will be within the grasp of the laboratory and field hydraulician. Some instrumentation and methods already are in the hands of a few, but still at the development stage, suitable for some applications, often with considerable effort, but not

---

[1]Research Engineer, Iowa Institute of Hydraulic Research, Iowa City, Iowa 52242.

[2]Supervising Hydrologist, U.S. Geological Survey, SSC, MS 39529.

yet sufficiently reliable for general unsupported use. Considerable promise exists for further developments based on principles presently just partially conceived.

Though significant advance in instrumentation and measurement methods have occurred, the development of instrumentation and methods has not always been a steady and orderly path toward progress. Commercial developers of instrumentation and data-acquisition software, and instrumentation users, have concerns and constraints that are not adequately realized or understood by the other group. To promote closer interaction between instrument suppliers and users, and to draw attention to instrument needs and possibilities, the organizers of this symposium decided to arrange a panel discussion between suppliers and users. The panel comprises three representatives from companies that commercially develop and market instruments, and three representative users, generally grouped as government agency, industry, and university. Each panel participant will briefly present their perspective of instrumentation and data-acquisition opportunities and concerns. Subsequently, a moderated discussion will be held between panelists followed by questions and comments fielded from the audience.

Some Issues

It is anticipated that the discussion will address the following, and other, issues:

- New directions instrumentation and data-acquisition software development.
- Unmet needs in laboratory and field instrumentation and data-acquisition systems.
- Constraints on commercial development of instruments and software.
- Interdisciplinary nature of instrumentation development.
- Economics of instrumentation and software development.
- Ballooning expense of obtaining and supporting contemporary instrumentation.
- Interaction of instrument developers and instrument users.
- Deficiencies in engineering education.
- Keeping up with developments.
- Acceptance of new instrumentation and data-acquisition methods.
- Instrumentation, data-acquisition, and industrial standards.

Concluding Remark

It is intended that a synopsis of the main points raised in the dialogue be documented in a paper to be submitted for publication and dissemination by ASCE's *Journal of Hydraulic Engineering*.

Subject Index
Page number refers to first page of paper

Accuracy, 509
Acoustic measurement, 1, 43, 129, 341, 351, 366, 530
Aeration, 268, 289, 425
Air flow, 289, 296
Air water interactions, 289, 296
Alluvial channels, 129
Alluvial streams, 233
Anemometers, 43, 159, 195, 223

Barges, 62
Bathymetry, 341
Bed load, 129
Bed load movement, 149, 233
Bed movements, 139
Bed roughness, 195
Boundary layer, 519
Bridges, 37, 104, 114, 366
Bubbles, 205, 278, 296, 454

Calibration, 119, 233, 268, 509
Cameras, 205
Cavitation, 301, 454
Channel erosion, 21
Channel flow, 311
Closed conduits, 278
Colorado River, 43
Computerized control systems, 396, 509
Corrosion, 301
Cracks, 416
Currents, 27, 242, 251, 341, 351, 376, 530

Dams, 43, 416, 472, 500
Data collection, 43, 62, 104, 530
Data collection systems, 296
Data processing, 53
Design, 86, 472
Digital techniques, 86
Discharge, 76
Discharge coefficients, 21
Discharge measurement, 27, 341, 376
Dissolved oxygen, 425
Doppler systems, 159
Drag, 331

Dredging, 119
Drop structures, 21
Ducts, 311
Dunes, 129, 139

Efficiency, 233, 445
Electrical conductivity, 289
Endangered species, 472
Equipment, 472
Erosion, 301, 472
Error analysis, 176
Estuaries, 1
Experimentation, 149, 396, 500

Fathometers, 366
Federal agencies, 37
Fiber optics, 386, 406
Field tests, 43, 268
Fish screens, 53
Flood plains, 311, 396
Floods, 104, 185
Flow characteristics, 1
Flow measurement, 27, 43, 251, 268, 289, 296, 351, 376, 425, 445, 490, 539
Flow rates, 445
Flow visualization, 213, 406, 490
Fluid flow, 53
Flumes, 205, 223
Force, 331
Friction, 76

Gas flow, 278
Grade control structures, 21
Gravity, 139, 331

Highways, 37
Hot film anemometers, 176
Humidity, 519
Hydraulic jump, 321
Hydraulic models, 43
Hydraulic performance, 233
Hydraulic properties, 396
Hydraulic structures, 86
Hydraulics, 104, 185, 301, 425, 539

Hydrodynamics, 139, 321, 454, 530
Hydroelectric power, 259
Hydroelectric power generation, 53
Hydroelectric powerplants, 416, 425, 445
Hydrogen, 205

Ice, 251
Ice cover, 242, 259
Ice flow, 259
Ice forces, 242
Image analysis, 86, 205, 406
Inflatable structures, 500
Inspection, 114
Instrumentation, 1, 37, 53, 76, 94, 104, 114, 176, 223, 251, 259, 268, 301, 351, 366, 509, 539
Interactions, 213

Laboratory tests, 205, 251, 331, 351, 376
Lakes, 1
Laminar flow, 311
Lasers, 1, 159, 195, 386, 406
Locks, 472
Low head, 445
Lysimeters, 519

Mapping, 53
Measurement, 1, 62, 76, 86, 94, 114, 149, 176, 223, 242, 268, 278, 301, 311, 331, 386, 406
Measuring instruments, 435, 519
Methodology, 176
Mississippi River, 62, 104
Model studies, 86, 445
Model verification, 86
Models, 259
Monitoring, 114, 366, 472, 539
Morphology, 119
Movable bed models, 119

Navigation, 119, 530
Nile River, 119
North Carolina, 366

Oceans, 1
Open channel flow, 76, 119, 159, 185, 195, 396, 406, 490
Open channels, 27, 331

Particle full velocity, 223
Particle motion, 149, 213
Particles, 406
Penstocks, 43, 53
Photographic analysis, 149, 213
Piers, 104
Piezometers, 259
Pressure measurement, 321, 454
Pressure measuring instrument, 509
Probability density functions, 149
Projects, 37
Prototypes, 482
Pump intakes, 482
Pumping plants, 435, 482

Recording systems, 205
Records management, 27
Remote sensing, 351
Research, 321, 454, 500
Reviews, 278
Riprap, 331
River flow, 376
River systems, 530
Rivers, 129, 259
Roughness, 76

Saltation, 149
Sampling, 94, 233, 296
Scale models, 482, 500
Scour, 37, 104, 114, 366
Sediment, 94
Sediment concentration, 301
Sediment deposits, 472
Sediment discharge, 94
Sediment transport, 104, 119, 139, 213, 223, 301
Sensors, 351
Separation, 435
Shear stress, 76, 195
Simulation, 251
Slabs, 321
Soils, 519
Sonar, 114
Sorting routines, 139
Standards, 509

Stilling basins, 321
Strain gages, 416
Stream gaging, 21, 27, 376
Streamflow, 242, 251, 376
Streams, 129, 139, 341
Stress measurement, 416
Submerged flow, 482
Surface waters, 27
Suspended sediments, 94
Suspended solids, 530

Tanks, 386
Technology, 539
Technology transfer, 37
Telemetry, 416
Testing, 435
Tidal currents, 366
Towing, 62, 386
Transducers, 454, 509
Turbines, 416, 425, 482
Turbulence, 1, 62, 159, 176, 185, 195, 213, 386, 406
Turbulent flow, 1, 159, 176, 185, 311, 490
Two phase flow, 86, 278, 289, 296

Unsteady flow, 185, 396

Velocity, 62, 76, 351, 376, 396
Velocity distribution, 43, 195
Velocity profile, 242, 268, 341
Vibration, 454, 500
Videotape, 205, 213, 482, 490
Voids, 278
Vortices, 482

Wastewater treatment, 435
Water, 519
Water flow, 159, 278, 289, 296
Water hammer, 435
Water levels, 530
Water quality, 425
Water supply systems, 396
Water surface profiles, 21
Weirs, 21
Wind tunnel test, 509

Author Index
Page number refers to the first page of paper

Abdel-Motaleb, Mohamed, 139
Ahmed, A. F., 119, 331
Almquist, Charles W., 445
Arndt, Roger E. A., 454

Balakrishnan, Mahalingam, 159
Bhowmik, Nani G., 62
Brown, Bobby J., 21

Cabrera, Ramon, 351
Chincholle, Lucien, 301
Christensen, B. A., 76
Chu, Chia R., 519
Crissman, Randy D., 259

Dancey, Clinton L., 159
Derrow, Robert W., II, 129
Diplas, Panayiotis, 205

Economides, T. A., 500
Ehler, David G., 268
Eichinger, William, 519
El-Gamal, F. S., 119, 331
Ellis, Christopher R., 454
Ettema, Robert, 242, 539

Fagerburg, Timothy L., 530
Frizell, K. Warren, 416, 482
Frizell, Kathleen H., 268
Fulford, Janice M., 251, 376

García, Marcelo, 213, 490
Gaweesh, Moustafa T. K., 233
Griffin, Michael S., 472

Hadjerioua, Boualem, 289
Hansen, David B., 445
Hardwick, J. D., 296
Head, Roger, 1
Hoisington, David, 435
Hsu, In-Song, 149

Imberger, Jörg, 1
Ivarson, W. R., 37

Kadota, Akihiro, 185
Kaehrle, William R., 376
Katul, Gabriel, 519
Knight, Donald W., 311
Kraus, Nicholas C., 351
Krylowski, T., 37
Kuck, Andreas J., 86
Kuhnle, Roger A., 129

Landers, Mark N., 104
Larsen, Johannes, 259
Latkovitch, Vito, 539
Laursen, Emmett M., 289
Lee, Hong-Yuan, 149
Lohrmann, Atle, 351
López, Fabián, 213, 490

Malinky, Stan E., 27
March, Patrick A., 445
Mason, Robert R., Jr., 366
Mefford, Brent W., 268
Miller, Jerry, 425
Mueller, David S., 104, 341, 472
Muste, M., 223

Nagao, M., 396
Naghash, Mahmood, 278
Nakagawa, Hiroji, 185, 406
Nezu, I., 396
Nezu, Iehisa, 185, 195, 406
Niño, Yarko, 213, 490

Oberg, K. A., 341
Orlins, Joseph J., 53
Otter, Paul, 435

Papanicolaou, Athanasios N., 205
Parlange, Marc, 519
Parthasarathy, R. N., 223, 386
Paul, Saurav, 454
Petersen, Margaret S., 289
Pinheiro, António N., 321
Pratt, Thad C., 530
Price, G. R., 114
Prinos, P., 176

Quintela, António C., 321

Raffel, David, 435
Ramos, Carlos M., 321
Rediniotis, Othon K., 509
Rhodes, David G., 311
Richardson, E. V., 94
Richardson, J. R., 114
Rizk, Tony A., 289

Saeki, Ken-ichi, 406
Schall, J. D., 37, 114
Schohl, Gerald A., 445
Sheppard, D. Max, 366
Soong, Ta Wei, 62
Stern, F., 386
Swenson, Lawrence J., 53

Teal, Martin J., 242
Thibodeaux, Kirk G., 376
Tominaga, A., 396
Tominaga, Akihiro, 195

Vermeyen, Tracy, 43

Wahl, Tony L., 425
Walker, D. A., 500

Young, Doug, 425

Zarrati, A. R., 296
Zufelt, Jon E., 259

TC 177 .F86 1994

Fundamentals and
advancements in hydraulic